The Foundation of Australia's Capital Cities

The Foundation of Australia's Capital Cities

Geology, Landscape, and Urban Character

Anthony Webster

LEXINGTON BOOKS
Lanham • Boulder • New York • London

Published by Lexington Books
An imprint of The Rowman & Littlefield Publishing Group, Inc.
4501 Forbes Boulevard, Suite 200, Lanham, Maryland 20706
www.rowman.com

6 Tinworth Street, London SE11 5AL, United Kingdom

British Library Cataloguing in Publication Information Available

Library of Congress Cataloging-in-Publication Data

Names: Webster, Anthony, 1961- author.
Title: The foundation of Australia's capital cities : geology, landscape, and urban character / Anthony Webster.
Description: Lanham : Lexington Books, [2022] | Includes bibliographical references and index. | Summary: "The Foundation of Australia's Capital Cities is the story of how the places chosen for Australia's seven colonial capitals came to shape their unique urban character and built environments, resulting in development patterns than have persisted today"—Provided by publisher.
Identifiers: LCCN 2021056753 (print) | LCCN 2021056754 (ebook) | ISBN 9781498597951 (cloth) | ISBN 9781498597968 (ebook)
Subjects: LCSH: Capitals (Cities)—Australia—History. | Cities and towns—Australia—History. | Urbanization—Environmental aspects—Australia—History. | Land settlement—Australia—History. | Urban geology—Australia. | Landscapes—Australia—History. | Australia—History. | Australia—Capital and capitol.
Classification: LCC DU110 .W43 2022 (print) | LCC DU110 (ebook) | DDC 994—dc23/eng/20220110
LC record available at https://lccn.loc.gov/2021056753
LC ebook record available at https://lccn.loc.gov/2021056754

To Josephine, Annette, Catherine, and Kathleen.
The best of sisters and friends.

Contents

List of Figures

Acknowledgments

The research for this book began several years ago, when I was working in a Pilbara iron-ore mining town in northwestern Australia. That is about as far as it is possible to get, within Australia, from the cities that form its subject matter. It would not have been possible to complete the work without the remarkable online digital resources that were then becoming available through State Libraries, archives offices, government departments, and university libraries. Digitized historic survey maps and charts, geoscientific information, artworks, and documents held by such institutions were the most important resources used for the project. There is, therefore, a strong bias to online sources in this book. I am sure that it would have been greatly improved if the undigitized materials had been examined too, but alas, to undertake such a task would have taken many more years.

Many researchers have collected and collated detailed historical information about the early settlements and they made the research for the relatively obscure beginnings of Australia's capitals a great deal easier. I am particularly indebted to the survey of the origins of Australia's capital cities edited by Pamela Statham, and to the remarkable historical research about the Brisbane convict era by the late Professor John Steele. The late Frank Bolt's daily records of events in early Hobart were also very valuable. There were many others.

Modern reconstructions of the pre-European vegetation were created over many years of painstaking observation and analysis by Doug Benson and Jocelyn Howell (1990) in Sydney and Darrell Kraehenbuehl in Adelaide. Ron Hateley recorded his deep knowledge of the Victorian bush, and Garry Presland provided a remarkable reconstruction of the once rich environments of pre-European Melbourne. These inspirational works, among many others, have been drawn on freely for this book.

The 3D topographic reconstructions were created using the remarkable Leapfrog Geo implicit 3D modeling software. Access to the software came through my role as an honorary researcher within the Centre for Ore Deposits and Earth Sciences at the University of Tasmania. Seequent provide their software to staff and students for research and teaching purposes. Leapfrog Geo is primarily designed for mining and exploration purposes but is superb for integrating historic maps in a 3D environment to reconstruct lost or modified topography, and then for modeling the subsurface geology. Unfortunately, the format of this book does not allow the models to be presented. The University of Tasmania and the University of Queensland are thanked for providing access to their rich library resources.

It is also important to acknowledge the importance to this project of Google Earth and Google Earth Street View. While all the places discussed in this book were visited, this remarkable free resource greatly increased the value of the field work by allowing locations to be "revisited," and details to be seen that had been missed. Google Earth also provided high-quality satellite imagery that was used in many landscape reconstructions, and to provide the broader contexts for the localities visited.

June Pongratz read the entire first draft of the manuscript and greatly improved the grammar. I accept responsibility for the many grammatical errors that certainly remain but thank June for there being many less of them. June is also that very rare professional in the modern world: a geological draftsperson. June provided a great deal of advice about the often-difficult process of preparing the complex figures for presentation in this book. The quality of individual chapters was also greatly improved by the early reviews of Ron Morland and Jane Capp. An anonymous peer reviewer of the original draft highlighted many of the scientific weaknesses that are hard to avoid in such a multidisciplinary work. Alas, no doubt a great many more remain, but there are now less than there might have been thanks to the thoroughness of this review.

And finally, I must particularly thank Kasey Beduhn, Acquisitions Editor at Rowman and Littlefield. Kasey recognized the potential for this book from a 100-word conference abstract and guided the preparation of the manuscript from concept to outline, to first review draft, and final submission. Thanks Kasey, I doubt this book would ever have been published without your support.

Tony Webster
South Arm, Tasmania
September 2021

Introduction

There are many historical accounts of the first British settler responses to Australian landscapes. Yet, there are no analyses of the ways in which the features within those landscapes influenced the decisions they made as they adapted to the new continent. This book is about the roles that the landscapes and geology played in the places chosen for the first permanent European settlements in Australia and in the land-use choices that were made as they adapted to their new environments.

The most historically significant sites of the first British settlements are now occupied by the modern central business districts (CBD) of Australia's largest cities. As a result, they are now intensely urbanized and heavily modified landscapes. Despite the effects of two centuries of development and landform modification, this book will show that the geology and pre-European terrains of these places had a profound influence on their early growth. It will also be shown that the effects of the original landforms remain deeply embedded in the modern urban landscapes.

The unique architectural character of each city is also a reflection of the local geology. Locally sourced stone was an important construction material in the first century of urban development. Unique types of stone were used in each capital to construct major buildings, and other distinctive architectural elements of their built environments, and now define the individuality of each city.

Purpose and Scope of the Study

The purpose of this book is to show the importance of local geological and landform influences on the establishment and early development of seven major Australian cities. All were founded on virgin sites, but their establishment and initial expansion was well documented in a variety of official records and personal accounts. The rich historical sources were used to remove centuries of development and landform modifications to reconstruct the original topography. An interdisciplinary approach was used that integrated the fields of 3D terrain modeling, geology, geomorphology, history, and archaeology through modeling software, CAD and GIS.

The landscapes encountered by the first Europeans were recreated as maps, and as three-dimensional digital terrain models (DTM) using a range of historic maps and charts, colonial-era landscape art, and modern topographic data. Information from these models was then used to evaluate how the first settlers interacted with the new landscapes through their activities. Those activities might include their choice of building sites, road and track alignments, places to farm or create cemeteries, the locations of quarries, or the locations chosen to construct jetties and wharfs.

The terrain reconstructions were then combined with modern geological mapping information, observations of surviving outcrop, and subsurface geological drilling data to evaluate how landforms reflected subsurface geology. The documented historic features of each settlement were then overlain on the reconstructed terrain using 3D and GIS software. The result was the recognition of the often surprising and subtle ways that small variations in geology and landform at each location had influenced settler land-use patterns. The largely preindustrial and unmechanized era of the late eighteenth and early nineteenth centuries was one when small landscape features could play an important part in land-use decisions.

Finally, the degree to which any evidence of historic land-use patterns might remain preserved in the modern urban landscape was evaluated by referencing the DTMs and maps to modern topographic information. Through understanding the landscapes present at the beginning of a settlement, the effects of lost features, such as wetlands and outcrops, could be recognized in the modern urban landscape.

It is not possible to provide in-depth geological descriptions of each city, but summaries are presented in each chapter. It is also beyond the scope of this book to discuss how geology continued to influence the later urban expansion of the settlements. They have all grown into great cities and expanded far beyond their original foundational footprints. Such a study would require multiple books of a comparable size. Comments are made about geological aspects of later urban expansion where needed, but they are generally brief.

Historical and archaeological records, supplemented with additional details from contemporary visual sources, were used to evaluate how the settlers interacted with those landscapes. To provide some historical context, summaries are also provided of early European exploration and the preludes to colonization, but the historic focus is on the formative years of each settlement. In general, this is the period from the first choice of the site and foundation of the settlement to the time when the modern street networks and land-use patterns of the modern CBDs were fixed in place by cadastral surveys, land ownership, and infrastructure placement. Once this stage had been reached, major modification became difficult. As a rule, this happened within ten to twenty years of the first settlement.

The architectural character of each city also came to be heavily influenced by the local geology. Stone was an important construction material in the first century of urban development and the most readily available sources were often used, even when not of the best quality. Characteristic building stones unique to each region were used in the construction of many historic landmark buildings and more humble structures. Surviving examples now form unique architectural elements that define the distinctive character of the built environment of each city. Sections about the built heritage and character of each city are included in each core chapter.

Separate core chapters are devoted to each capital city (chapters 1–7), arranged chronologically according to the date of their foundation. The geographic focus of each discussion is the foundational footprint, the site where settlement first occurred and where most urban growth happened in its first two decades. The areas discussed are small in areal extent (e.g., Sydney is approximately 10 km^2 and Hobart is approximately 10.5 km^2). In all cases these places have remained the focus of most of the commercial, administrative, and cultural activity of their respective states since their beginning. This is not meant to imply that all the early activities of each settlement only occurred within these boundaries because they were porous, and constantly and rapidly expanding. Tendrils of exploitation reached across much wider areas from the first days as parties ranged outward to explore, collect building materials, and to gather food. These broader influences are discussed when necessary.

To the author's knowledge, this integrated study presents the first comprehensive reconstructions of the presettlement landforms of the CBD of each Australian state capital. It is also one of the first studies to evaluate the part that geology and landform play in the early growth of cities. It is hoped that by presenting these seven historical Australian examples, new insights will be gained into broader issues of urban geology and the development of cities. Some of these broader themes are discussed in chapter 8.

HISTORICAL BACKGROUND

The British colonization of Australia was the result of the slow evolution of a more enlightened legal system, and the American Revolution. In the late eighteenth century, British judges had become increasingly reluctant to execute criminals for minor offences (e.g., Hughes 1988, 35–6). Many of the convicts had then been transported to the North American colonies, but after 1776, the newly independent United States would no longer accept them (Martin 1989, 39–41). The results were an explosion in the British prison population. Adapting decommissioned warships as temporary prison hulks was used as a stopgap measure but a more permanent solution had to be found. There began a worldwide search for a new and permanent place to transport convicts.

After potential penal colony sites in places such as Africa were considered (e.g., Kenneally 2006, 8–19), a decision was made in 1786 to create a penal settlement on the east coast of Australia—on the far side of the world. The result was the foundation of the first permanent British settlement on the coastline of this vast continent. It was followed by other coastal settlements, which led to the slow but inexorable displacement of the Indigenous people by Europeans arriving from multiple entry points (figure I.1). The most important settlements became the modern Australian state and territory capital cities, and its largest population centers.

Australia has six state capitals: Sydney, Hobart, Brisbane, Perth, Melbourne, and Adelaide; and one territorial capital, Darwin (figure I.1). All but Darwin were founded during a fifty-year period of British colonial expansion between 1788 and 1836. Despite their often similar historical and cultural origins, each city has a unique "feel" that is strongly influenced by its landscape setting and built environment. Sydney is folded into the sandstone coves and headlands of Port Jackson (chapter 1). Hobart looks out from its position at the foot of Mount Wellington to the drowned graben of the River Derwent Estuary (chapter 2). Brisbane is defined by its river pocket peninsula setting (chapter 3). Perth sparkles beside "Perth Water" and undulates across the sandy limestone coastal plain to the Indian Ocean (chapter 4). Melbourne's center is still subtly delineated by the Yarra River, on a site where the volcanic plain meets the ancient Silurian bedrock and the edge of the Yarra Delta (chapter 5). Adelaide is focused on the Torrens River but extends across an ancient and deeply weathered plain from the Mt Lofty Ranges to Gulf St Vincent (chapter 6). Darwin occupies its steep-sided peninsula on the shore of Port Darwin (chapter 7). Each city is made unique by its landscape setting.

Yet despite the importance of their situation to the modern character of each city, their locations were often chosen for relatively short-term reasons. Arthur Phillip, the first governor of New South Wales, chose the site for Sydney, in preference to Botany Bay, because it had a reliable spring-fed freshwater source, and a large and sheltered harbor with deepwater anchorages close to shore. The location of Melbourne was mainly dictated by "the falls," a natural rock ledge that spanned the Yarra River, marking

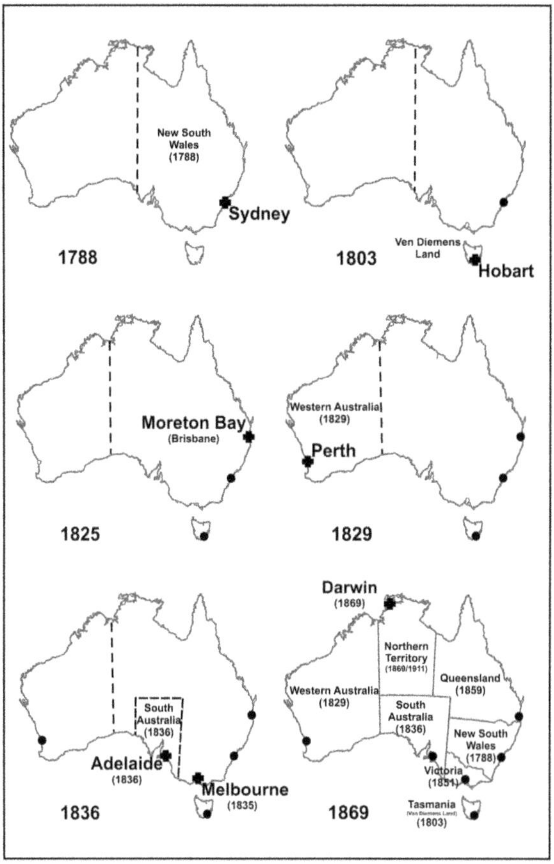

Figure I.1 Settlement Locations and Foundation Dates. The evolution of the British colonisation of Australia. Each panel shows the foundation year of each colony and the location of the principal foundation settlement discussed in this book (shown as crosses). The boundary of the area claimed by Britain at that time are shown as a dotted line. This figure is based on the front and endpaper maps of Statham (1989) and a topographic base from Geoscience Australia. Please refer to the extended caption at the rear of the book for additional information.

the place where the tidal influence ended, and the river became freshwater. Most of the original reasons for choosing these sites for settlement are now obsolete but become more apparent when the historic landscapes are reconstructed.

It is often unclear if the settlement founders had any long-term vision of the way that these places would develop. William Light (Adelaide) and perhaps James Stirling (Perth) seem to have foreseen a longer-term picture of how the settlements they founded might develop, with perhaps Light the most forward-looking. Arthur Phillip may have also had a longer-term urban design in mind for Sydney (Medland, published in Phillip 1789), but it was only implemented at the later inland settlement of Parramatta. John Batman and John Fawkner do not seem to have had any plan for how their village might develop, and it was not until the arrival of the first government superintendent (later lieutenant governor), Charles La Trobe, that urban planning began with the city grid and a ring of parklands. It is probably fair to suggest that none of the founders could have envisioned these small settlements becoming the centers of major cities with populations in the multimillions.

Once chosen, any shortcomings of the sites had to be overcome to make these places work, whether to meet the short-term needs of the initial settlement, or for any future needs of the developing towns. Occasionally it was decided that the geological shortcomings of a site were too great to establish a permanent presence, or it was abandoned after considerable work had been undertaken. Examples include Botany Bay in 1788, Sullivan's Bay (Port Phillip) in 1803, and Escape Cliffs in 1868.

Servicing the major towns and cities that grew from the small foundations often required considerable effort and landscape modification to provide the facilities that growth required (e.g., Melbourne and the Yarra River canal or Brisbane's rail and road transport corridors). Given that all of the seven settlements discussed have grown into cities with populations in the millions (and continue to grow), and each city remains the commercial and administrative center of each state, many of the shortfalls of the early site choices are still being rectified.

British Settlement of Australia, 1788–1869

Two phases of British settlement occurred in Australia during this period of rapid colonization. Each phase had a characteristic style of settlement and urban impact.

Phase 1. 1788–1829: Military Camps as Penal Settlements.

The first settlements were established on military lines, as penal establishments, including Sydney, Hobart, and Brisbane. The primary purpose of these establishments was as centers of punishment and exile, and they were primarily designed to service the needs of the establishment and its administrators. They were serviced by convict labor. There were also overtones of colonial rivalry, with these new settlements serving the secondary purpose of creating a British presence in places that might potentially be claimed by colonial rivals such as France.

The foundation settlements were usually followed quickly by others during this period, including Parramatta, Newcastle, and Port Macquarie; Georgetown then Yorktown, and finally Launceston at Port Dalrymple (in northern Tasmania). An outpost of New South Wales was established at King George Sound, now Albany (Western Australia) in 1826, prior to the foundation of Perth directly from Britain (see figure I.2).

Several attempts were also made to establish military outposts in northern Australia at this time, including Port Essington, on the Cobourg Peninsula. Most were short-lived, but one remained long enough to greet HMS *Beagle* on its second voyage to Australia in 1838, during which Darwin Harbor was discovered and surveyed (see chapter 7).

Phase 2. 1820s–1860s: Civil (Commercial) Settlements

By the late 1820s, the convict transportation system had fallen out of favor and free settlement was becoming preferred, with entrepreneurs beginning to see commercial opportunities in Australia. Coupled with an ongoing perception of French territorial ambitions, this led to the establishment of new settlements in Southern and Western Australia that would not be dependent on convicts. The new settlements would be free and formed by land speculators. So, the second phase of British settlement of Australia was mostly a civilian undertaking, rather than military, and it was not intended to use convict labor. Three settlements would be founded during this phase, Perth, Adelaide, and the initially unauthorized venture that became Melbourne. Brisbane was also refounded as a free settlement at this time.

The high ideal of being convict-free would rapidly change in Perth and Melbourne, where convicts were eventually requested by the free settlers, but it persisted in Adelaide. The settlements created at this time would be generally better-planned and therefore take on a more organized urban character than those of the more ramshackle earlier phase.

Brisbane was the last military settlement of note to succeed. Although founded as a military outpost, it would effectively be refounded when opened to free settlers in the 1840s, following the winding down and closure of the penal settlement. By the 1860s, even the street pattern of the military phase would be mostly erased (see chapter 3).

Darwin was different again. Begun as a South Australian colonial initiative, it was established as a government outpost, with many of the features of the earlier military camps, though there was no military presence (see chapter 7). The camp was the base for the land surveyors who were sent to mark out the landscape for sale to investors, using a similar model to that used when Adelaide was created in the 1830s.

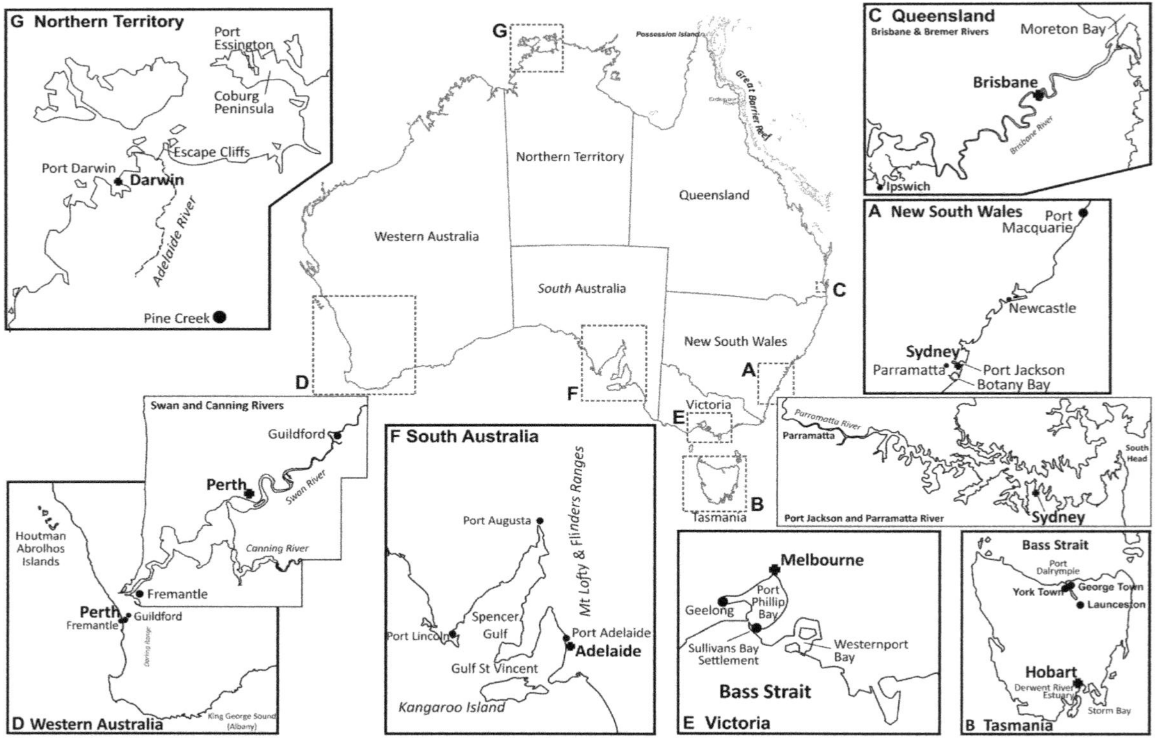

Figure I.2 Places Mentioned in Text. Many places beyond the main settlements were important during the first years of each Australian colony. The locations of the most important places in each colony mentioned in the text are shown on the inset maps.

It is also important to note that by the 1820s, settlements were being founded by locally born Australians, or by people raised in Australia. Sydney and Hobart were settled by fleets sent directly from Britain (except Risdon Cove), but those founded in the 1820s could call upon experts who had been born, or had spent most of their lives in Australia, such as John Oxley (Brisbane) and John Roe (Perth). By the 1830s, it is no longer accurate to describe founders of settlements as British because they were local enterprises. Melbourne was founded by the Tasmanians John Batman (born in Parramatta) and John Fawkner (arrived as a child in 1803); and Darwin (Palmerston) was founded by South Australians. They may have had British ancestry, but they were culturally Australian.

ABORIGINAL LANDSCAPE INFORMATION

Australian landscapes were inhabited for thousands of years before Europeans arrived. Any discussion of the sites on which cities were founded that did not include the first inhabitants would be incomplete without recognizing this fact. The information about Aboriginal land use provided in this book is a brief overview, focused on historically documented locations where people were living or working within the study areas around the time of European arrival; or where physical evidence for their activities was subsequently discovered during archaeological investigations of inner-city developments. Many local Aboriginal place names were recorded at the time of first contact and were also collated.

The readily available information was compiled as maps and evaluated within the context of the reconstructed landscapes to suggest how Aboriginal people interacted with the sites of each CBD. The approach taken was like that used to evaluate the European use of landscapes. Unfortunately, figure constraints prevent a more comprehensive presentation of this aspect of the work in the accompanying map figures.

Sections are included in each chapter at the interface between the descriptions of the pre-European environment and the period of European development. They are not comprehensive reviews. To do full justice to this aspect of the landscape history of each city would require a separate book.

WIDER APPLICATIONS

Australia was settled at a time of massive European expansion across the globe and many major settlements were founded by the former colonial powers. Despite the examples discussed here all being Australian, they may provide useful insights into how other contemporary cities evolved in other parts of the world (particularly other European settler cities).

There are likely to be applications of this study beyond the fields of history and landscape archaeology, and particularly in the disciplines of town planning, urban geology, and geohazard management. Australia's capital cities provide excellent test-cases for the study of geological and landscape influences on urban development because the transition from natural landscapes to intensely urbanized centers was rapid, but well documented. All cities were founded on virgin sites in coastal or riverine locations, but in very different landscape and geological settings. The impact on historic urban growth of a diverse range of geological features and landform types can be recognized for each city. This information may provide insights into how similar conditions might affect the growth of other urban centers if they were encountered. Perhaps most importantly, the examples described here illustrate how significant relatively subtle geological and landscape features can be during the rapid development of virgin sites into major cities.

The landscape reconstructions may have applications to problems of modern city planning and geohazard management in Australia through the better definition of former wetlands and drainage lines that could be potentially reclaimed by flood events. The landscape reconstruction methodologies used may also have applications internationally, to create better flood-prediction maps (as an example). The potential wider applications of the study are discussed briefly in chapter 8.

Chapter 1

Sandstone Harbor

Sydney, New South Wales, 1788

The vast east coast of Australia was settled by the British with almost no knowledge of the place. What information they had was from James Cook's pioneering visit of 1770, while in command of HMS *Endeavour* (e.g., Cook 2005). Cook encountered the coast on 19 April and made some brief visits to the shore, but it was not until he saw a sheltered bay that he decided to lay anchor and water the ship. The *Endeavour* spent seven and a half days at a place Cook eventually named Botany Bay (see figure I.1 for location). He explored the local area with Joseph Banks and Daniel Solander (his botanist passengers), focusing mainly on the southern shore, near the anchorage opposite modern Kurnell. He charted the bay to the mouths of two rivers that enter it from the west, and briefly explored the north shore. Cook then sailed northward along the coast.

Ever observant, Cook noted the probable entrance to a harbor just to the north of Botany Bay, and named it Port Jackson, but did not enter it. He later explored the shore at various anchoring places further north but only made one other extended stay in Australia (over six weeks), when the ship was holed on the Great Barrier Reef, at a place Cook named Endeavour River (Cook 2005, 157). Cook named the entire Australian east coast "New South Wales," at Possession Island, in the Torres Strait, just before leaving Australia and heading to Java (see figure I.2 for location). Cook may have chosen the name long before reaching Possession Island, because of the visual similarity between part of the coast of Wales and the near-horizontal sedimentary strata that formed the rugged sea cliffs near Botany Bay.

When the British government decided to establish a permanent penal colony in New South Wales in 1786, Botany Bay was chosen as the location. Then in 1787, just seventeen years after Cook's visit, and with no further data, a convoy of eleven ships was dispatched to create a settlement at Botany

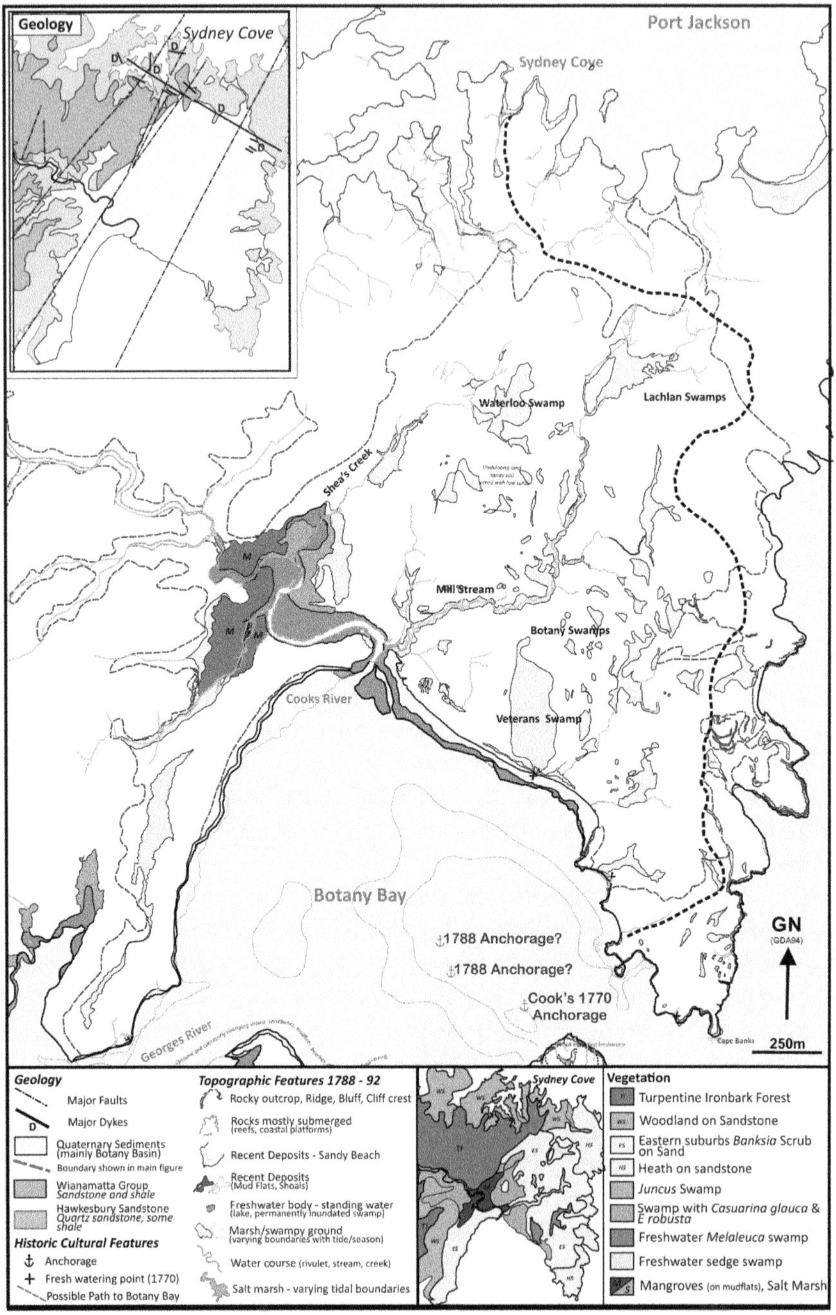

Figure 1.1 Sydney Region. Topographic reconstruction of the landscape context of Sydney Cove c1770-1792, with an emphasis on the region between the southern shore of Port

Figure 1.1 (Continued)
Jackson (Sydney Cove) and Botany Bay. It was in this relatively small area that most important historical events took place during the first year of European settlement. **Figure 1.2 Inset 1**. The geology of the Sydney region, simplified from Brunker and Rose (1967), with additions from Chesnut (1978) and Och et. al. (2004). Note the significant expanse of Quaternary sediments immediately south of Sydney. **Figure 1.2 Inset 2**. Reconstruction of the pre-European vegetation of the Sydney region, modified after Benson and Howell (1990). Please refer to the extended caption at the rear of the book for additional information.

Bay, with a foundational group of over 1,400 convicts, military guards, and administrative personnel on board. The task of establishing the settlement and making it work fell to Arthur Phillip, the Royal Navy officer who led the expedition and who would become the first governor. Within three days of arrival, Phillip decided that despite Sir Joseph Banks' strong support for Botany Bay, it was unsuitable for a permanent settlement and decided to move to Port Jackson (figure 1.1).

Phillip and his staff were the first Europeans to enter Port Jackson and decided to relocate there after exploring this gigantic new harbor for a day. The site of the first permanent European settlement in Australia, and of the future nation's largest city, was moved from the one site where there was some detailed information, to Sydney Cove, a place about which almost nothing was known.

No scientists accompanied the First Fleet, so the records of the natural environment of Sydney Cove are scant. Contemporary sources briefly describe the first camp and the surrounding harbor (e.g., Tench 1789, 1793; Collins 1798, 1802) but systematic scientific observations were not made until many years later (Benson and Howell 1990, 28–29). The lack of scientific investigation is understandable because the priorities of 1788 were very different to Cook's more scientifically focused 1770 stopover (Cook 2005). The new colony was to be permanent and there were over a thousand people to sustain on limited resources. The immediate priorities were to locate safe anchorages, reliable freshwater, arable soil, and building materials (e.g., Collins 1798, 1802; Tench 1789, 1793; Phillip 1789). The result is that we know very little about the pre-European natural environment of Sydney Cove.

Within days of the arrival of the First Fleet, the French ships *Astrolabe* and *Boussole* arrived in Botany Bay, under the command of Jean-François de Galaup, comte de Lapérouse. This was a scientific expedition and stayed on for several weeks after the British fleet moved to Port Jackson. Contact was maintained between the two groups by boat and overland, probably using Aboriginal coastal tracks (figure 1.1). This was the first significant overland travel by Europeans in Australia. Alas, the two French ships sailed on to the north and were subsequently wrecked, with the loss of all hands. If the expedition had returned home safely, the scientific collections made during the extended stay at Botany Bay would now represent a priceless record of pre-European Australia.

Despite the lack of scientific data from 1788, it is still possible to learn much about the 1788 landforms and environments because the mainly military leadership were generally skilled at mapping, charting, and landscape sketching. Charts and plans were included with official reports and correspondence with the home government and there were many private journals and artworks, some of which were published. And the native vegetation was used for building materials and as a food resource. It is from such records that the original landscape of the CBD of modern Sydney can be reconstructed.

GEOGRAPHIC SETTING AND STUDY AREA BOUNDARIES

Sydney was founded on the southern shores of Port Jackson, at latitude 33.52S, longitude 151.12E (Sydney GPO, Google Earth). Like Botany Bay, this deeply incised eastward draining estuarine inlet was formed as one of many river valleys cut into the Triassic sedimentary rocks of the Sydney Basin. They were subsequently drowned during Holocene sea-level rise. The eastern ends of these once major river valleys were inundated, leaving their headwaters occupied by minor rivers and creeks.

The focal point of this discussion is an area of approximately 10 km^2, incorporating the site of the first camp at the head, and along the western and eastern shores, of Sydney Cove. This site is now the core of the modern CBD and includes The Rocks and the Domain-Hyde Park green corridor. The landscape features marking the approximate boundaries of this area are Potts Point and Garden Island to the east, Pyrmont and Darling Harbor to the west, and the north shore of Port Jackson. To the south, the boundary is marked by a ridgeline that forms the drainage divide between Port Jackson and Botany Bay and is now occupied by Oxford Street. The Botany Wetlands lie to the south and were important influences on later settlement growth and land use, and as the source of the first engineered water supply.

Impact rapidly spread outward from the central camp to Farm Cove, Garden Island and Rushcutters Bay to the east, Cockle Bay to the west, Careening and Sirius coves to the north, and the brickfields to the south. The 1780s footprint eventually occupied an approximately 18 km^2 land area that is roughly contiguous with the modern municipal boundary of the City of Sydney. Some reference is also made to later phases of urban development and the evolution of the wider road networks and transport infrastructure.

The time frame of the discussion is the first two decades of settlement, from January 1788 to the appointment of Governor Lachlan Macquarie in 1809. Many of the key events of the development of Sydney occurred during these formative decades. Botany Bay was abandoned, and Sydney

Cove was chosen for permanent settlement; brick clay deposits were discovered, leading to the development of the Brickfields village, and Parramatta was established as the first planned settlement. The second fleet arrived in June–July 1790, bringing the newly raised regiment, the New South Wales Corps (Johnson, in Flannery 2009, 99; Hughes 1988, 105), which was followed by the third fleet in 1791 (Hughes 1988, 105). The colony would also become largely self-sufficient in food production during this time. There were five distinct phases in the urban development of Sydney during this period.

Phase 1 The first camp was formed wherever clear space allowed huts to be built or tents pitched.

Phase 2 The land was cleared, and the camp was laid out more regularly, with the establishment of the administrative divisions of the settlement to the east and west of the Tank Stream. Phillip's urban design concept was recorded by Dawes et al (1789) but never implemented, except in Parramatta (see Karskens 2010, 75–76).

Phase 3 The Brickfields outstation was established.

Phase 4 Parramatta was founded in October 1788.

Phase 5 Meehan's 1807 survey fixed the formal street grid in place (figure 1.2).

Greater Sydney is now the largest city in Australia, with a population of over 5.3 million (Australian Bureau of Statistics 2021). The site of the original eighteenth-century settlement remains the cultural, administrative, and commercial center of the modern state of New South Wales. It is also the location of the most internationally recognized landmarks of modern Australia, including the Sydney Opera House and the Harbor Bridge.

ON THE SHORE OF PORT JACKSON: THE GEOLOGY OF SYDNEY COVE

Sydney Cove occupies a fraction of the southeastern margin of the onshore component of the c. 36,000 km^2 Lower Permian to Middle Triassic Sydney Basin, comprised of mainly terrestrial sandstone, shale and mudstone laid down in fluvial, alluvial and paludal environments (e.g., Mayne et al. 1974; Herbert and Helby 1980). The Sydney Basin extends about 380 km along the east coast of Australia, south of latitude 32S. It is a major structural basin bounded by the New England Fold Belt (east) and Lachlan Fold Belt (west). It forms a part of the larger Sydney-Bowen Basin that reaches into

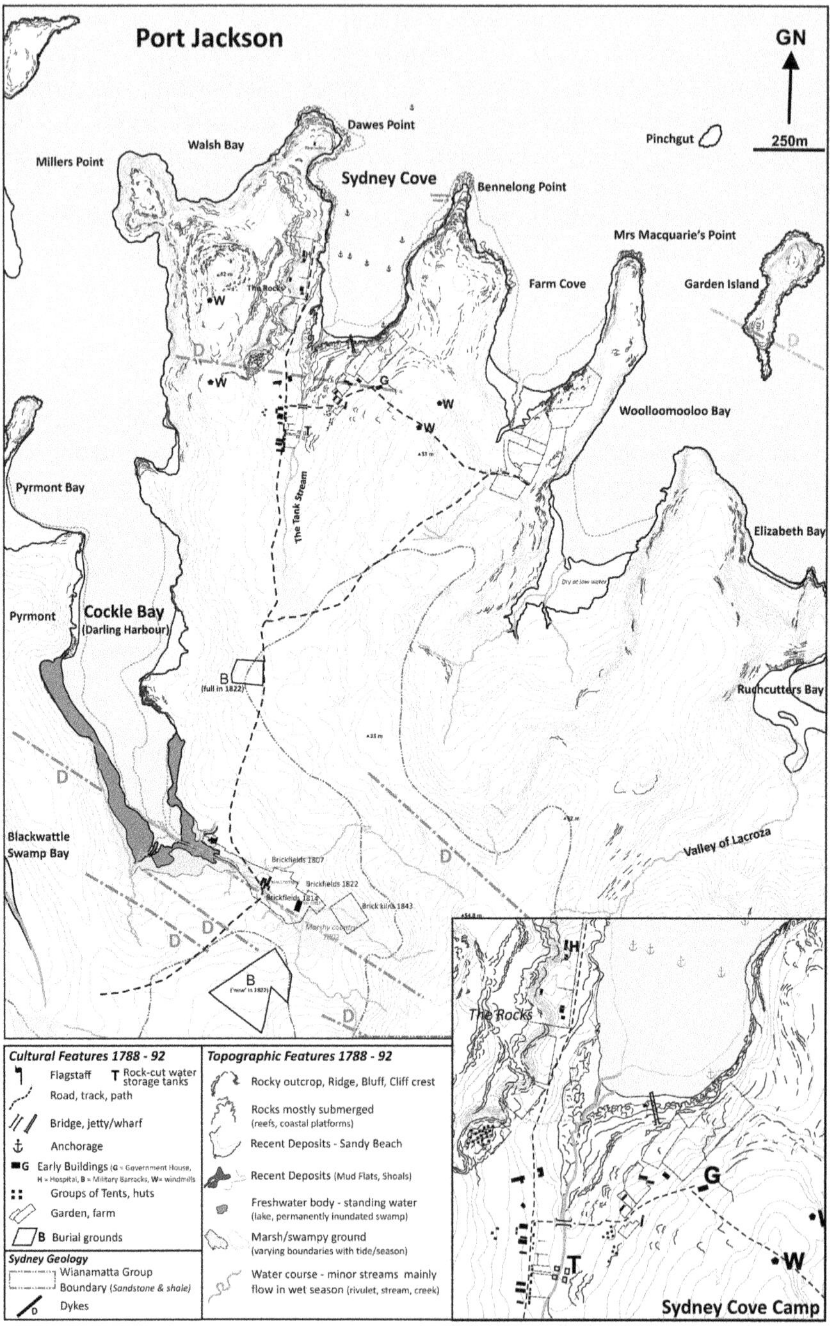

Figure 1.2 Sydney Settlement. Topographic reconstruction of the Sydney Cove landscape circa 1788-1792. The settlement was established on a hand-shaped peninsula with

Figure 1.2 (Continued)
five northern promontories; Millers Point to the west and Potts Point (and its northern continuation of Garden Island) to the east. The camp and the first farms were formed at the heads of the Sydney Cove and Farm Cove inlets. The topographic contours are at 2 m intervals and are derived from the 3D terrain reconstruction model. **Inset.** Detail map showing the location of Hawkesbury Sandstone bluffs and outcrops originally present on the slopes, and at the shoreline of Sydney Cove in 1788. The modern road network preserves the geometry of many of these original outcrops. Please refer to the extended caption at the rear of the book for additional information.

Queensland. The western margin is depositional or erosional and onlaps the rocks of the Lachlan Fold Belt. The southeastern margin is the continental slope and is possibly a rift boundary. The northern boundary is defined as the Mount Coricudgy Anticline—a syn-depositional basement ridge and the northeastern boundary is a zone of major faulting—the mid-Permian to Triassic Hunter Mooki Thrust System, with the rocks of the New England Fold Belt thrust over those of the basin. Extensive Cenozoic warping has emphasized the outline (Herbert and Helby 1980, 5).

Hawkesbury Sandstone

Approximately 4,500 m of Triassic strata underlie the landscape of Sydney Cove and the surrounding bays and headlands (figure 1.1 inset). It is largely comprised of upper Hawkesbury Sandstone; a mainly massive sheet facies and massively cross-bedded sandstone with some shale lenses (Herbert 1980). The Hawkesbury Sandstone was deposited during the breakup of Pangaea (250–160 Ma), as Gondwana separated from Laurasia (Blewett 2012, 108). The higher ground in the southern part of the study area is formed by a remnant of the overlying basal Wianamatta Group, represented by Ashfield Shale and thin occurrences of Mittagong Formation (Chesnut 1983).

Sydney Cove and the first settlement site are mostly underlain by bedded Hawkesbury Sandstone with sheet, massive and mudstone (<5%) facies that reaches a maximum thickness of 275 m (Chesnut 1976, 3). It is typically composed of detrital quartz grains (50–75%), lithic fragments (24%), clay matrix (mainly kaolinite and illite), secondary quartz (4–10%), and siderite (Pells 2002). It is dominated by massive or coarsely cross-bedded sandstone strata but contains approximately 5% of interlayered shale lenses (Branagan 2000a). Ripple marks in the sandstone always dip north, suggesting a southern source (Flannery 1999, 8–9).

Extensive exposures of Hawkesbury Sandstone occur throughout the northern part of the study area, particularly on the coastline and in road cuttings. In most excavated exposures the sandstone is a massively bedded, medium-grained, and generally cross-bedded sandstone. Bedding ranges in thickness from around 1 m to >2.5 m. Most massive beds are internally layered by cross-bedding (see figure 8.1, 1E).

Weak regional deformation is manifested as a gentle warping and tilting of the almost horizontal strata, resulting in generally south or southwest dipping bedding ranging from 0–2 degrees and locally up to 10 degrees. Easterly dips are known. There is a regional southwest dip of 10–20 m per km (Chesnut 1976). Shale layers acted as slippage planes during deformation (localizing strain) so can be sheared (e.g., Chesnut 1979).

The sandstone is variably weathered (moderate to strong), with regions of pale-white friable low-strength material surrounding lenses and blocks of hard, intact rock. Weathering is focused on approximately horizontal bedding planes and other discontinuities and the regions of friable material tend to be preferentially eroded. The irregular, pock-marked texture of some rock faces suggests a significant but variable carbonate content in the sandstone matrix (siderite?). Dissolution of matrix carbonate during weathering may have significantly weakened the rock mass.

Mafic Dikes

Late Triassic to Cenozoic dolerite and basalt dikes are widespread throughout the Sydney region, particularly on the eastern edge of the Sydney Basin. They form a dominantly west-northwest to east-southeast trending swarm that is closely related to jointing (e.g., Pittman 1903, Willan 1925; Branagan 2000a, b; Och et al. 2004; Och et al. 2009). Dikes average around 1 m wide, rarely exceed 3 m but occasionally reach up to 6 m (Herbert 1980, 103). Three significant dikes transect the study area (Pittman 1903; Branagan 2000b; Och et al. 2004): one near the modern Cahill Expressway; the Oxford St Dike and the Great Sydney Dike (Och et. al. 2004; Pittman's 1903 dike LVI), just to the north of Central Station (figure 1.2).

Though relatively rare near Sydney Cove, dike occurrences are important to modern civil construction projects because they can be weathered to clay to depths exceeding 50 m, and they form planes of geotechnical weakness in excavations (e.g., Branagan 2000b; Work Cover Authority 2006). The near-vertical orientation makes them difficult to detect during modern site assessment drilling (Chesnut 1978; Och et al. 2004).

The influence of the dikes on the 1788 landscape was significant, particularly in the south, around the brickfields area and the head of Cockle Bay (modern Darling Harbor). Dikes are highly susceptible to weathering and their association with jointing created linear topographic depressions (Branagan 2000b), such as the example coincident with the surface trace of the Great Sydney Dike, one of the most persistent intrusives. The topographic low associated with this dike hosted the northwest draining creeks and wetlands at the head of Cockle Bay and was the site of the first brickfields. The topographic depression coincident with this dike also influenced the narrowing of the southern end of the Pyrmont Peninsula (figure 1.2).

Faults and Joint Swarms

Large-scale displacement faulting is uncommon in central Sydney, though normal and reversed faults with displacements up to 10 m are known. More typically, faulting and shearing occurs as zones of closely spaced joints associated with breccia, clay pug, and slickensided faces up to 10 m wide (Chesnut 1979).

Several near-vertical, north- to northeast-trending brittle faults and joint swarms traverse Sydney Cove (figure 1.1 inset). The Woolloomooloo and Luna Park Faults are the most significant (Och et al. 2009), but smaller structures, such as the Martin Place Joint Swarm and the GPO Fault, are present between the main faults and are important at local scale (Chesnut 1976; Pells et al. 2004; Och et al. 2004, 2009). The timing of fault formation is constrained to between 120 and 50 Ma by K-Ar ages obtained from fault gouge (a thermal event at c.120 Ma) and K-Ar ages of 51 +/– 1.1 Ma obtained from dikes intruding the north-northeast fracture systems (Och et al. 2009).

Published information about the geometry of the faults and joint swarms is limited and existing interpretations are simplified. During this study, historic topographic information showing lost or obscured outcrops and now obscured topographic lineaments was combined with modern imagery of coastal outcrops (Google Earth and Nearmap) to provide additional information about the geometry of brittle structures that traverse Sydney Cove. The work suggests that the traces of such structures are more sinuous than depicted in published mapping representations and that there is greater geometric complexity in faults and joint swarms (Pells et al. 2004; Och et al. 2004, 2009).

Quaternary and Recent Sediments

Quaternary and recent sediments mantle the basement rocks and form beaches, mud and mangrove flats, salt marshes and wetlands, particularly on the margins of Botany Bay (figure 1.1). There are only small occurrences of Quaternary sediments mantling the Permo-Triassic basement rocks in the inlets, coves, and bays of Port Jackson (figure 1.2). Sydney Cove lies at the point in the harbor where the style of coastal sedimentation in these embayments changes in character from sand-dominated in the east, to mud-dominated in the west. This observation was first recorded by William Bradley (1802b) after completing his survey of the harbor (with John Hunter) on February 9, 1788. Bradley (1802a) also illustrated the clear sandy beaches of Sydney Cove in his contemporary watercolor. Archaeological excavations at the former Gateway Plaza site exposed a section of the original Sydney Cove shoreline, which were found to be sandy, and was interpreted to have been a clean sandy beach. No evidence for mudflats or mangroves was found (Higginbotham 1986).

Inlets to the east are sandy and shelving at their heads while generally steep-sided and rocky. Those to the west, toward the head of Port Jackson and the Parramatta River, are dominated by broad mudflats and mangroves (e.g., Darling Harbor). Sydney Cove was the only inlet on the southern shore of Port Jackson that provided the near-shore, deep water anchorages that Phillip wanted. Those to the west, where shale formed much of the bedrock, were too shallow for anchorages and were occupied by mudflats. For the immediate needs of the new settlement, Sydney Cove seemed to be well-chosen.

Recent dune sands and marine sediments dominate the landscape beyond the drainage divide to the south of Sydney Cove and underlie most of the land bordering the north shore of Botany Bay (figure 1.1). They are the dominant geology underlying the industrial areas, airport, and Port Botany infrastructure. Extensive wetlands occupy these younger deposits and are known as the Botany and Lachlan swamps.

Quaternary and Recent Sediments and Early Farming

As mentioned, Quaternary and recent sediments are less common on the shores of Port Jackson and mostly occur as small beaches and mudflats at the heads of inlets and coves (figure 1.2). At these locations, recent sediments often hosted wetland systems formed where creeks and streams entered the harbor. The small expanses of relatively moist and fertile soils formed on these sediments provided the sites for the first European-style farming in Australia—at the heads of Farm and Sydney Coves. While largely unsuccessful, these patches of good agricultural land were one of the few local contributors to food production at Sydney Cove in the first two years of the colony. Reservation of Farm Cove as part of the Government Domain led to its preservation as the modern Botanic Gardens.

THE SYDNEY COVE LANDSCAPE IN 1788

Sydney Cove occupies a central position in a landscape of high sandstone bluffs and ridges, headlands, and inlets formed by the sandstone strata, which together form a multi-promontoried, hand-shaped peninsula (figure 1.2). The land reaches an elevation of over 60 m on the western shore of the cove and 154 m to the south and southwest. Ashfield Shale caps the high ground of this elevated ridge, which creates a drainage divide. It separates the deeply incised, northward drainage into Port Jackson from the southward flowing water courses that feed the low-lying and waterlogged landforms of the onshore Botany Basin (Griffin 1963).

The interactions between the shallowly dipping sandstone strata, and northeast-trending faults, such as the Luna Park Fault Zone (Och et al. 2009), produced many inlets and headlands forming the hand-shaped peninsula on which Sydney developed. From west to east the finger-like peninsulas include Pyrmont, Millers (Barangaroo), Dawes, Bennelong (including the Domain), Mrs. Macquarie's Point, and Potts Point—Garden Island to the east. Each peninsula has a central north-south trending linear ridge or "spine" of high ground between which lie coves and inlets.

Influence of Strata on Landforms

The landforms of Sydney Cove are the product of the differential weathering and erosion of small-scale and localized variations in the outcropping Hawkesbury Sandstone. Small differences, such as the dip of massive sandstone units (including bedding planes and surfaces); the locations of intercalated shale layers within the sandstone sequence; and the interaction of the strata with jointing, faults, and dikes, could produce influential landforms.

Massive sandstone units were particularly important influences on the details of Sydney Cove topography because of the resistance of some layers to weathering and erosion. This strongly influenced landforms that were important at local scales for infrastructure placement. Important early examples include the following.

- The topography of the floor of the cove created deep water close to shore where large ships could moor, and jetties could be economically constructed (Phillip 1789).

- Sandstone shore platforms on the southeastern shore of the cove, which were often bedding surfaces, provided a convenient landing place, and led to the placement of the first flagstaff, a location that is still marked near the former Customs House. The first jetty at this site (and another on the western shore) probably exploited existing sandstone outcrop bedding surfaces.

- The water-holding tanks formed on the Tank Stream in 1792 to increase water storage capacity were cut into sandstone bedding surfaces on the banks (Collins 1802, unpaginated)

- The first bridge across the Tank Stream was constructed at a narrow point between the sandstone outcrops forming the banks.

The orientation of resistant massive sandstone units was also important to the topography of Bennelong Point and Dawes-Millers Points, the two headlands bordering Sydney Cove. The influence of the favored shallow east to south-east dip of strata results in both peninsulas having strongly asymmetric topographic profiles, with more gently sloping eastern sides that are more strongly influenced by the inclined bedding plane surfaces. The ridgelines forming the spines of these headlands slope more gently on their eastern sides and are steeper and more prominently bluffed on their western sides. Gentle east-dipping bedding can be observed on Bennelong Point (Tarpeian Wall) and Millers Point (Hungry Mile) (see figure 8.1, 1C and 1D). This asymmetry is the result of the undulating dip of strata in the Hawkesbury Sandstone. The eastern ridge (the "spine" of modern Sydney) was chosen as the site for the main colonial hospital and other administrative buildings (now the State Parliament), and then Hyde Park and the ANZAC Memorial.

The broad, gentle sandstone slopes of the upper western shore of Sydney Cove may have been important to the choice of this area for the first tent encampments in 1788, because they were flatter, and the thinner vegetation created natural clearings (refer to Dawes et al. 1789; figure 1.2 inset). The gently sloping ground of this shore may also have influenced its choice as the main landing site for offloading heavy stores. The eastern shore of Bennelong Point became the site of the first farming at the head of Farm Cove. The steeper topography of the western sides of both points was more strongly bluffed and less amenable to settlement.

Northeast-southwest trending, near-vertical faults and joint swarms are more susceptible to weathering and erosion. Selective erosion of such structures produced low, linear (often deeply incised) valleys and drainage lines, and prominent finger-like headlands, of which the Tank Stream Valley and Sydney Cove are the most significant. The preferential weathering and erosion of dikes was also important. The southeast-northwest trending late Triassic to Cenozoic dikes are readily weathered to clay. The locations where they traverse the sandstone have also become topographic lows that are now the sites for minor northwest-trending valleys and inlets, such as the example already mentioned at the brickfields (see figure 1.2).

The drainage system formed in the fracture system of the Hawkesbury Sandstone was invaded by the sea during the postglacial marine transgression between 20,000 and 6,000–7,000 years BP, producing the harbor and inlets (Harvey and Caton 2010, 83). Most of the important landscape features of Port Jackson, the middle and outer harbors, and Botany Bay were formed at this time.

Wetlands and the Drainage System

Small freshwater wetlands backing saltmarshes were present at the heads of the inlets along the southern shore of Port Jackson, including Farm Cove, Cockle

Bay, and Woolloomooloo Bay. It was uncommon for the small streams draining the higher ground to the south to flow directly into the harbor (as the Tank Stream did). They usually fed into a freshwater wetland behind a saltmarsh flat and then exited into the harbor from an outlet of the saltmarsh.

Wetlands were places where important building materials could be sourced (such as casuarina for shingles and cabbage palms for easily worked structural timbers. They were also sources of thatching (rushes)). The moister and more fertile soils of the small wetlands at the heads of Sydney Cove and Farm Cove became the locations for the first farming of the new settlement.

The Tank Stream and Marsh

The most historically significant wetland in early Sydney was the small marshy area that formed the head of the Tank Stream. The reliable freshwater source for the new settlement was a rivulet rising from springs at the contact between the Hawkesbury Sandstone and the overlying Wianamatta Group (Ashfield Shale), expressed as a broad marshy area at the head of the Tank Stream Valley (e.g., Collins 1798; Karskens 2010). The rivulet flowed northward through a well-defined valley that follows the approximate trace of the "GPO Fault Zone" (Och et al. 2004) and was also fed by a series of small runnels that drained the adjacent sandstone ridges (figure 1.2). The stream acquired its name after water storage cisterns were cut into the sandstone banks to hold fresh water for the settlement. The course of the stream divided the military (western) from the administrative (eastern) sector of the first camp (Phillip 1789). Government institutions still dominate the eastern side of the modern Sydney CBD.

In 1788, George Worgan (unpaginated) noted that the other freshwater courses in the settlement also arose from springs associated with wetlands. Several small rivulets that drained into the head of "Cockle Bay" (modern Darling Harbor), which became important for brickmaking, were also probably fed by springs rising along this shale boundary.

The Botany and Lachlan Swamps

Wetlands that became known as the Lachlan and Botany swamps occupied most of the area south of the drainage divide now occupied by modern Oxford Street and extended south to the shores of Botany Bay. The broad expanses of wetlands and marshes on the northern shore of Botany Bay impressed James Cook and Joseph Banks in 1770, and they suggested that it could be easily and rapidly turned over to agriculture (Cook 2003, 125–126). However, it was these same wetlands that were at least partly responsible for deterring Arthur Phillip from establishing his first settlement there and moving it to Sydney Cove.

The Botany Wetlands and Lachlan Swamps played a role in some of the significant historical events of the early colony. On Saturday, February 9,

1788, George Worgan (1788, unpaginated) recorded that a party traveled overland to visit the French ships still moored in Botany Bay. They encountered swampy and rocky ground and found the going hard. On arrival, they were treated well by La Perouse and his officers but discovered that the terrain had not prevented several convicts from already finding their own way back to Botany Bay, probably via Aboriginal tracks, in an attempt to escape with the French (Phillip 1789; Karskens 2010).

In 1789, a punitive expedition sent against the Botany Bay Aboriginals was reluctantly led across this country by David Collins and Watkin Tench, to seek revenge for the murder of Arthur Phillip's game keeper. The Botany Bay mudflats and swamps impeded their progress, and they found the going difficult, becoming mired in mud while trying to cross the creeks and mudflats (probably the Cook River). No one was captured (Tench 1789). In this instance, the wetland and salt marsh landscape probably prevented an early escalation in hostilities.

1788 Vegetation and Environment

There were no botanical collections made at Sydney Cove in 1788 (Benson and Howell 1990; Macphail 1999; Karskens 2010), but several descriptions and artworks provide information about the density, undergrowth style, height and girth of trees, and naturally clear areas (e.g., Tench 1789; Collins 1798). This evidence shows that on January 26, 1788, Sydney Cove was densely timbered, with close-spaced trees that were only occasionally interrupted by clear places or bare rock outcrops. There was also an understory of flowering shrubs, "weeds," "flowers," or underwood. The first tents and huts had to be built wherever naturally clear places existed. So, the first camp was very "irregular" (Phillip 1789, June–July 1788). The spacing of the trees was less dense at the cove than elsewhere in the harbor but they were very large (Phillip 1789, 26 January).

More detailed information about the pre-European vegetation and environment of Sydney Cove comes from several sources, including contemporary charts and maps, general written descriptions, and modern archaeological investigations. Reconstructions of the vegetation based on modern knowledge of the soil and topographic preferences of species have been published by Benson and Howell (1990, 42–46) (figure 1.1 inset). They suggest that the higher ground, which is underlain by Ashfield Shale, was probably occupied by Turpentine-Ironbark Forest comprised of Turpentine (*Syncarpia glomulifera*) and Grey Ironbark (*Eucalyptus paniculata*), while the lower sandstone slopes were probably inhabited by a woodland of Scribbly Gums (*Eucalyptus racemosa*), reaching a height of 10 m, with an understory of *Leptospermum*, *Banksia* and *Callistemon*. Blackbutt (*Eucalyptus pilularis*); Red Bloodwood

(*Eucalyptus gummifera*), Peppermint (*Eucalyptus piperita*), and smooth-barked *Angophora costata* occupied the sandstone on the slopes. Large Forest Red Gums (*Eucalyptus tereticornis*) with a ground layer of ferns probably followed the line of the Tank Stream Valley (Benson and Howell 1990, 42). The trees were very large and difficult to fell, and to remove. Some of the trees on the sandstone ledges were of such a size that Arthur Phillip found it hard to understand how they managed to find enough nourishment to reach such dimensions (Phillip 1789, February 7, 1788).

Small seasonal rivulets drained the sandstone bluffs and ridges into Sydney Cove and were environmentally significant at local scales. Some may have influenced the locations of the first camp gardens. A small eastward draining creek, which became known as Hospital Creek, flowed from the sandstone ledges and ridges of the western shore of the cove. Plant microfossils were recovered from probable pre-European soil deposits at 200 George Street, near the original course of the rivulet. They suggest that there were deeper, well-watered organic soils creating a microhabitat that supported a casuarina swamp forest and ground ferns in the lower reaches, near the Tank Stream estuary (Owen and MacPhail 2018, 19; 23).

No contemporary records show that Phillip considered the better-quality timber of the Turpentine-Ironbark Forest as one of his reasons for choosing Sydney Cove, but this vegetation type was heavily exploited for timber in the colonial period and is now endangered (Department of Environment and Climate Change NSW 2008). It would have been a significant source of timber in 1788 (though considered too poor for construction). Subsequently, the timber was found to be resistant to marine organisms and rot and it became an important timber for constructing wharf pylons, bridges, railway sleepers, and mine supports.

Swamp oak (*Casuarina glauca*) was an important source of roofing shingles from the first days of the camp (Worgan 1788, 11). Shingling parties were noted at work in the forest to the south of Sydney Cove by Fowkes (1789), in an area that Benson and Howell (1990, 42) interpret to have been occupied by swamp forest in 1788.

Cabbage palms (*Livistona australis*), which reached a height of over 20 m (70–80 feet), were abundant around Sydney and probably occurred along the marshy areas of the Tank Stream Valley. They were also sourced from down-harbor to the east. This soft and easily worked timber was an important early building material and used to frame the first wattle and daub huts. The top of the plant was an important food source. It was preferred as timber because it was easy to work with the poor-quality axes available to the settlers (Worgan 1788, 11).

Rushes were important as roof thatching and were gathered from marshes and swamps, such as those at the heads of Rushcutters Bay and Cockle

Bay (modern Darling Harbor). One of the first violent interactions between Europeans and Aboriginals occurred between a party of rush cutters and Aboriginals in Cockle Bay in May 1788 (Karskens 2010, 361), though recent evaluation of a contemporary chart of Port Jackson suggests that the attack may have occurred at Iron Cove, some way further up the harbor (Gapps 2019).

Mangroves

Mangroves were not present in Sydney Cove, Farm Cove, or Woolloomooloo Bay in 1788. They were not mentioned by Arthur Phillip (Benson and Howell 1990, 42) or recorded in contemporary charts and artworks (e.g., Bradley 1788). Only traces of samphire pollen and possible mangrove pollen were identified in archaeological deposits by Owen and MacPhail (2018, 24), supporting the information available from the historic sources. Mangroves were probably discouraged by the sandy intertidal zone, the steep rocky shorelines that fell away rapidly into deep water, and the stronger marine influence. There does not seem to have been any significant salt marsh either. The Tank Stream flowed directly into the cove at a sandy beach bordered by rocky shelves. This clear and firm shoreline may be one of the reasons why Arthur Phillip selected Sydney Cove for settlement.

Mangroves were common up-harbor to the west and Sydney Cove lay at the point where coastal sedimentation in the coves and bays of Port Jackson changed from sand-dominated to mud-dominated. William Bradley noted this on February 9, 1788, when he wrote that "those Coves above where the ships lay were surrounded by Mangroves & had mud flats at the bottom, those below had sandy beaches most of them" (Bradley 1802b). Coastal rocks and mangroves on the shores of Port Jackson were thickly covered in oysters, which became an important early fresh food source, and Bowes Smyth (1790, 92) recorded that he often brought back mangrove branches to the ship that were loaded with oysters.

Mangroves were widespread and well established elsewhere in the estuarine reaches of the river systems entering Port Jackson and Botany Bay in 1788 (Cook 1770; Tench 1789; Bowes Smyth 1790; Bradley 1802b). Lieutenant Watkin Tench, a Marines officer, was involved in many parties that explored the region around Sydney Cove and developed a detailed knowledge of the landscape. He considered it "strikingly singular" that three "such noble harbors" as Botany Bay, Port Jackson, and Broken Bay all ended in "shallows and swamps filled with mangroves" fed by only minor rivers such as the Parramatta River at the head of Port Jackson, and the Georges and Cooks rivers at Botany Bay (Tench 1998 / 1789, 53).

Mangroves seem to have been particularly common on the northwestern and southwestern shores of Botany Bay. On May 6, 1770, James Cook described the vegetation lining the shores of Botany Bay, which he considered rather monotonous, but specifically noted that there were "mangroves about the head of the harbor" (Cook 2003, 129). Cook made no detailed chart showing the location of the mangroves he mentions. However, from the locations of his soundings of the bay, which mark the track of his boat during the 1770 survey (Cook 1893 chart), it seems that he only explored the Georges River as far upstream as Kogarah Bay and only sounded the outer margin of the shoals at the outlet of the Cooks River (figure 1.1). Given that Cook hardly explored the Cooks River, which he considered a creek, it seems likely that Cook's "head of the harbor" was the estuary of the Georges River and that mangroves were present at its mouth in 1770. In 1788, John Hunter spent ten days surveying Botany Bay, exploring the Georges River as far upstream as possible. He then only briefly explored the Cooks River, because like Cook, he decided it was not worth the time or effort to survey, recording it as a sketch quality chart only (subsequently surveyed in more detail by Lieutenant Bradley) (Hunter 2003 / 1793, 134). It seems likely that any mangroves seen were ignored during the 1770 and 1788 surveys because more detailed work was not considered worth the time and effort. Subsequent cadastral surveys of Point Sutherland dating to the 1830s (e.g., NSW Land Registry Services c 1830s; 1868–1874; Irish and Ingrey 2013, their Figure 1.2) record mangrove distributions along the southern shores of the bay similar to the present, and plans of the mouth of the Cooks River show that mangroves were well established on Sheas Creek and surrounding the estuary of Cooks River in the 1850s and 1880s (e.g., Willis 1859; Higinbotham and Robinson 1888). The historic records support the pre-European mangrove distribution reconstructed by Benson and Howell (1990).

It has been suggested that the extent of mangroves in Botany Bay and Port Jackson has greatly increased since 1788, and particularly in the twentieth century, at the expense of salt marshes, which are invaded on the landward side on the Georges River (Saintilan & Williams 1999; Haworth 2010). This expansion is attributed by these authors to human-induced global warming, sea-level rise, sedimentation, and the storm-water flushing of estuaries.

While the relative proportions of saltmarsh and mangrove may have changed, the seaward extent of surviving stands of mangroves seems to have remained relatively constant in Botany Bay since European settlement. Only at the western end of Towra Point, to Killigalere Point, has there been a significant seaward expansion of mangroves, which have colonized sandbanks and shoals accumulating around former oyster farms frames, and have enveloped Carters Island.

The mangrove stands at the mouth of Cooks River have largely been destroyed by airport development, but surviving maps suggest that their extent remained constant throughout that nineteenth century. This also seems to be the case for the surviving mangrove stands in the coves on the southern shore of Botany Bay.

The Aboriginal Landscape

The people inhabiting the southern shores of Port Jackson in 1788 knew themselves as the Gadigal, one of twenty-nine clans within the broader language and cultural grouping now known as the Eora. The Eora were a mainly coastal people who harvested shellfish and fished from small, skillfully built and handled watercraft. Inland to the west (modern Parramatta), far beyond the head of the harbor, were the Dharug, to the south of Botany Bay were the Dharawal, and the Guringai probably lived along the coast to the north, beyond Port Jackson. The Eora spoke a variety of closely related dialects around the coastal areas, but there was greater language diversity inland. When Colbee and Balladerry accompanied Arthur Phillip's exploratory journey to the Hawkesbury and Nepean rivers in 1791, they could understand the language of the inland people, but it was a distinct dialect with many different words (e.g., Tench 1793, 84; Attenbrow 2010, 31; Karskens 2010, 42). Such contemporary observations suggest that a significant linguistic change occurred beyond the head of the harbor and toward the Hawkesbury and Nepean rivers further west. This change may possibly reflect how the type of natural resources used by the people changed from marine and estuarine to riverine and terrestrial and how different groups had become specialized in their exploitation.

The people of the coastal regions were also bound by common material cultures and technology, initiation rites (e.g., male right front incisor avulsion, which the inland people did not do), body adornments, hairstyles, and ceremonial practices. Movement of groups was known from north to south, but there is no evidence for the east-west movement of peoples (Heiss and Gibson 2009).

Aboriginal occupancy of the Sydney region extends back at least 6,000 years, into periods preceding the modern climatic conditions. Evidence for this includes the butchering marks identified on a dugong skeleton excavated in the Alexandra Canal in the late nineteenth century (Etheridge, Edgeworth David and Grimshaw 1896, 170–173). Dugongs are only rarely observed as far south as Sydney but may have once been more common (Haworth 2004, 52; Attenbrow 2010, 66).

Aboriginal Landscape Feature Names

The Aboriginal names for many key landscape features and places were recorded for the southern shore of Port Jackson. Lieutenant William Dawes formed a close friendship with Patyegarang, a young Aboriginal woman, and through her learnt and recorded the first written record of an indigenous Australian language (the Gadigal language). It may also be through this friendship that the Aboriginal names for many of the landscape features around Sydney Cove came to be known and documented. Several place names were recorded as annotations during the drafting of maps and charts (probably by William Dawes), while others were included in local language word lists. Several Gadigal landscape feature names were also published in the works of Phillip (1789) and Collins (1798). This information has been collated by Attenbrow (2009, her Table 1.1).

Aboriginal clan groups often derived their names from the places they mainly lived. For example, the Cemeeragal were the people who lived in the bay of Cameera while the Cedigal (Gadigal) lived around the bay of Cad-I, while Europeans were identified as Bereewolgal, the people who came from far away (Tench 1793, 134).

Some Aboriginal place names never fell out of use. The small bay to the east of Sydney Cove known as Woolloomooloo, seems to have never lost its original Gadigal name. It was adopted as the property name by one of the first European occupants and is now the name of the modern suburb.

Campsites

There is widespread archeological evidence of Aboriginal occupation throughout the study area, which is remarkable given that for over two centuries, it has been one of the most intensely developed and urbanized places of modern Australia. Sites predating European settlement have been discovered at the original land surface, or beneath the earliest European occupation deposits during archaeological investigations undertaken for redevelopment (e.g., Attenbrow 2009). Many of the sites represent long- and short-term camping or food preparation sites containing shellfish and fish remains, charcoal, and stone tools of a variety of types and materials.

Watkin Tench (1789, 38) noted that most campsites were simple features, with shelters built from wood and bark. However, he felt that the most important living places were not these small open sites but the many caverns and rock shelters around Port Jackson. Such shelters may once have been present on the shores of Sydney Cove (particularly on the eastern shore).

Aboriginal Use of Geological Materials

The Hawkesbury Sandstone and other strata exposed around Sydney Cove are not good sources of the fine-grained lithologies that are useful for manufacturing sharp-edged tools or ground-edged axes. Consequently, flaked tools are rare in sites around the harbor (Lampert 1980, 2). Raw materials for such tools had to be transported in or obtained through trade. Luckily, much of the staple food supply of local coastal people came from the sea and shoreline, so fishing equipment and canoes were the key technologies.

Stone artifacts recovered from several large and small archaeological camp/midden sites around Sydney Cove include flakes and reworked tools that were mostly manufactured from river pebbles and cobbles that were probably transported in from some distance away (probably sourced from water courses and coastal areas). At the Moore's Wharf midden campsite, flaked artifacts comprised small flakes of silcrete, quartz, quartzite and chert. Cores found with the flakes showed that most had been derived from small pebbles that may have originated within the Hawkesbury Sandstone (conglomeritic units?). At the Angel Place site, raw materials included silicified tuff, indurated mudstone, silcrete, and quartz (McDonald 2004, 15–16). In addition, several ground-edged hatchets manufactured from hard and dense igneous or metamorphic rocks have also been discovered and may also have originated as pebbles or cobbles (Attenbrow 2010, 100).

Rock Engravings and Paintings

Large-scale engraved petroglyphs are widespread around Port Jackson, located mainly on the sandstone ledges and natural pavements of the higher ground (e.g., Stanbury et al. 1990) with over 2,000 sites recorded (Attenbrow and Kononenko 2017, 161). They were created by abrading, cutting, hitting, or drilling into the surface of the sandstone with stone, shell or wooden implements, usually by drilling initial holes and then linking them. Designs are varied but usually comprise outlines of animals, weapons, or people. The sites chosen for placing engravings were often on high points with panoramic views, or locations with special environmental significance (Stanbury et al. 1990, 1–2). The purpose of the rock engravings is uncertain, but they may have been associated with male initiations. District-scale patterns of stylistic differences may reflect group boundaries (Attenbrow 2010, 149). Others may have indicated territorial boundaries or teaching places (Irish and Goward 2019).

There are two locations within, or adjacent to the study area where petroglyphs once occurred. One was described on a sandstone outcrop to the south of the current Sydney Cricket Ground (Mann 1883, in Attenbrow 2002, 11) but has since been lost. It was a flat sandstone outcrop near the Association Ground, Sydney Common, and was covered with the representations of kangaroo, possum, fish, and boomerangs (Attenbrow 2002, 11). Another was reported to have occurred on flat sandstone surfaces at the tip of Dawes Point (Irish and Goward 2019).

LANDSCAPE, GEOLOGY, AND THE EARLY DEVELOPMENT OF SYDNEY

When the First Fleet arrived at Botany Bay in January 1788, Arthur Phillip and his officers immediately explored the area to find the best site for a large, permanent settlement. Within three days they had decided that the bay was unsuitable. They recognized that any sheltered anchorages were shallow, and far from the shore, while deepwater moorings, such as off the southern shore, would be directly exposed to ocean swells. The shore was mostly comprised of sand, low-lying wetlands, salt marsh, mangroves, and mudflats, which offered little potential for significant agriculture and might prove unhealthy. And there was no freshwater source that could support a large population (Phillip 1789).

Phillip decided to abandon Botany Bay for mainly geological reasons. The subdued topography of the bay is mostly underlain by the Quaternary and recent sediments of the Botany Basin, which overlie a paleo-topography formed on the Permo-Triassic basement (Griffin 1963). The bay is also very shallow, shoaling rapidly to the west where it is entered by the Cooks and Georges rivers, and their recently deposited estuarine sand flats and mudflats. The result is a largely flat and relatively featureless western and southern shoreline. The low-lying sediments extend inland from the shores of the bay, where they form extensive freshwater aquifers and wetland systems, particularly on the northern and southern shores. In 1788 they were host to successive rings of mangroves, salt marshes and extensive freshwater wetlands (figure 1.1), including what were later known as the Lachlan and Botany Swamps (Benson and Howell 1990). This diverse natural landscape did not provide any places for a significant European settlement because the mangroves, salt marshes and wetlands prevented agriculture, and the shallow bay did not provide secure, deepwater anchorages. Arthur Phillip chose to move to Port Jackson and Sydney Cove because there was a reliable, spring-fed source of fresh water, it provided a huge and well-sheltered harbor, there were deepwater anchorages close to shore, and there was (apparent) good soil and plentiful timber (Phillip 1789).

The early settlement occupied less than 18 km^2 of the 36,000 km^2 onshore extent of the Sydney Basin, and it was underlain by only a narrow stratigraphic interval of the Hawkesbury Sandstone and Ashfield Shale. Sydney Cove is located to the immediate north of the stratigraphic boundary between the Hawkesbury Sandstone and the eastern-most outcrop extent of the overlying Wianamatta Group. This boundary was an important focus of springs, such as those that fed the Tank Stream. This spring-fed freshwater rivulet arising at the contact of the Hawkesbury Sandstone and Ashfield Shale was the primary reason for the choice of Sydney Cove as the settlement site. The mangrove-free shoreline and deepwater anchorages were also important.

Although the boundary between shale and sandstone was important as a source of water, sandstone predominated around the shores of the cove. It was the fine details and character of this rock that most defined how Sydney developed in the first decades after 1788. Seemingly insignificant variations in the local geology could be influential, including the dip of strata, how massive those strata were, or whether good agricultural soils formed. Almost every decision that was made by the leadership of the new colony was influenced by the geology of the chosen site and the character of the Hawkesbury Sandstone.

Small-scale and localized variations in the thickness, composition and orientation of sedimentary strata, their interaction with relatively minor faults and joint swarms, and their variable resistance to erosion produced small topographic variations that were of the greatest importance. In other circumstances, such minor features might have been considered insignificant, but to the mostly preindustrial European culture that arrived in Sydney in 1788 they were of great consequence. The dependence of that transplanted society on manual labor, without access to working animals, amplified the effects of the geology and landscape on behavior, and on the patterns of their early exploitation. The landscape could not be modified, so the settlers had to fit into it. The result is that the geological foundations of Sydney Cove set in place the street grid of the modern Sydney CBD, which still reflects the original bluffed outcrop pattern of massive sandstone units of the Hawkesbury Sandstone. This is particularly well preserved in The Rocks (see figure 1.2 inset).

Landforms dictated the layout of the first camp and its administrative divisions. High ground was preferred for buildings of government authority, defense (gun batteries) or science (astronomical observatories), while expanses of naturally clear and flat ground were chosen for camping areas. Early road alignments followed the outcrop pattern of sandstone units and the north-south trend of the Tank Stream Valley formed subparallel to faulting.

The location of the first bridge to link the east and west administrative divisions of the settlement was chosen because it was the narrowest point between the sandstone outcrops forming the banks of the Tank Stream. These early tracks became locked into the fabric of the settlement and fixed in position as adjoining land tenure was formalized and the land usages from the formative years became regularized.

Wianamatta Group Strata and Agricultural Soils

Early farming at Sydney was not very successful, and it was quickly realized that the sandstone soils were too poor for agriculture. Quaternary alluvial sediments at the heads of Farm Cove and Sydney Cove provided the first arable land, but such areas of richer soils were too restricted for significant farming. Searches for better farmland led to the recognition of higher-quality soils to the west, formed on Wianamatta Shale. The identification of good soils led to the foundation of Parramatta in November 1788 (Kass et al. 1996), which shifted the center of agriculture and urban expansion westward. Phillip hoped that Parramatta would develop into the model settlement of the colony, while leaving the poorly planned Sydney as the port (Karskens 2010).

Sydney Cove lies at the eastern extremity of the outcropping Wianamatta Group (figure 1.1 inset), represented in the area by the basal Ashfield Shale (Mittagong Formation). While originally ranging in thickness from 5 to 15 m in the city area, the Ashfield Shale has been largely removed by development and is now mainly observed in drill core and excavations (Chesnut 1979, 184). The most significant undisturbed area is on the eastern side of modern Hyde Park (Chesnut 1979, 184).

The change from Hawkesbury Sandstone- to Wianamatta Group-dominated geology has a marked influence on the topography and landforms of the southern shore of Port Jackson. It coincides with the change from the angular and steep-bluffed sandstone landforms surrounding the Inner and Middle Harbors (east) to the gentler and subdued topography of the Parramatta River (west). Sydney Cove lies at this topographic transition.

The distribution of the Wianamatta Group also had a profound influence on the natural environment of 1788 Sydney. There was a close relationship between changes in soil quality and the transition from sandstone- to shale-dominated bedrock near Sydney. This transition coincided with a marked change in the type of native vegetation (Benson and Howell 1990). Reconstruction of the pre-European vegetation shows that the settlement was placed at the eastern extremity of the range of the shale-adapted Turpentine-Ironbark Forest (Benson and Howell 1990). Agriculture spread west across the better, shale-derived soils and a settlement developed at the "Kangaroo Grounds" between Parramatta and Sydney in 1793 (Karskens 2010).

East of the cove, eucalypt woodland predominated, which was better-suited to the poor sandstone-derived soils. Further east, in the region of strong oceanic influence, the forest rapidly gave way to coastal *Banksia* scrubland and sandstone heaths (Benson and Howell 1990). These eastern areas of Sydney would remain largely undeveloped until the late nineteenth century.

Places underlain by shale had deeper soils and may therefore have been easier for digging graves. This factor may have been important to the choice of the site for the first main cemetery (figure 1.2), a site now occupied by Sydney Town Hall.

Brickmaking Clay and the Brickfields

Brickmaking was the first industry developed in Sydney, beginning the day after landing at Sydney Cove (Worgan 1788, unpaginated). It is not recorded where the first clay came from, but sources of good-quality brick clay were rare in the largely sandstone dominated terrain.

Shale-derived soils were probably an early source of brickmaking clay, particularly where transported material accumulated in water courses. Deposits of clay suitable for brickmaking were soon found along Cockle Creek during land clearing, about 1.6 km south of Sydney Cove (Worgan 1788, unpaginated). The location became the center of brickmaking in the colony and the small village that evolved around the clay deposits became known as the "brickfields." Brickfields was the first settlement developed outside of Sydney and was the destination of the first road. It remained a site of clay extraction and brickmaking well into the 1830s and was the source of bricks for all of Sydney's earliest buildings, before being overwhelmed by the urban expansion of Sydney Town.

Reconstruction of the location of the brickfields as they developed from 1788 to the 1830s (figure 1.2) shows that the clay resources and associated kilns closely followed the trend of the "Great Sydney Dike." Clay was exploited in pits and small pockets in the area (e.g., Casey and Lowe Pty Ltd 2011), and its extraction was never on a large scale, suggesting restricted deposits of good-quality material. Deeply weathered dike and adjacent wall rocks (clay-rich Ashfield Shale) were the likely source of these first clay deposits. "Stiff fissured clay" encountered during excavation of the Eastern Distributor Tunnel at depths of around 10 m was part of the 5–7 m wide Great Sydney Dike (Justice 2000).

Clay may also have been sourced from weathered shale, and the Quaternary sediments derived from it, which filled the watercourses draining the shale ridges into the head of Cockle Bay. Fresh water for brickmaking probably came from spring-fed streams rising at the contact between the Ashfield Shale and Hawkesbury Sandstone.

Lime Sources

There was no limestone found for the manufacture of lime mortar in the rocks near Sydney. Contemporary writers frequently comment on the abundance of good building stone, including high-quality freestone (e.g., Phillip 1789; Worgan 1788), but there are persistent references to the lack of limestone for mortar (e.g., Collins 1798). Expeditions were sent out in search of limestone but had no success (Tench 1793). The lack of lime restricted the type of structure that could be built and slowed the construction of public buildings (Phillip 1789). So, by necessity, the first buildings were made of timber and daub. Later, some were built from brick or stone with mud mortar. While a small quantity of limestone was imported from Norfolk Island (Collins 1798), this problem persisted for several decades.

In 1788, lime could only be produced by burning intertidal shell deposits collected in neighboring coves (Phillip 1879), some of which may have been Aboriginal middens. Female convicts were often employed picking up shells around the shoreline of Cockle Bay (modern Darling Harbor), which was a particularly rich source. The oldest surviving buildings in Sydney are built using such shell mortar (e.g., Hyde Park Barracks).

The Road Network

Once the camp at Sydney Cove was established in January 1788, parties radiated outward to explore the region and to search for resources, particularly good farming land and limestone. The locations of the resources discovered led to the placement of additional infrastructure and roads that set broader patterns in place as the settlement expanded. Tracks developed to the sources of brickmaking clay, and to other building material resources (e.g., to rush-filled wetlands, timber sources, and shell beds or middens). Tracks also developed to the places where arable land occurred, and to burial grounds, the locations of which were possibly chosen for largely geological reasons (deep and easily dug soil).

The new arrivals could not significantly modify the land surface and so had to fit into it as best they could, a fact well illustrated by the layout of The Rocks, Sydney's oldest suburb. This largely residential area developed along the rocky western shore of Sydney Cove and evolved from the main convict camp. It grew organically between the mostly immovable sandstone outcrops and the residents worked around them. Huts were often placed on natural rock shelves, while an interconnecting network of paths developed between the sandstone ledges and approximately parallel to the trend of prominent outcrops (e.g., Karskens 1999). The paths eventually became the modern road network of this part of the Sydney CBD, and still record the geometry of the original sandstone outcrops, which have long since been removed (figure 1.2 inset).

A track developed southward along the level ground of the Tank Stream Valley to the Brickfields village and this road was eventually continued through to Parramatta. Most of the earliest commercial development of Sydney took place along the western shore of the cove and then progressed southward along the western side of the Tank Stream Valley. A similar track developed northward from the camp to a small observatory operated by Lieutenant Dawes (and a later gun battery) on Dawes Point. These two tracks set in place the main north-south axis of the settlement that evolved into George Street.

In October 1788, a bridge was built over the rocky course of the Tank Stream to link the governor's residence and farms with the main camp. This was achieved by the convict gang rolling logs across the narrowest point in the often rocky and steeply bluffed banks formed in the sandstone (Collins 1798, unpaginated). The tracks that developed to the bridge set in place the east-west axis of the settlement and would become Bridge Street.

Significant infrastructure elements were placed on the high ground created at the bluffed ridgelines, including the current Government House, an observatory, defensive sites (gun batteries), hospitals, barracks, jails, and the later windmills. The strongly bluffed outcrops on the steeper slopes of the western sides of the headlands were less amenable to residential development and became the locations of early stone quarrying (e.g., "The Hungry Mile" and the Tarpeian Way (Bennelong Point), particularly as the port infrastructure began to grow. Later, the Pyrmont Peninsula would become a center of quarrying, and a major source of the high-quality sandstone used to build many of Sydney's most significant buildings.

THE BUILT ENVIRONMENT

Sydney is "founded and excavated in sandstone" and has long been recognized as a place where geology defines its urban character (McNally 2000). As an originally cheap and abundant, high-quality building material, the Hawkesbury Sandstone has had a major influence on the built environment of Sydney and remains a key element of the modern urban character.

The ruggedness of the sandstone landscape of Port Jackson has resulted in the survival of extensive outcrop in the modern Sydney CBD, and outcrops and cuttings of Hawkesbury Sandstone are the defining character elements of the modern city. Some coastal outcrops and shore platforms survive that seem largely unchanged since first European settlement (e.g., Mrs. Macquarie's Point and the western shore of Woolloomooloo Bay). Extant outcrops, bluffs, and platforms can be identified on early charts and probably look much as they did in July 1788 (refer to Dawes et al. 1789). There can be few other cities in the world where the rocks that underlie a CBD still play such an important part in its unique urban character.

The Rocks was the first Sydney suburb and developed along the western shore of the cove from the first convict camp. The suburb gains its name from the influence of the underlying geology on its urban character. Many of the buildings there rise out of the sandstone that is immediately beneath them.

By Australian standards, this is a very old place of European settlement and the buildings there seem to have been fitted into their native bedrock. Many are also built from rock that was probably sourced nearby, if not from the actual site.

The landforms of Port Jackson are formed on outcropping Permo-Triassic rocks, which creates a relatively steep and incised landscape of headlands and inlets. This landscape is distinctly different to the relatively flat plains formed by the Quaternary sediments that underlie Botany Bay and most of its shoreline. The contrasting geology produces a major change in character of the inner-City landscape southward, from Centennial Park to the airport and the shoreline of Botany Bay.

The poorly drained land underlain by the Quaternary sediments of the Botany Basin was mostly unsuitable for residential development, except in elevated areas, so was turned over to market gardens and industry. Those industries tended to be of a more noxious variety, such as tanneries, wool washeries, and chemical manufacture.

Later in Sydney's history, a part of the Botany Wetlands was developed into water reservoirs to replace the Tank Stream as the freshwater supply for the growing town. They were connected to a distribution point at the north end of Hyde Park, via the Busby's Bore water supply tunnel, from where it was distributed to households by water carts. This area is now occupied by sporting facilities, recreational and industrial areas, and the airport. The most significant remains of the wetland system now form the Centennial Park lakes (Benson and Howell 1990).

Mudflats at the mouth of the Cook River became one of the first sewerage farms. The Alexandra Canal was developed from the former Shea's Creek, a northern tributary of Cooks River, to create a navigable canal through the mangrove flats, salt marshes and wetlands, to service the growing industries on the marshy flats. Much of the low-lying ground became a focus for industry, particularly along the course of the canal. In the 1930s, the broad and relatively flat expanse of land of the marshy north shore of Botany Bay provided a relatively undeveloped and flat area near central Sydney where an airport could be developed. It would become the site of Sydney's international airport and is now also the location of the main container port. Industry and transport infrastructure are still the dominant land uses of the former wetland region underlain by the sediments of the onshore component of the Botany Basin.

The geometries of the once extensive wetlands are now ghosted by the pattern of many streets that once flanked their shores or developed to service the inhabited higher ground. Such patterns record the historical suburban development and preserve a remnant of the drainage pattern.

Historic Landmark Buildings and Features

The historic built environment of central Sydney is mostly constructed from Hawkesbury Sandstone (see figure 8.1 for examples). One of the most important sources of high-quality sandstone for Sydney buildings was the Pyrmont Peninsula, on the western border of the study area. It produced "yellow block sandstone" from extensive quarries, mostly located at the northern end of the peninsula. Important buildings constructed from Pyrmont sandstone include the Sydney Town Hall, the University of Sydney, St Mary's Cathedral, St Andrew's Cathedral, the Sydney General Post Office, the Great Synagogue, the Art Gallery of NSW, the former Customs House, and the Australian Museum. Stone sourced from quarries on Goat Island was used to construct the Government House extensions and the Darlinghurst Jail. Other landmark sandstone buildings of the Sydney CBD include parts of Parliament House and the Treasury Buildings, which originally comprised wings of the Rum Hospital. Other key buildings include Victoria Barracks, the Australian Museum, the Queen Victoria Building, the Sydney Observatory, Fort Dennison (Pinchgut), and the remains of the Dawes Point Battery.

Minor structures and ornamental features built from sandstone occur throughout Sydney, including sea walls, monuments, retaining walls, and gate pillars. Steps have been cut into the rock faces and road cuttings tower over major streets and motorway on-ramps. Many of the cuttings once also served as quarries (see figure 8.1).

Quarries

Former quarries are significant elements of the modern urban landscapes of the city. Important examples have extensively modified the shorelines of Bennelong Point and Millers Point, particularly on the more steeply bluffed western slopes where the best sandstone outcrops were. Other examples include the northern margin of Observatory Hill, the Argyle Street Cut, and the Hungry Mile bluff. The Tarpeian Wall of Bennelong Point has also been extensively modified by quarrying, and there were several smaller quarries on the western shore of Farm Cove. The former quarries of modern Pyrmont, where yellow block sandstone was won, are now significant aspects of the built urban and recreational environments of this inner-city precinct. Most of the green recreational spaces are former quarry workings and its northern promontories are defined by the extensive rock-cut sandstone faces.

Chapter 2

On the Edge of the Graben

Hobart, Van Diemen's Land 1803–1804

As Australia broke away from Antarctica during the final stages of the dismemberment of Gondwana (late Cretaceous, from 95 to 85 Ma), crustal thinning and seaway development created sedimentary basins in the Gippsland and Bass Strait areas. Deep rifting scars were left in one of the last remaining links between the two developing continents. This small crustal fragment would become the island of Tasmania (Blewett 2012, 109).

Throughout the Cenozoic, the rift graben scars became the focus of drainage from highland glaciers, which created deeply incised river valleys that were eventually drowned during the last postglacial marine transgression between 20,000 and 6,000–7,000 years BP (Harvey and Caton 2010, 83).

From the late eighteenth century onward, the exceptional sheltered estuaries formed by these flooded grabens would become the focus of intense European political, geographic, and scientific interest as British and French colonial ambitions began to compete in the region. Ultimately, this led to the establishment of a permanent British settlement on the western shore of one of the estuaries in 1804 that would be named Hobart. The site of that settlement is now the commercial and administrative center of the modern state of Tasmania (see figures I.1 and I.2 for location).

GEOGRAPHIC SETTING AND STUDY AREA BOUNDARIES

The City of Hobart lies in the southeastern corner of the island of Tasmania, at latitude 42.88°S, longitude 147.33°E (Hobart GPO, Google Earth). The modern urban area is focused along the shores of the River Derwent Estuary.

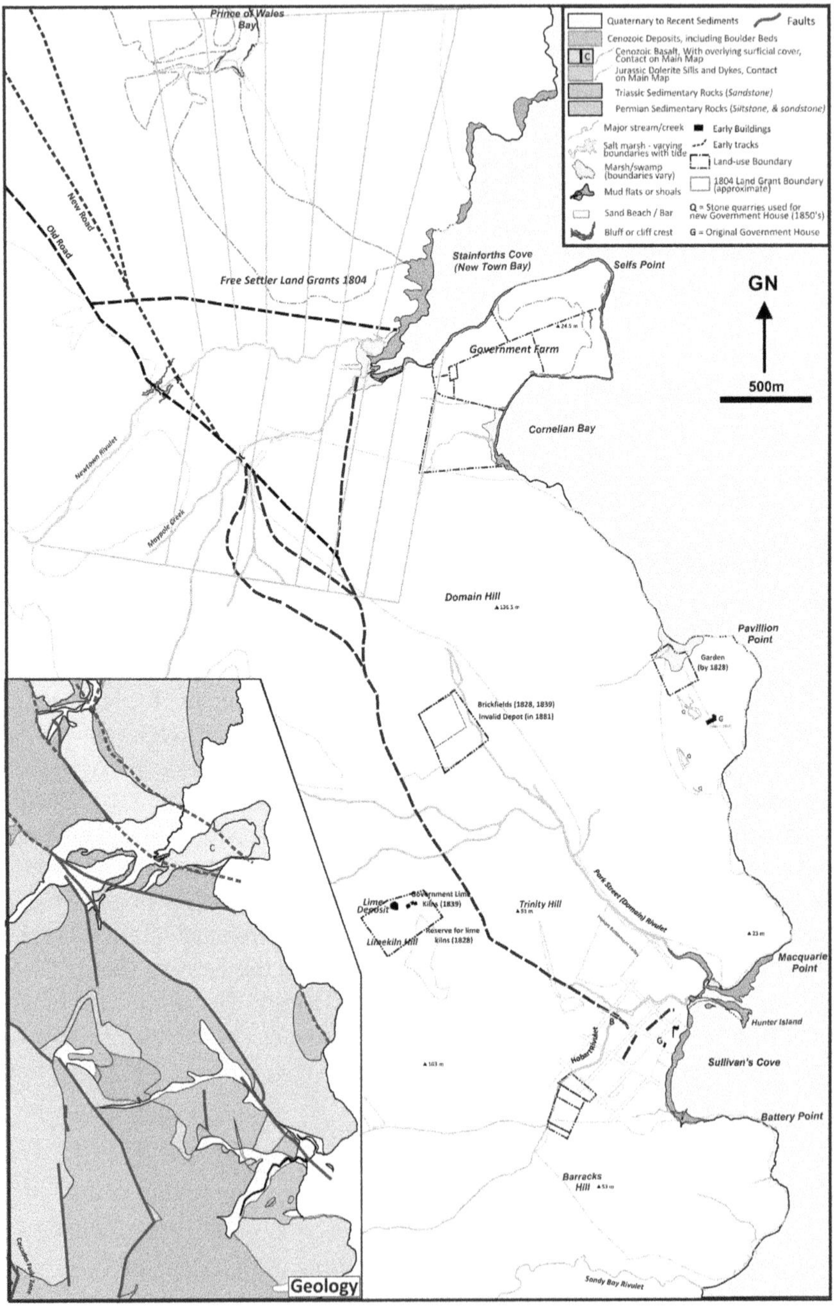

Figure 2.1 The Hobart Region. Reconstruction of the topographic and cultural features located between the 1804 settlement at Sullivans Cove, and the free settler land grants at

Figure 2.1 (Continued)
New Town Bay (Stainforth's Cove). The need to travel from the main settlement to the free settler's farms resulted in the development of an early track and road network that eventually evolved into Elizabeth Street and Main Road. The approximate boundaries of several of the 1804 New Town free settler land grands are shown, which still largely control the modern street network of New Town and Moonah. **Inset.** The geology between Sullivan's Cove and New Town Bay. Hobart spread northwards along a lower-lying valley formed largely on Triassic sandstone. The valley was bounded on either side by tilted blocks of Jurassic Dolerite. This map is based on Forsyth and Clarke (1999). Please refer to the extended caption at the rear of the book for additional information.

The CBD is on the western shore and focused on the original 1804 settlement site at the head of Sullivans Cove (figure 2.1).

The study area measures approximately 3 × 3.5 km (10.5 km^2) and is contained within the municipal boundary of the modern City of Hobart. The eastern boundary extends to the Derwent Estuary shoreline and incorporates the Domain Rivulet, Queens Domain Hill, Royal Tasmanian Botanical Gardens, and Macquarie Point. The northern boundary extends to the approximate position of the former brickfields (now North Hobart Oval), but important sites further north are also discussed, including New Town Bay (Stainforth's Cove), the former New Town Coalfield, and the former government farm (Cornelian Bay Cemetery). The southern boundary approximates the course of the Sandy Bay Rivulet and includes Battery Point and Barracks Hill. To the west, the boundary is marked by the steeply rising ground of Knocklofty and the cascades on the Hobart Rivulet (figure 2.1). Within this small area, the land rises to approximately 140 m on Domain Hill and 54 m at Barracks Hill. To the west, Mount Knocklofty reaches a height of 374 m.

The focus of this chapter is on the landforms of the River Derwent Estuary and how they influenced the formative years of permanent British settlement, from 1803 until the 1830s. During these first decades, the Risdon Cove settlement was established by Lieutenant John Bowen (1803), the first survey traverse across the Sullivans Cove area was completed by James Meehan (1803–1804) and the Collins expedition arrived from Britain, establishing a permanent Hobart convict settlement. A concurrent, mainly agricultural, settlement was established at Stainforth's Cove (modern New Town Bay) by land grants to free settlers that arrived with Collins. Throughout these first decades, Hobart developed in a largely unplanned way, spreading to access available resources, and in deference to the landscape elements that could not be easily modified. By the 1830s, the modern road network of central Hobart had been established and many of the land-use patterns of the modern city created. The development of Hobart occurred in several distinct, but often overlapping, phases:

Phase 1 1803–1804: The Risdon Cove Settlement.

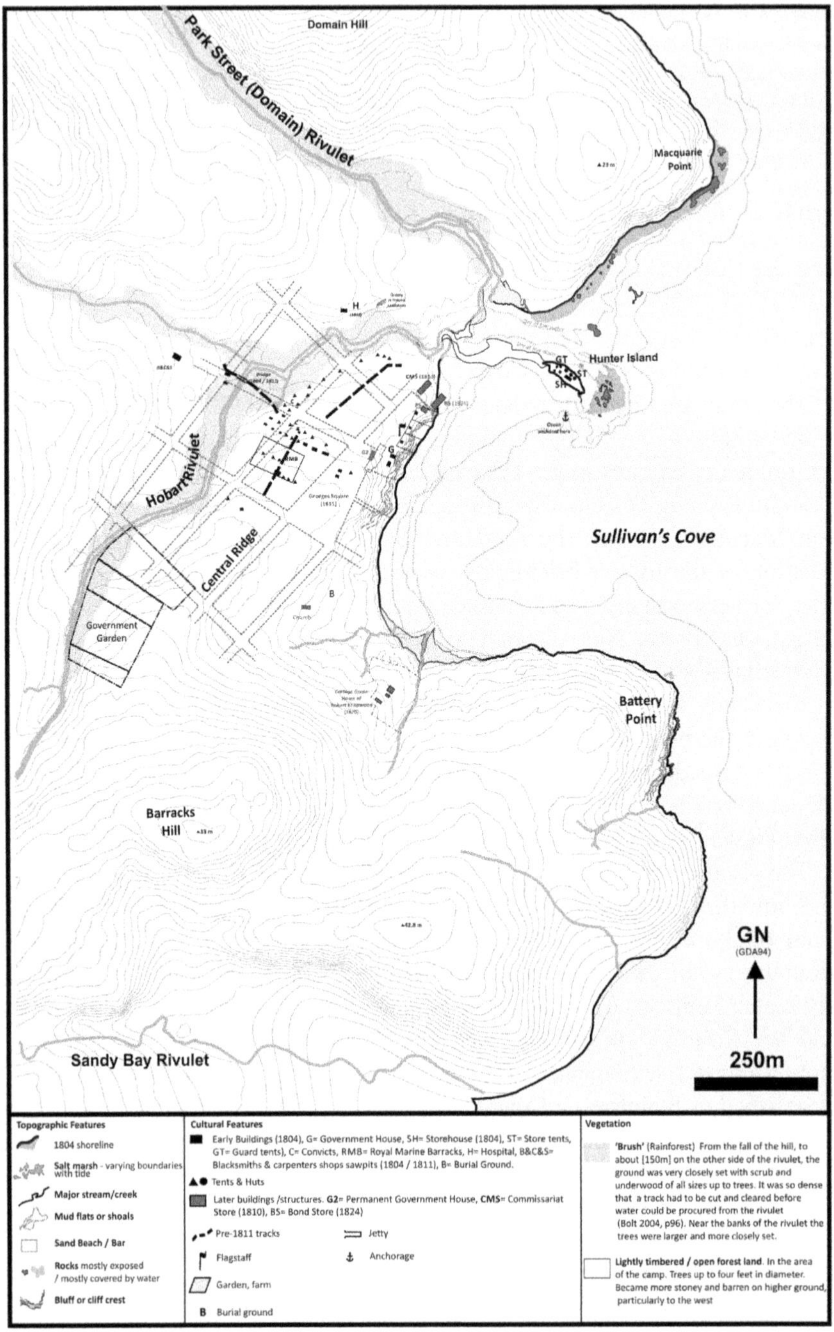

Figure 2.2 Hobart Settlement. Topographic reconstruction of the 1804 landscape of Sullivan's Cove, with 2 m contours derived from the 3D topographic terrain reconstruc-

Figure 2.2 (Continued)
tion. The layout of the initial camp is shown, and is based on Harris (1804) and Bolt (2004) with additions from Mault (in Walker 1889). The first camp was focussed along a lightly timbered and rocky central east-west ridge mainly formed by sandstone. The woodland became drier, stonier, and more sparsely timbered as the land rose gently to the west. A belt of thick "brush" (scrub and remnant temperate rainforest) occupied the deeply incised course of the rivulet between the steep sandstone banks. James Meehan's 1811 surveyed street grid is shown as an overlay. Please refer to the extended caption at the rear of the book for additional information.

Phase 2 1804–1805: The first camp at Sullivan's Cove.
- The layout of the first camp is influenced by the rising ground of the Central Ridge, the steep banks of the rivulet and the cliffs of the north shore of Sullivans Cove,
- Land grants for free settlers at Stainforths Cove (New Town Bay) are laid out normal to the New Town Rivulet. The government farm is established on the basalt-derived soils forming a peninsula north of Cornelian Bay.

Phase 3 1804–1812: The Collins Settlement, a military camp,
- First settlement develops in an unplanned way until Meehan's town grid survey.
- Stainforth's Cove (New Town Bay), the location of the free settlers, becomes an important early farming district.

Phase 4 1812–1825: Macquarie's orderly settlement.
- Meehan's 1811 street grid survey (figure 2.2), ordered by Macquarie, regularizes the streets and establishes the modern grid (Bolt 2004).
- The first town planning (Craig 1945).
- First suburban development—Battery Point and New Town.
- The existing land-use and tenure, and the earlier camp layout influences Meehan's street survey and prevents a true street grid being established (Bolt 2004).
- Greater settlement permanency with the first significant public works and buildings.

Phase 5 Arthur's appointment of Survey Commissioners in 1825—the layout of the central part of the settlement.

Phase 6 1841 Onward: Post Macquarie, Sprent's map (Sprent 1846) and modern Hobart.

Some reference is made to subsequent development of the town, and the evolution of the wider road networks and transport infrastructure.

EUROPEAN DISCOVERY AND EARLY EXPLORATION

On the November 24, 1642, Abel Tasman, explorer, and employee of the Dutch East India Company, sighted several prominent mountain landmarks on the west coast of a place he named Anthony Van Diemenslandt (Heeres and Coote 1898, 12), now known as Tasmania. Tasman's expedition continued south and eastward, following the rugged shoreline and several islets and peninsulas were documented. They eventually sighted the entrance of the River Derwent Estuary but bad weather in the place they named "Storm Bay" prevented any close examination, so they proceeded up the east coast and anchored in Blackman Bay. After a brief land exploration, and the planting of a flag on shore by the ships carpenter, Tasman's small expedition steered to the east, anchored in Frederick Henry Bay and Marion Bay, and then left Tasmanian waters.

Two prominent west coast mountain peaks first seen by Tasman would later be named "Zeehan" and "Heemskirk," after his ships, by British explorer Matthew Flinders (1814, December 11, 1798 entry) during the wave of exploration of the southeast Australian coastlines in the late eighteenth and early nineteenth centuries.

In March 1773, Lieutenant Tobias Furneaux, in command of HMS *Adventure* had become separated from James Cook's ship *Resolution* as they traversed the southern Indian Ocean. He decided to shelter in a bay first sighted by Abel Tasman on the southern shore of Van Diemen's Land, which he named Adventure Bay. After spending nine days there he proceeded on to create the first chart of the coastal features of southern Tasmania (Turnbull 2006, 14–15). Furneaux's confirmation that Adventure Bay provided good supplies of wood and water triggered several further visits over the next two decades. James Cook anchored HMS *Resolution* in Adventure Bay to collect wood and grass during his third discovery voyage in 1777. While there, the bay was charted, creating the first accurate map of an Australian landscape (Turnbull 2006, 49). William Bligh was Cook's sailing master on the *Resolution* in 1777 but returned in command of HMS *Bounty* in April 1788. Bligh visited once again in 1792, with Midshipman Matthew Flinders, on board the *Providence* (Turnbull 2006, 18). Despite these visits, it was not until the scientific voyage of French explorer Bruny D'Entrecasteaux, in the *Recherche* and *Esperance* that it was recognized in 1792 that Adventure Bay was on the east coast of an island, subsequently named Bruny Island, and separated from the mainland by the D'Entrecasteaux Channel.

D'Entrecasteaux was the first to enter the large estuary to the north of the island in early 1793 and named it Rivière du Nord. It was charted for the first time in April 1793 by Lieutenant John Hayes, in the *Duke of Clarence*, who named the upper reaches the River Derwent, after a river in Cumbria, England (Tardif 2003, 19). He named various geographic features in the estuary, including Risdon Cove and Stainforth's Cove, Prince of Wales Bay, and the 300-acre King George's Plains lying behind (Flannery 2006, 30). However, Hayes left no written descriptions and the primary record of his visit is an annotated chart held by the Admiralty (Thomas 1991, 52).

The French exploration of the southern and southeastern coastlines of New Holland in the 1790s, began to worry the British government (Robson 1989, 80). Britain had a foothold on the east coast at Sydney, but the rest of the vast continent was largely unsettled by Europeans. Concern was raised that this region might become the site for French territorial ambitions and so both the local colonial and home governments reacted to the perceived threat by dispatching expeditions to establish a permanent British presence in the southeast.

In early 1798, Matthew Flinders and George Bass were tasked with confirming reports from sealers working in the Furneaux Group that a strait existed between the mainland and Van Diemen's Land (Owen 2010, 15). Entering from the eastern end in the *Norfolk*, by December 11, they had confirmed that the strait existed, and that Van Diemen's Land was an island (Flannery 2006, 26). The recognition that Tasmania was an island may have weakened any British claim to it because it was now recognized that it was not a part of the east coast claimed by Cook in 1770. Flinders continued down the west coast of Tasmania and explored the River Derwent. He climbed Mount Direction to take a series of survey angles that provided the base for his subsequent charting of the area.

It was the descriptions of landscapes and estuaries of these earliest visitors, and their assessments of future settlement potential, that ultimately led to the choice of the River Derwent as the location for the first British settlement in southeastern Australia. This was not an expedition from Britain but from Sydney and was a local response to the perceived territorial threat from France.

THE GEOLOGY OF SULLIVANS COVE

The geological relationships between the island of Tasmania and mainland Australia are uncertain, despite being just 240 km apart. This small fragment of continental crust (64,519 km^2) has a distinctive geological history and possesses features not observed on the nearby continent (see discussion in Rawlinson et al. 2010, 382–384). Proterozoic rocks outcrop in western Tasmania, but none occur in the Lachlan Orogen, which underlies most of nearby mainland Victoria (Rawlinson et al. 2010, 382). The oldest Tasmanian rocks, the Rocky Cape Group, have no apparent relationship with time-equivalent rocks on the mainland and are correlated with rocks exposed in the lowest levels of the Grand Canyon, United States, based on age and lithology (Mulder et al. 2018). The Oonah Formation, exposed on the northwest and west coasts, differs from the time-equivalent Adelaide Rift Sequence on the mainland but is similar to the Pahrump Group (Death Valley, United States) and the Cobham Formation (Transantarctic

Mountains), suggesting a connection between Tasmania, East Antarctica, and southwest Laurentia within the supercontinent of Rodinia (Mulder et al. 2018). The Ordovician-Devonian Mathinna Group sedimentary rocks of northeast Tasmania are similar in provenance and metamorphic history to the Siluro-Ordovician rocks of the Lachlan Fold Belt (Patison et al. 2001), and also host orogenic gold deposits, but their structural history is different (Rawlinson et al. 2010, 382). The Cambrian volcano-sedimentary sequence that forms an arcuate belt from the west to north coasts has few outcropping correlates on the mainland. Tasmania may have more in common geologically with far distant Antarctica than it does with its near-neighbor Australia.

Geology of the 1804 Sullivan's Cove Settlement Site

None of the most ancient rocks outcrop in southeastern Tasmania. The geological history of the Hobart region begins in the Permian, with the deposition of the thick marine and freshwater sedimentary sequences of the Parmeener Supergroup (Forsyth and Clarke 1999). The study area is mostly underlain by the downfaulted, younger freshwater Triassic sandstone, siltstone (with some mudstone) of the Upper Parmeener Supergroup. The Triassic rocks can be fossiliferous, including *proterosuchian* reptile fossils, which were recovered from early Triassic rocks exposed in the Arthur Street quarry on Knocklofty (Camp and Banks 1978). The high ground to the west, forming the foothills and lower slopes of Mount Wellington, is underlain by older Permian glaciomarine siltstones and sandstones (with some argillaceous limestone) of the Lower Parmeener Supergroup (figure 2.1 inset). Strata beneath central Hobart lie in a fault-bounded block and are mostly tilted to the west, toward Mount Wellington, except on the upper mountain slopes where the dips reverse (Harrison 1865, 493). Dip angles range from 5 degrees to 50 degrees but are generally between 15 degrees to 25 degrees (Forsyth and Clarke 1999).

The Triassic sandstones were an important influence on the 1804 landscape. They underlie most of the low ground of the floor of the linear Hobart Valley, which parallels the trend of the downfaulted block of sandstone on the western margin of Lower Derwent Graben. In the settlement area, Triassic sandstone formed low cliffs and the steep banks of the Hobart Rivulet, as well as stepped waterfalls and pavement outcrops in the stream bed further west, which became known as the "cascades" (figure 2.1 inset).

The Triassic sandstone was also the source of the freestone used to build some of the most historically important landmark buildings of the city (see Built Heritage section below). Historically important sandstone quarries in the study area were developed in a small roof cap overlying the Jurassic dolerite that otherwise dominates the geology of the Queens Domain. Freestone from this erosional remnant was used to construct the current Government House. The former quarries are now the Government House lakes and the site of the former Beaumaris Zoo. The Rossbank Magnetic Observatory was established near the sandstone quarries in 1840, with the site on thick sandstone chosen to remove any doubt about interference in the magnetic field measurements from the underlying dolerite (Savours and McConnell 1982, 531). Several views of parts of the quarries were incorporated into watercolor paintings of the Observatory made by Francis Simpkinson DeWesselow, who worked there in the 1840s (e.g., Plate 5 in Angus 1984).

The highest hills of the 1804 settlement area are almost invariably formed by Jurassic dolerite, intruded into the Permo-Triassic sedimentary sequence as sills and dykes. There are two or three major sills in the Hobart area, ranging in thickness from 1 m to 300–400 m. The sedimentary rocks are contact metamorphosed for a few meters beyond the sill margins, with the greatest effects in the roof (Leaman 1975, 176; Stacey and Berry 2004, 156). Dolerite forms the linear ridge of the Queen's Domain and Macquarie Point to the east, and Battery Point to the south (the site of Hobart's first significant suburb). Jurassic dolerite also formed Hunter Island, the place where the stores were first offloaded from the ships in 1804.

Dolerite is hard and resistant to erosive forces such as wave action, so forms many of the spectacular coastal cliffs of the Derwent Estuary. Despite its strength, the dolerite has many weaknesses resulting from the internal tensions that occurred as the molten magma slowly cooled and contracted within a confined space. This process generated columnar jointing, and layered zones of differing strengths, which were etched and weathered out to form often-spectacular columns. Columnar coastal landforms were noted by many early navigators along the coast, such as Fluted Cape and Cape Pillar and were named for this geological feature.

The sequence of rocks beneath Hobart is heavily dissected by late Mesozoic faulting associated with the western margin of the Lower Derwent Graben. The most important fault system in the study area is the Cascades Fault Zone, which defines the western margin of the graben (figure 2.1 inset). This fault system forms the structural boundary between the older upthrown sequence of Lower Parmeener Supergroup rocks (forming the rising ground to the west of the fault system), and the younger Triassic rocks that underly the relatively low-lying area of central Hobart (Forsyth and Clarke 1999).

The Parmeener Supergroup and dolerite is mantled by Cenozoic boulder deposits and talus slopes comprised of varying proportions of poorly consolidated boulders and cobbles of sedimentary rock and dolerite in a sand to silt matrix. The boulder deposits have been shed from the flanks of the higher ground to the west, including Mount Stuart and Knocklofty, and probably formed during periglacial conditions of the last ice age. The boulder beds underlie most of the western side of the study area, including the modern suburbs of West Hobart and South Hobart (Forsyth and Clarke 1999).

Isolated occurrences of Cenozoic basalt and related volcanics are present, mainly to the north, in New Town Bay, where they form Self's Point and underlie the Cornelian Bay Cemetery. Basalt also occupies the river valley under the modern Tasman Bridge (Leaman 1977). An isolated, deeply eroded eruption center lies to the south, in modern Lower Sandy Bay, where it was first recognized by Charles Darwin in 1836 (Banks and Leaman 1999).

Quaternary Geology

An extensive icecap developed in the Tasmanian central highlands in the late Cenozoic (Kiernan 1991), with the first, and most extensive in the early Pleistocene. The ice cap discharged southward via a major glacier that occupied the upper Derwent Valley, and which may have extended 70 km down-valley, to as low as 230 m ASL. The effects of the ice cap extended further south, with probable periglacial landforms being created by the effects of freeze-thawing action, such as that identified in the Baskerville area northeast of Hobart (Vicary and Mazengarb 2019, 17).

The islands and peninsulas of the lower Derwent Estuary have very different seaward and lee shores. The high cliffs of the erosional seaward shorelines are mostly formed by wave-carved dolerite while the depositional lee shores are occupied by white sand beaches, sand-spits, and mudflats. Dolerite is resistant to the relentless wave action of Storm Bay and tends to form cliffs. However, the north-flowing coastal currents can no longer transport their sediment loads as they pass the sheltering islands and peninsulas and deposit it as sand, silt, and mud. Most of the present depositional features were formed during the Pleistocene, at a time when the sea level was about 5 m higher (Scanlon et al. 1990), but they have continued to accumulate over the last 6,000 years. One of these forms the South Arm Neck near Hobart, with the rhythmically rippled mud and sand flats of Ralph's Bay lying behind. The more famous is Eaglehawk Neck, on the Tasman Peninsula, which once isolated the Port Arthur convicts from the rest of Tasmania and was occupied by the "dog line." The sand neck joining North and South Bruny is occupied by a significant colony of Fairy Penguins. Many of these sand deposits became important sources of building materials (and still are). Sand sheets at Glenfield and Old Beach began to be exploited for concrete making in the 1930s (Sigleo 1978, 146) while the South Arm Neck remains one of the most important sand sources of modern Hobart.

Raised shell beds occur from 3 to 13 m above the modern high-water mark along the shoreline of the Domain and are higher elsewhere in the estuary. They may have once been more extensive in the Hobart area, before removal of some for early lime-making. They were particularly large on the shore of Ralphs Bay, on the east side of the Derwent, which became the main source. Most shell beds are the result of land uplifts since the Last Interglacial Stage, with one lying 24 m above sea level being dated at 125,000 years BP (Banks and Leaman 1999, 47). While most shell beds are natural deposits, some are Aboriginal middens.

THE DERWENT ESTUARY LANDSCAPE

The River Derwent and its estuary occupy a structurally complex belt of faulted rocks that have been extensively deformed and downdropped, creating a stepped graben (Stacey and Berry 2004). The rocks in such structures are often deformed and weakened by the faulting and so more easily weathered and eroded. It was such a belt of weakened rocks that was exploited by erosion to form the deeply incised River Derwent Valley.

The Hobart region is dominated by two physiographic features, the Mount Wellington-Mount Marian plateau, and the Derwent lowlands (Leaman 1976, 10). The elevated parts of the landscape are generally formed on more erosion-resistant dolerite, forming linear topographic highs between narrow valleys that result from step faulting within the graben. The general topographic asymmetry observed in the hills and ridgelines bordering the Derwent Estuary is a consequence of the gently west-dipping strata (and sills) in the fault-bounded tilt blocks of the stepped graben. The gradients of the western dip slopes are gentler because they generally follow the strata, but the eastern hill slopes are more steeply bluffed where they abut faults. They create the characteristic skyline profiles surrounding Hobart. Some of the more prominent hills, such as Mount Direction and Knocklofty, were used as landmarks and survey reference points by early explorers such as Matthew Flinders, James Meehan, and Charles Grimes. Later, several high points became the sites for the semaphore stations (an early long-distance visual communication systems) that transmitted information between Port Arthur and Hobart (e.g., Mount Nelson, which was later developed as a harbor signal station).

The ocean invaded the lower reaches of the ancient river valley, forming the Derwent Estuary between 20,000 and 6,500 years BP (Harvey and Caton 2010, 83), leaving the harder and less deformed surrounding rocks as the coastline, hills, ridgelines, and islands on either side. The drowned riparian landscape results in steep topography that usually extends to the high-water mark. This old topography had been enhanced by the strong tidal scouring and the Derwent Estuary is deep. At the site of the modern

Tasman Bridge, which crosses the estuary from the west side at Pavilion Point, in the Queen's Domain, to Rosny Point on the eastern side, the channel is 35 m deep.

The Aboriginal Landscape

People arrived in what is now Tasmania, then attached to the mainland, at least 35,000 years ago, occupying the periglacial uplands where they subsisted on a range of terrestrial animals. Their use of marine and plant resources at this time is unknown (Cosgrove 1999, 395), probably because the coastal landscapes that they occupied were inundated as the sea level rose.

As the last glacial maximum waned, heralding the current warmer epoch, sea levels rose and many of the large river systems of southeastern Australia and Tasmania were drowned. The Aboriginal populations of Tasmania, King Island, and Flinders Island were slowly isolated from their mainland relatives by progressively narrower and marshier links. Eventually, the Bass Strait seaway had formed by about 12,000 years ago. The Tasmanians developed a unique, insular, and simplified material culture and the populations of the smaller Flinders Island and King Island went extinct around 4,700 years ago (Flannery 1995, 263). It would be several thousand years before there would be any further contact with the outside world. Genetic evidence shows that the closest mainland relatives live in Victoria (Presser et al. 2002, 36).

Aboriginal society must have developed in tandem with the fluctuating Tasmanian environments, as they changed from an ice age to a more temperate and then drier climate. The changing landscapes and newly evolving environments of the island must have been continually colonized and adapted to. Radiocarbon dating of charcoal from sites at the Botanical Gardens and Bedlam Walls shows that the Hobart region was inhabited for at least 5,000 years before European arrival (Kerrison and Binns 1984). Tasmanian Aboriginals witnessed the evolution of the modern landscape following the waning of the last ice age and the evolution of the modern estuary. During that long occupation of the River Derwent Estuary, the linguistic divisions, tribal affiliations, and hearth groups that were encountered by Dutch, British, and French explorers from the seventeenth century onward developed.

Archeological Evidence of Aboriginal Occupation of the Derwent

Direct archaeological evidence for Aboriginal occupation of central Hobart is rare but rich. In their review of archaeological studies undertaken in the Derwent Estuary, Stockton & Wallace (1979) found seven previous studies. Subsequent work was done by Lister (1979), Officer (1980), and Gaffney and Stockton (1980). Officer (1980, in Kerrison and Binns 1984) documented 416 shell middens (118 on the west bank of the Derwent and 298 on the east), and eleven quarry or tool manufacturing sites, and rock shelters along the shores of the estuary from New Norfolk to a line connecting Blinking Billy Point to Trywork Point.

Several rock shelters were probably occupied in the "Bedlam Walls" area of the eastern shore, opposite New Town. Mussel shells found eroding from their floors led Vanderwal (1977) to excavate the Shag Bay Rock Shelter, located 15 m above the level of the river, where evidence for a 700-year occupation was found (Stockton and Wallace 1979, 82)

Raised shell strandlines are common around the shores of the Derwent Estuary and were once present in the settlement area (see Quaternary Geology section above). Most of these deposits are natural, including many recorded by Stockton and Wallace (1979), and are a record of tectonic processes that have raised and tilted the land surface. However, some also preserve evidence of Aboriginal activity. A surface scatter of shell and charcoal occurs close to the shoreline near the northeastern corner of the modern Botanical Gardens. It was excavated by Kerrison and Binns (1984) to reveal a midden. Radiocarbon dating of charcoal at the base of the midden indicated a first occupancy of 5,000 years ago. A variety of stone artifacts were found within the midden strata, mixed with various shell species, ash, and soil. The most common raw materials used were quartz hornfels, chalcedonic rock, and fine-grained basalt. Fine-grained granite, devitrified volcanic glass, and schistose quartzite were also used (Kerrison and Binns 1984, 59). Other sites are likely to have been destroyed or obscured since European settlement, due to urban development (Kerrison and Binns 1984) and the collection of shells as an early source of building lime.

Aboriginal Natural Feature Names

In their compilation of Tasmanian Aboriginal place names, Plomley and Goodall (1990) list several in the study area that were recorded by early settlers. They include,

- Nib.ber.loon.ne / Niberlooner, Ling.He / Lineg.he: the environs of Hobart Town
- Le.bra.la.wag.ge.na or Le.bra.la.wa.gen.na: Hobart Town
- Unghanyahletta or Pooranattere: Mount Wellington
- Bur.nang.ye: The Big Hill, Mount Wellington
- Go.nun.ye / Gur.nang.ye: Mount Wellington
- Cree.wer.lae.dy: Mount Nelson
- Lide.wane.ne: Sandy Bay Point
- Kree.wer: Little Sandy Bay (large village)

Aboriginal Use of the Hobart Landscape

The natural environments and landscape features of the diverse bioregions of Tasmania heavily influenced the range and exploitation patterns of the approximately fifty Aboriginal bands identified from historical ethnographic information (Jones 1974, in Gilligan 2007, 556). The movement of groups were seasonal, with coastal regions favored in winter and no groups remained inland all year round (Gilligan 2007, 556). The group inhabiting the modern Hobart region west of the River Derwent utilized the marine resources of their extensive coastline and did not range far inland. The Oyster Bay people, who occupied the eastern shore ranged far inland on a seasonal basis (Ryan 1996, 16–19).

Landscape features also influenced linguistic and clan boundaries. The River Derwent and estuary is wide and deep and formed a barrier that could not be easily crossed, except in its upper reaches. It was therefore a major geographic boundary between groups and speakers of different language dialects (Bolt 2004; Tardif 2003; Bowern 2012; Cosgrove 2015). The River Derwent also formed a boundary between different types of raw material exploitation. The people of the eastern shore tended to use more diverse stone types to manufacture implements, including hornfels, quartzite, and siliceous breccia while those of the western shore used mainly hornfels (Sutherland 1972, 15).

In 1803, the largest populations in the Derwent Estuary seem to have been focused on the eastern shore, which opened out onto the east coast and midland plains while the western shore mainly offered access to the more densely forested country to the west. In May 1804, a group estimated to have been around 160–200 people (Johnson and McFarlane 2015, 97) was moving southward, from the broad basin-shaped valley behind the Meehan Range (and the midland plains beyond), toward the Derwent Estuary. They were following a steep-sided, narrow valley drained by Risdon Brook, which then turned west, entering the narrow Risdon Valley, which provided an outlet route to the river. They unexpectedly encountered British settlers at Risdon Cove, who were unnerved by their large numbers. Mutual misunderstanding of intent resulted in the death of at least three Aboriginals and the wounding of at least another when the settlers felt the need to respond to apparent aggression (Owen 2010, 21). The people of the eastern shore became wary and more aggressive after confrontations with the colonists (Owen 2010).

Despite some early confrontations, such as an aggressive group of men throwing stones at James Meehan's survey party in 1803, the settler relations with the Aboriginal inhabitants of the western shore were usually friendly (Owen 2010, 3–4).

Aboriginal Use of Geological Materials

The use of earth resources was a significant component of Tasmanian Aboriginal material culture, and their exploitation of earth resources was diverse and widespread. Earth materials were used as raw materials for tool manufacture, adornment, weapons, and artworks.

Eighty-one pieces of stone were recovered during the midden excavation at the Botanical Gardens, of which sixty were flakes that had been produced during stone tool manufacture or repair. They included quartz hornfels (derived from various sedimentary rocks, including well-sorted siltstones and sandstones, poorly sorted sandstones, sandstones with clay, and siltstones with Permian fossils), chalcedonic rock, and fine-grained basalts. These materials were likely sourced locally from Cenozoic, Triassic, and Permian aged rocks. Nonlocally sourced materials included fine-grained granite, sheared quartzite, pieces of devitrified volcanic glass and a piece of schistose quartzite (Kerrison and Binns 1984, 59–60).

The chalcedonic rocks were sourced from the volcanic complex at Lower Sandy Bay, while the other materials were largely sourced from hornfelsed contacts of dolerite sills and dykes with both Triassic and Permian sedimentary rocks. Given the abundance of such contacts in the Hobart area, it is difficult to accurately identify the exact sources of the material. The quartzitic and granitic materials are not from the local area and may have originated as glacial dropstones, or "erratics," which are common in the Permian sedimentary rocks of the area, though their origin as trade items cannot be completely discounted (Kerrison and Binns 1984, 60).

Tasmanian Aborigines generally preferred siliceous materials for tool manufacture because of their hardness and conchoidal fracture, which assisted flaking (Sutherland 1972, 2). In his review of the materials used to create the Tasmanian Aboriginal stone implements held in the collection of the Queen Victoria Museum, Sutherland (1972) found that the materials used were hornfelsic rocks, quartzites, quartz, chalcedony, opal, impure chert, chert, spongolite, and silicified breccia. Basalt and dolerite were also occasionally used.

Eleven quarry or tool manufacturing sites were recorded in the Derwent Estuary area, but none have been recognized in the study area (Officer 1980, in Kerrison and Binns 1984). However, chalcedonic artifacts excavated from the Botanic Garden midden were sourced from the volcanic complex in Lower Sandy Bay, 7 km to the south (Kerrison and Binns 1984, 59).

Ochre, in the form of red hematite, was produced by heating iron oxides such as limonite in the fire (Noetling 1910, 31), which produced a durable red pigment when ground down. This was used to decorate bodies, usually by men, who mixed it with grease and spread it through their hair and beards (e.g., Noetling 1910, 30). It was also used, along with charcoal, to decorate bark sheets on the insides of huts and on burial structures (Flood 1997, 229). Pieces of ochre were found by Noetling (1910, 30) at a former Aboriginal camp site at Old Beach, on the eastern shore of the Derwent, who recognized that it was a highly valued resource. Ochre sources were rare, and no quarries are known in the study area, so trade in the material was likely (Noetling 1910). Material was probably traded across the island from the well-documented Toolumbunner site in the Gog Range of northern Tasmania (Sagona 1994).

BRITISH SETTLEMENT OF THE RIVER DERWENT ESTUARY

British settlement occurred twice within twelve months on the River Derwent Estuary. Both settlements were responses designed by home and colonial authorities to counter perceived French ambitions in areas claimed by Britain since Cook's 1770 visit, but not yet settled. French explorers had been assessing the region for over a decade, building on the preliminary work of Cook, and earlier Dutch explorers (e.g., Robson 1989). To counter the potential territorial interest of France, John Bowen was ordered by Governor King to establish a small outpost of NSW there in 1803. In 1804, David Collins would arrive with a small fleet from Britain, having been sent for the same reason. Collins was originally ordered to establish a base on the mainland shore of Bass Strait, and the newly discovered Port Phillip Bay was chosen (in the modern state of Victoria). The Collins settlement was meant to reinforce territorial claims on Bass Strait and southeastern Australia (see chapter 5). Originally, only the Bowen settlement was meant to be established at the Derwent.

The foundation of Hobart was followed soon afterward by settlements on the north coast at Port Dalrymple, at the mouth of the River Tamar and facing Bass Strait. They included Georgetown (November 1804), York Town (March 1805), and finally Launceston (1805) (see figure I.2). The foundation of these settlements consolidated Van Diemen's Land as a British possession.

The Risdon Cove Settlement

The first British settlement in Tasmania was established at Risdon Cove, on the eastern shore of the Derwent Estuary in 1803. It was a direct result of the earlier discovery of the Dutch, but particularly the more detailed coastal charting of French and British explorers at the turn of the nineteenth century. The settler group consisted of a convict, military, and small free settler group dispatched from Sydney by Governor King, so was a local enterprise. The site had been recommended by Hayes in his earlier exploration reports, and

Mathew Flinders had also visited. Based on their information, the Governor directed John Bowen to form his settlement there. However, Risdon Cove was probably never the best location for the establishment of a permanent settlement. The stream was unreliable and generally dry, and the anchorage was poor. Ships could only moor at a considerable distance from the only significant landing place.

When Collins arrived at Risdon Cove on February 15, 1804, he inspected the Risdon settlement, and the "general opinion" was that the location was unsuitable for a "town" and the water source was poor (Knopwood 1977, 43). The Risdon "River" was only comprised of a "few pools of dirty standing water" and was no longer the running stream that Bowen had encountered in the previous September (Owen 2010, 21). Even Flinders had found it unsuitable as a watering place when he anchored there in December 1798, being unable to enter the creek with a ships boat (Flannery 2006, 30).

The state of Risdon Brook must have contrasted sharply with Meehan's discovery of the still flowing Hobart Rivulet (Owen 2010, 21), which he observed shortly before Collins' arrival. Collins inspected the new western shore location with Robert Knopwood on the day after he arrived and immediately made the decision to form the new settlement there (Knopwood 1977, 43). Collins was acting under different orders to Bowen and had the freedom to choose his own site. The new location on the Hobart Rivulet was chosen quickly, probably because of James Meehan's advice, a contribution for which he has not specifically been acknowledged (Bolt 2004, 75–76).

Ultimately, the decision to evacuate Risdon Cove was made by Collins because it lacked reliable freshwater, poor arable land, and because there were no deepwater anchorages. However, there was also a political element to the decision. The lines of authority were not clear when Collins unexpectedly arrived at the Derwent in February 1804. Bowen remained in charge at Risdon, and so the two settlements continued in parallel for several months. Ultimately, Collins was confirmed by Governor King as the one in authority and this confirmed the Sullivan's Cove settlement as the sole seat of government. The Risdon Cove settlement was eventually closed and mostly abandoned.

The 1804 Landscape of Sullivan's Cove

On January 24, 1804, James Meehan—convicted Irish rebel—transported political prisoner and surveyor—began a survey of the western shore of the Derwent Estuary for Lieutenant Bowen, the Commandant at Risdon Cove. Using the Risdon settlement as a base, his party undertook a reconnaissance survey over the next few weeks that would cross Sullivan's Cove, Sandy Bay, and Stainforth's Cove (New Town Bay) (Tardif 2003).

Meehan began at a landing site on Cartwright's Point, to the south of Sullivan's Cove (in modern Taroona), and progressed northward (Bolt 2004, 74–75). Late on the January 25, 1804, Meehan and his surveying party decided to move camp a further mile north along the western shore of the Derwent because the initial campsite they had chosen lacked freshwater (Tardif 2003, 108). They came to a small cove with a flowing stream and decided to camp the night. Next day, they began a survey of the place that would become the center of modern Hobart from their landing place at the confluence of the Hobart and Domain Rivulets, where he placed a survey mark on a tree. In 1811, he would again use this same survey base as the beginning of the street grid layout of the modern city (Bolt 2004, 74). He traversed up a gently rising ridge as he followed the southern bank of a flowing freshwater stream to the west of its outlet at the cove. He described the country as he went, making general notes about the nature of the country and soils which he recorded in his survey books and as annotations on the chart he produced. His track is recorded on his 1804 map of the Derwent where, above the site of the camp, in the approximate locations of South Hobart and Cascades, his annotation on the Central Ridge reads: "The land on each side of this stream is high and barren" (Meehan 1804).

George Harris later surveyed the course of the Hobart Rivulet from the mouth to the Mount Wellington foothills and documented many details of the vegetation and banks, including the stand of rainforest in the bed of the stream and rocky outcrops along its lower course (Harris 1804; Bolt 2004, 97). While not the best of surveyors, Harris was a good artist who made the earliest record of the Hobart landscape at the time of European settlement and provides the only contemporary record of the layout of the first camp (Harris 1804). Some of the details he recorded include the geography of Hunter Island and the landing place, the strand lines at the mouth of the rivulet and the low rock outcrops and cliffs forming the banks of the rivulet, and islands and water falls in its channel (figure 2.2).

The Hobart Valley

The 1804 Hobart settlement was established at the southern end of a broad, gently rising, valley that extends approximately 2.6 km northward from Sullivan's Cove, toward New Town Bay (figure 2.2). The valley is approximately 1 km wide at the cove but narrows markedly to the north, where it is less than 100 m wide at its head. At this point, a low pass crosses a drainage divide into the low-lying and marshy ground at the heads of Cornelian Bay and New Town Bay (now largely reclaimed). The pass is now occupied by the Brooker Highway. The floor of the valley forms a broad "plain," seen by

Robert Knopwood on Friday, February 17, 1804 (Knopwood 1804, 43). It is bordered by the high ground of the Queens Domain (dolerite) to the east and by the rising ground formed mainly by the fault-bounded dolerite, Triassic sediments, and slope boulder deposits forming Knocklofty and Mount Stuart to the west, with mainly Permian sedimentary rocks underlying the rising ground beyond.

The geometry of the valley is fault controlled. It occupies a fault-bounded downthrown step block on the western side of the Lower Derwent Graben, with high ground to the east and west representing horsts. The linearity of the Hobart Valley locally parallels the Cascades Fault Zone (Stacey and Berry 2004). The rounded but asymmetric hills formed on the horsts are underlain by more resistant Jurassic dolerite. The Triassic sandstones and siltstones forming most of the valley floor occupy a fault-bounded graben and Sullivan's Cove is the drowned southern end. North of the drainage divide, around New Town Bay and Cornelian Bay, the floor of the graben is infilled by Cenozoic basalt and some sediments, which form the cemetery peninsula and extend northward into modern Moonah (Forsyth and Clarke 1999).

The primary drainage line of the Hobart Valley is the Domain Rivulet, which occupies the deepest point of the valley for its entire length. Its southward course follows the faulted boundary of the dolerite tilt block forming the Domain Hill, and the eastern wall of the valley.

The Central Ridge and the Sullivan's Cove Coastal Cliffs

In 1804, Sullivan's Cove was an almost semicircular inlet of the Derwent Estuary. It was backed by moderately to steeply rising ground of the Central Ridge and there were two "headlands"; Battery Point to the south and Macquarie Point (and its associated reefs) to the north. Both headlands are topographic highs formed on dolerite (figure 2.2). They became the locations for the first defensive gun batteries of the settlement and the oldest is now the site of the Cenotaph. The northern shore was shallow and shelving, with sand flats extending further outward near the mouth the rivulet, where a small dolerite outcrop named Hunter Island was located. The cove became deeper west of this island and fully laden ships were able to anchor close to shore there.

Hunter Island was associated with low rocky reefs on the seaward side. Similar but more extensive rock platforms were also present at the tip of nearby Macquarie Point. Over the next half century, the wharf areas that developed around Hunter Island would be extended westward along the shore of the cove through reclamation and it would become a focus of commercial activity and industry. The Collins Street Canal allowed the rivulet estuary to be filled in and the place became a market square. The surrounding low-lying area remained flood prone and became industrial “Wapping” and a focus of several noxious industries, including tanneries and slaughterhouses. It also became a place of poor working-class housing and a red-light district. The market site is now occupied by City Hall. The Wapping district of central Hobart is now being revitalized as a trendy suburb and education precinct.

The most important landforms in the settlement were an approximately east-west trending central ridge, on which the first camp was laid out, and the low cliffs on its southern flank. They are underlain by Triassic sandstone (Forsyth and Clarke 1999, figure 2.1 inset). The cliffs formed the north shore of Sullivan’s Cove and created a backdrop to the otherwise gently shelving shoreline. The Hobart Rivulet and the sandstone cliffs along its banks defined the northern flank of the central ridge. These landforms were the site of the first camp and their topography controlled its layout. Locations chosen at the time the camp was first established remained in place for many years afterward and were influential when Meehan surveyed the street grid in 1811 (e.g., Bolt 2004, figure 2.2).

The original steeply bluffed northern shoreline of Sullivan’s Cove remains interwoven with this early colonial fabric but has been heavily quarried and modified. The face is now further north than in 1804. The lower reclaimed land of the wharf area preserves a ghost of the reefs and sand flats that once joined Hunter Island to the mainland, and which could only be walked at low tide. Hunter Island remains beneath Hunter Street and the IXL factory precinct as a recognizable geophysical feature (Leaman and Richardson 1981).

Important Hills in the Settlement Area

The generally smooth and rounded hills forming the high ground at the peripheries of the study area are mostly formed on tilted, fault-bounded blocks of Jurassic dolerite (figures 2.1 inset and 2.2). Wintle (1864) recognized that dolerite capped most of the high points near Hobart and important examples include Domain Hill, Macquarie Point, and Battery Point. As Hobart grew, it incorporated further high points within the urban area, such as the Barracks Hill, Hospital Hill, Trinity Hill, and Limekiln Hill. More distantly, the dolerite massif of Mount Wellington is a legacy of earlier attempts at rifting (Blewett 2012, 109). To the northeast, Mount Direction was an important observation and survey point for Matthew Flinders in 1798.

Drainage System and Wetlands

The River Derwent is one of the largest river systems in Tasmania. It drains a catchment of around 89,000 km^2, with 80% of the freshwater part of the catchment located in the central highlands. The headwaters are at Lake St Clair, at an elevation of 735 m, and it then flows southeast for 190 km before entering the Derwent Estuary at New Norfolk (Eriksen et al. 2011, 2). The estuary extends for over 50 km before entering the Tasman Sea at Storm Bay and has an approximate area of 195 km^2.

All the water courses of the central Hobart area drain into the Derwent Estuary. They are fed by rain and snow falls on Mount Wellington and from the multiple springs that exit the boulder beds enshrouding the lower slopes of the mountain and underlying the western suburbs of modern Greater Hobart. The most significant water courses in the study area are the Hobart, Domain, New Town, and the Sandy Bay rivulets, of which the Hobart and Domain Rivulets are of most significance to this discussion. The Hobart Rivulet is ultimately sourced from such springs, which rise at the head of the rivulet at the Springs on the mid slopes of Mount Wellington (e.g., Bonar 2011). The alluvial flats at the mouths of some of these rivulets were important agricultural land in the first decades of Hobart (e.g., on the shores of New Town Bay). Many of the rivulets remain flood prone during exceptional rainfall, such as in 2018, when the Hobart Rivulet once again found its original course and entered the Derwent Estuary at Sullivan's Cove.

Hobart Rivulet and Estuary

The Hobart Rivulet comprises one of several moderately incised east-flowing water courses that drain the southeastern foothills and slopes of Mount Wellington (figure 2.1). The catchments of most of these streams are underlain by Permian freshwater to marine sediments and talus slope deposits. Each of the major drainage lines enters the Derwent Estuary in a cove embayment and it is likely that they existed as major drainage lines and tributaries of the early River Derwent, prior to the inundation of the primordial Derwent Valley by the ocean at the end of the last glacial maximum between 20,000 and 6,000–7,000 years BP (Harvey and Caton 2010, 83).

Unlike most other east-flowing drainage lines in the study area, the Hobart Rivulet does not follow a straight course to the estuary but is sharply deflected to the southeast at its confluence with the southeast flowing Domain Rivulet. This is caused by the drainage lines having formed along the fault lines bordering the tilted blocks at the western edge of the Lower Derwent Graben. The Domain Rivulet follows one such fault and is now the alignment of the Brooker Highway. The persistent flow of freshwater in the Hobart Rivulet was a key reason for David Collins' choice of Sullivan's Cove as the site for Hobart. James Meehan saw that it was still flowing in January 1804, despite the dry season and this information was probably passed on to Collins. The persistent flow of the Hobart Rivulet contrasted with the poor flow observed in the Risdon River by Flinders in 1798 (Flannery 2006, 30) and by Robert Knopwood, when he inspected the location with David Collins in February 1804 (Knopwood 1977, 43).

In 1804, the rivulet was only navigable to its mouth in Sullivan's Cove and the high and often rocky banks could only be bridged in one place. This first log bridge linked the main settlement to the saw pits to the north, where the best timber was located, and to the surgeon's camp and hospital (figure 2.2). It became an important hub for the development of the northern road network that fanned out from the camp, and the track that developed became Elizabeth Street.

The lower reaches of the rivulet have been extensively modified and are now mostly underground. The earliest modification connected it to the Domain Rivulet by a canal along Collins Street, to free up land for reclamation as a market square. Subsequently, both the Domain and Hobart rivulets were connected to the estuary by a tunnel excavated under Macquarie Point, where the outlet portal still exists south of the Regatta Ground. The Collins Street Canal was mostly covered over in the 1950s and the location is marked by an anomalously wide section of the street at the east end of the modern CBD. The upper reaches of the rivulet still flow through an open channel. The site of the "cascades," a series of rocky ledges formed within the Cascades Fault Zone, and the water from the rivulet determined Degraves' choice of the site for his brewery, now the Cascade Brewery (Leaman 2009b, 89).

The Hobart Rivulet also remains but is now largely channeled within a subterranean drain beneath the city. However, the alignment can still be clearly seen in any Google image of the modern CBD, by the configuration of buildings along its course. Unlike Sydney's Tank Stream, it is not completely buried by development and there are many places it can still be seen in the central city, such as on Harrington Street, below the 1844 sandstone bridge. It is also still possible to walk the banks of the upper reaches, in the linear park at the western end of the city center. The original mouth of the stream, above Hunter Island, is lost because the rivulet now debouches into the estuary via a drainage adit tunneled through Macquarie Point dolerite in 1916.

Domain Rivulet (Park Street Creek)

A major tributary of the Hobart Rivulet followed the western foot of the Domain Hill and drained most of the Hobart Valley. It merged with the Hobart Rivulet just before debouching into Sullivan's Cove at a small sandy delta near Hunter Island. This significant stream became known as the Domain (or Park Street) Rivulet and the steep-sided valley it occupied came to define the eastern edge of Hobart Town, as it did for more than a century before it was hidden beneath the alignment of the Brooker Highway.

The Domain Rivulet is now built over, occupied by the alignment of the Brooker Highway, where it continues to define the eastern edge of the settlement, as it has done since 1804. Only the suburb of Glebe, which was developed on land originally reserved to financially support the local clergy, spills across to the other side. The north-trending linear valley with its steep eastern bank was extensively quarried.

Other Significant Watercourses

The course of the Sandy Bay Rivulet, to the south of the study area, is sharply diverted to the east at the point where it encounters the Battery Point dolerite hill. The Sandy Bay Rivulet became the focus of the first engineered water scheme for Hobart, needed to replace the heavily polluted Hobart Rivulet and is now the focus of the historic Waterworks Reserve. The upper reaches of the valley of this rivulet are now occupied by the northern section of the modern Southern Outlet motorway.

Other important drainage lines in the area include the North Hobart Rivulet, which was deflected parallel to a fault line marking the western margin of the Domain Hill, where it merged with the Domain Rivulet. The New Town Rivulet, with its upper tributaries of Brushy Creek and Pottery Creek, was the main water supply for the free settlers established at Stainforth's Cove. The heavily modified lower reaches of the rivulet are now contained within a concrete culvert that directs its flow through the extensive areas of reclaimed land at the entrance to the estuary. The Humphrey Rivulet, to the north of New Town Bay, was the location of the 1872 Glenorchy mass flow (Mazengarb et al. 2007; Stevenson et al. 2016) but later became the site of the Lower Glenorchy Reservoir.

Wetlands

There were few significant wetlands within the boundaries of the study area in 1804. The most significant was a coastal marsh on a small delta at the mouths of two creeks that entered the southwest corner of Sullivan's Cove. The largest water course drained the steep-sided valley of Montpelier Retreat and the smaller runnel drained the higher ground of St David's Park. The salt marsh was an important place for acquiring wild fowl in the early days of the colony. The location is now occupied by the reclaimed land of the Parliament House lawn. Another small swampy area and saltmarsh existed at the confluence of the Hobart and Domain Rivulets, near the shore of Sullivan's Cove.

1804 Vegetation and Environment

In January 1804, James Meehan described the land around West Hobart and the Cascades as "high and barren," however Robert Knopwood, who visited Sullivan's Cove twice in February 1804, before the camp was established, was more impressed. He found the size and quality of the timber trees to be excellent and saw the potential of the "plain" for future agriculture (Knopwood 1977, 43, 44). Soon afterward, George Harris (1804) made some particularly valuable observations of the Sullivan's Cove native vegetation by documenting the distribution of the dense rainforest remnants in the lower course of the Hobart Rivulet. When combined with Meehan's more accurate, but broader-scale survey traverse and the brief descriptions and survey map annotations, Harris' chart provides detailed information about the vegetation of Sullivan's Cove in 1804.

Where the land was underlain by sandstone and dolerite, in the areas now forming South Hobart and the Cascades, both Meehan (1804) and Harris (1804) described the vegetation as sparse, and the land as "barren" and rocky. Further to the north on Knocklofty, on similar geology at the same approximate height, Wintle (1864, 466) noted that the sandstone forming the lower slopes was almost bare of vegetation while the steep-sided valleys supported dense underwood and the dolerite crest sustained very dense growth.

The linear stand of rainforest "brush" mapped by Harris occupied most of the stream bed between its mouth and the approximate position of modern Molle Street, where the steep-sided banks and stream bed were formed in Triassic sandstone. Harris also documented smaller stands of "brush" with tree ferns higher up in the course of the rivulet to the west, beyond the Cascades, which also occurred where it crossed sandstone. The vegetation was poor where the rivulet crossed boulder beds or sandstone and dolerite forming the high ground. The forest became dominated by "blue gums" beyond the Cascades.

Such contemporary observations suggest that the vegetation surrounding the first camp was comprised mostly of mature eucalypts on the higher ridgelines and clumps of rainforest remnants in the valleys of the rivulets. The vegetation growing on the relatively small areas of Triassic sandstones in the valley of the rivulet was thicker than that occupying the surrounding boulder beds and dolerite hills, particularly where sandstone formed the bed of the rivulet and the small flats near its mouth. The floor of the rivulet, near its mouth, was a tangle of fallen logs and dense vegetation (Bolt 2004, 97).

These early descriptions show that the 1804 environment of the shoreline and hinterland of Sullivan's Cove was dominated by ecological systems developed around the permanently flowing Hobart Rivulet and the drier ridges bordering it. Small details of the geology had a significant influence on the distribution of vegetation types, density, and distribution.

The densest vegetation was focused in water courses such as the rivulet. The banks of the rivulet were generally steep sided and there were sandstone cliffs where the Triassic sandstone underlay the western end, around the modern Cascades. These banks confined the "brush"—remnant temperate rainforest—within this deepest course of the stream. The steeper rising higher ground to the north and south of the rivulet course was much drier, "stony," "clear of brush," and "highly barren" (Meehan 1803; Harris 1804). It was occupied by dry, open eucalypt woodland and the hill crests were probably quite thinly vegetated.

LANDSCAPE, GEOLOGY, AND THE EARLY DEVELOPMENT OF HOBART

There were several key influences on the layout of the first settlement. They include the positions of the landing place beach (just inside the mouth of the rivulet), Hunter Island and the first secure store, and the east-west trending high ground of the Central Ridge with its subtler high-ground features used as the sites of the Marine Barracks and governor's tent. On a broader scale, the north-west trend of the fault-controlled Hobart Valley "plain" and the several prominent hills around the camp would later become the focus of important tracks and timber felling, and as sites for administrative and defense infrastructure.

The camp was formed on gently to steeply rising ground that extends northwestward from the landing place at the mouth of the rivulet, following a ridge developed on contact-metamorphosed Knocklofty Formation forming a thin veneer on the underlying Jurassic dolerite (Forsyth and Clarke 1999). Dolerite is occasionally exposed in excavations in the central city area. The orientation of this ridgeline would determine the location and layout of the first camp (also the access to the freshwaters of the rivulet) and soon afterward, many of the most important government buildings of the developing settlement, including the homes and gardens of key officials, and of the first Government House.

The landscape of central Hobart draws the eye upward and to the northwest. From the narrow sandy beach that provided the first landing place inside the mouth of the rivulet, the land rises gently to steeply up a southwest-trending ridgeline (the Central Ridge) toward Barracks Hill and the foothills of Mount Nelson. This gentle ridge is the spine of the modern CBD and the site of the first settlement camp (figure 2.2).

In 1804, the ridge was clothed by dry but thick bush; to the north was steep-sided gully full of thick temperate rainforest (brush) and edged by low rocky bluffs on the southern bank. A small freshwater stream trickled through the gully but was probably mostly obscured by the thick vegetation. The lower reaches were covered by a dense tangle of fallen dead wood, rushes, and shrubby vegetation until just upstream of the sandy beach and low marshy ground at the mouth of the stream (Bolt 2004, 87, 97). The height of the central ridge increases to the southwest and the banks of the gully become steeper, occasionally forming low cliffs formed in Triassic sandstone. The steep banks of the rivulet gully and thick tangle of brush were difficult to cross until the first bridge was built, and the convict timber getters had to do this routinely while they were collecting the building materials for the new settlement. The southern side of the ridge formed a steep slope and cliffs that dropped to the shelving shoreline and sandflats of the cove. Reaching the shoreline from the ridge crest would have involved negotiating the almost cliff-like bank to reach the water's edge. Later, steps were added on Argyle Street to help pedestrians negotiate this slope. There is a saddle-like position with a patch of mostly level ground at the midpoint of the ridge spine and this was where the first encampment was placed.

The geology of the ridgeline spine of central Hobart is underlain by bluish-gray (pale yellow when weathered) freshwater Triassic sedimentary rocks in which plant and animal fossils are found (Leaman 1976, 26). Much of the rest of inner Hobart is underlain by dolerite (Macquarie Point, Battery Point, The Domain) and a thick blanket of talus (rubble and debris) shed from the flanks of the hills and mountains to the west, which has accumulated on the western slopes and floor of the Hobart Valley. Central Hobart is an island of older rocks surrounded and mantled by large areas of younger intrusive rocks, poorly consolidated debris, and recent sediments (Darwin 1836, in Banks and Leaman 1999; Krause 1883; Forsyth and Clarke 1999).

Reasons for the Choice of Sullivan's Cove

The primary reasons for David Collin's choice of Sullivan's Cove were that it provided one of the few significant, sheltered deepwater anchorages with a reliable freshwater supply in the Derwent Estuary, and there was better agricultural land potential than at Risdon Cove. Northward, the estuary narrows and becomes fresher (Flannery 2006, 30) but the bays in this sector, such as Risdon Cove (on the eastern shore), Stainforth's Cove (New Town Bay), and Prince of Wales Bay, are too shallow and shelving to provide good harbor facilities. They also lack reliable freshwater sources during the warmest months.

The geological elements of the Sullivan's Cove site were instrumental in its choice. The reliable freshwater flow of the rivulet was provided by the rain- and meltwater-fed talus slopes on the flanks of the nearby slopes, which acted as groundwater reservoirs. A small dolerite outcrop in the cove, named Hunter Island, provided a secure site for early food storage (an important consideration in view of food shortages that Collins experienced in Sydney). The deepwater anchorage close to shore minimized manual handling when offloading ships.

Early Phases of Settlement Development

Despite its simple beginnings, the settlement of Hobart rapidly developed into a complex urban center after passing through several distinct development phases. Each phase influenced succeeding stages to varying degrees, which has resulted in aspects of even the earliest settlement being preserved in the modern urban fabric of the central city. The landscape imprints of the pre-European landscape that formed the context for the earliest camp remain embedded within these patterns.

1804–1812: The First Camp and the Collins Settlement

At first light on February 20, 1804, several boatloads of convicts and soldiers left the *Ocean* and, passing a small dolerite island on their left, entered the mouth of a small rivulet near the northeastern corner of Sullivan's Cove. Just negotiating the narrow and shoal entrance of the small stream would have been difficult, at least for the first time. The shallow estuary of the rivulet did not allow the boats to travel very far upstream and after turning sharply left, they landed at a small sandbar/beach. Clearing ground began immediately (Bolt 2004, 91, figure 2.2).

The first phase of establishing Hobart involved the choice of sites for the convict and free settlers, and then the rapid establishment of temporary housing for the military and civil staff. The convicts were kept together in one area and pitched their tents in lines along the south bank of the rivulet (Bolt 2004, 91, figure 2.2). The first temporary camp was organized on military lines and comprised of tent lines with some more solid structures built of local timber. It took advantage of the smaller and local landscape features of the Central Ridge, the initial focus of the first settlement, including access to water, level ground for housing and higher ground for defensive and authority establishments (e.g., the Marines Barracks and the lieutenant governor's tent, and then temporary house). The early growth of the camp was then largely "organic," building on these central foundational arrangements.

The site chosen for the main camp was on higher ground, which was occupied by more open (dryer) woodland than the course of the rivulet, which was occupied by this "brush," and there was no flat and easily cleared ground near the rivulet. The Marine Barracks was formed on higher ground at the west end of the camp and the convict tent lines and was surrounded by a fence. This was probably a good defensive position. Government House and the flagstaff were placed on the high shoreline, in a prominent position at the southern end of the main convict tent lines. When a more permanent residence was built, it was oriented parallel to cliff top (figure 2.2).

There was a need for some geographic separation of different levels of society, such as free settlers and convicts, and the civil and military administration. Government House was placed in a prominent position above the cliffs near the mouth of the rivulet and next to the Commissariat Store and was one of the most important permanent structures of the first settlement. The other civil officials built their dwellings on blocks to the east and west of Government House, following the rising spine of the ridge backing the coastal cliffs and northern shoreline of the cove. The surgeon chose a healthy location north of rivulet to place a hospital tent. The small group of free settlers was granted land around Stainforth's Cove (modern New Town Bay) 4 km to the north of the cove where they were well away from the convict settlement.

High points overlooking the heads of the cove were developed as defensive batteries. The first, on the hill overlooking Macquarie Point is now the approximate site of the Hobart Cenotaph. The second, created around 1818, at the foot of the steep hill on the southern side of the cove would give its name to one of the oldest Hobart residential suburbs, Battery Point. Further defensive gun emplacements were built around the estuary in the following century, including the Alexandra Battery, on an eminence formed by the erosional remnant of a Cenozoic volcano in Lower Sandy Bay. Important defense sites utilizing prominent topographic positions include,

- The Mulgrave Battery—the first defensive structure built and the feature from which Battery Point gained its name.
- Macquarie Point—where the later Queens Battery was constructed. The site was later redeveloped as the Cenotaph.
- Anglesea Barracks—was established on the high ground of Barracks Hill in 1811, with a panoramic view of the settlement and estuary, and
- A line of semaphore stations forming a rapid communication system. The example on Mount Nelson later became a shipping signal station.

Most of these fortifications became important reserves and parks (including the Hobart Cenotaph, Battery Point Park, the Alexandra Battery Park, and Kangaroo Point on the eastern shore). Anglesea Barracks remains the headquarters of the Australian Defense Force in Tasmania.

Many of the buildings housing government institutions in central Hobart occupy sites that were originally at the Triassic sandstone cliff top, in places that were chosen by the first administrators to pitch their tents and build huts in 1804. These prominent sites grew into fenced plots with gardens and were eventually consolidated into the grounds of first Government House. The plots were regularized by Meehan's 1811 street survey and the establishment of the more formal grid of the CBD. They are now occupied by the most important government institutions in the state, including the Executive Building at 15 Murray Street, the HCC Service Centre (formerly the Hydro Building), and the Tasmanian Museum and Art Gallery (TMAG) Complex. The Hobart Town Hall (including the Maritime Museum) occupies the site of the original flagstaff. The southwestern end of the former Georges Square is filled by the Department of Treasury and Finance, while Franklin Square occupies the northeastern end, thus retaining its two-century role of public open space.

The Royal Hobart Hospital is still in the approximate location where the camp surgeon established the first convict hospital in 1804, on a site just north of the rivulet (figure 2.2). This location remains the center of health care in Tasmania and the modern Royal Hobart Hospital has grown to fill an entire city block.

The thick vegetation in the steep-sided rivulet gully hindered easy access to the stream, so a watering place was cleared on the bank near the camp. Sanitation was an immediate issue, with convicts discharging their "filth" into the rivulet. One convict was assigned to choose the site for a privy and to ensure that the other convicts used it. The first bridge across the stream to the carpenter's workshops was built near this spot (Bolt 2004, 105).

At Sullivan's Cove, the first agriculture began as the vegetable gardens established upslope of the Marine Barracks and in the grounds of Government House (Harris 1804; Mault, in Walker 1889; and Bolt 2004). Each government official had their own garden. This rapidly proved inadequate and the first significant government farm was established 4.8 km north (Semmens 1988, 20), on a basalt peninsula on the north side of Cornelian Bay (then known as Farm Bay), which later became the Cornelian Bay Cemetery.

The layout of the first temporary camp created a general structure that remained in place for many years. Several key land-use "anchor points" were created at this time that would determine the types of land use of the places into the future. Of most importance was the location of Hunter Island and its use as the first store's depot: the first permanent Commissariat Store that was built on the shore nearby. The first jetties were developed on Hunter Island and the adjacent shore, and they serviced the deepwater anchorage just off the island. Hunter Island was progressively developed into the primary harbor facility of Hobart and the surrounding area became the focus of industrial activity. The general alignments of the tents and temporary barracks would become more firmly fixed as more permanent huts and other structures were built. Once Meehan began his street survey in 1811, the existence of such structures and land uses had to be accommodated in the road network, which prevented the establishment of a perfect right-angle grid.

Stainforth's Cove (New Town Bay)

Stainforth's Cove was chosen by Collins as the location of the land grants that would be offered to the first free settlers of Tasmania, to keep them at a distance from the convicts at Sullivan's Cove (figure 2.1). The original land grants were surveyed by George Harris and James Meehan and shown in Meehan's 1804 survey plan. The area did not flourish as a commercial center but later came to be a favored rural retreat for Hobart's elite. New Town Bay also later became the site of a Female Convict Depot, based on moored prison hulks such as the *Anson* and the *Duke of York* (Williams 2005). The hulks were later replaced by a building complex constructed on the former brickfields (and now the site of North Hobart Oval).

Later, Trinity Church would establish a burial ground just south of the New Town Bay. In one of the interesting convergences in the early history of Australian cities, Captain Henry Miller, first commandant of the Moreton Bay Penal Settlement and the founder of Brisbane was buried here. Later, when the burial ground fell into disrepair, Miller's remains were disinterred and moved to the Cornelian Bay Cemetery, originally the site of the first government farm in Hobart.

The modern street pattern of eastern Moonah and New Town is still heavily influenced by the geometry of the free settler blocks surveyed by George Harris and James Meehan along the New Town Rivulet, which were originally laid out normal to the course of the stream, presumably so that each block had access to freshwater.

Meehan's 1811 Survey to Arthur's Appointment of the Commissioners in 1825

In 1811, Governor Lachlan Macquarie visited Hobart and decided that the settlement was too disorganized, and he ordered James Meehan to lay out a more orderly street grid (Bolt 1981, 2004, 74). Meehan began his survey from the tree he had blazed as a survey marker at the rivulet mouth in 1804 (Bolt 1981, 2004, 74). As Meehan worked, he seems to have made concessions to existing landholders and boundary features and did not impose a rigid grid. The locations of several earlier buildings and landholdings already set in place by the more organic growth and layout of the first camp, and influenced by the natural landscape, would affect the shape of the arrangement of the new streets. Accommodating some of the existing landholdings into the survey meant that a true street grid could not be imposed on the settlement. The result is that modern Hobart streets are not a true 90° grid and elements of the original organic camp layout continue to influence the slightly irregular modern core of the Hobart CBD. Features such as the gap originally left between the convict hut lines and the palisade of the first Marine Barracks became the alignment of Murray Street (Bolt 2004, 212; figure 2.2). And the alignment of modern Franklin Square still retains an orientation that is a "ghost" of the grounds of the first Government House and the crest of the original steep shoreline. Meehan's 1811 survey could never quite eradicate the disorganizing elements of the earlier camp and settlement layout.

In 1825, Governor Arthur appointed three "Commissioners of Survey," with responsibility to not only direct the general survey of the island but to also make recommendations about ways to improve the layout of Hobart Town. This was the first significant town-planning effort in Australia (Craig 1945, 99). The consolidation of all these early phases of town development were documented in Sprent's 1841–1846 town survey (Solomon 1967) and remain in the modern streetscapes of the Hobart CBD.

The Road Network

The first tracks to develop beyond the camp were to the north of the cove, where convict gangs were accessing timber resources in the valleys and dragging logs down to the saw pits and carpenter's shops established on the north bank of the rivulet. The best stands of high-quality timber trees stood on the talus slopes and boulder deposits of the eastern slopes of Knocklofty and Mount Stuart (both foothills of Mount Wellington), to the northwest of the settlement, and particularly in the deep valleys formed by southeast and eastward flowing streams. These early tracks became the beginnings of Elizabeth Street, Liverpool Street, and Augusta Road (Bolt 2004).

To travel from the cove to the northern timber areas required crossing the deep and steep-banked rivulet, so a bridge was soon required. The first bridge was placed across the narrowest point of the lower rivulet (see Harris 1804), where many people gathered to collect water. It was also located at the foot of a run of timber that extended up the ridge and valley to the northwest (Bolt 2004, 106). Once this log and deck bridge was in place, it allowed the carpenters and saw millers to access the stands of timber they needed (figure 2.2) and became one of the most important influences on the northern development of the layout of the settlement. The track from the bridge eventually continued further northward as the convict gangs accessed more distant stands of timber, gradually evolving into Elizabeth Street. North of the rivulet, both east and west tracks were formed by the convict gangs as they dragged the timber back to the carpenter's shop, which eventually became Liverpool Street. At the top of the rise, this track veered to the north, toward the free settlers in Stainforth's Cove (New Town Bay), where a surviving section is still known as Bay Street (Bolt 2004, 106) (figure 2.1).

With the establishment of the free settlers at New Town and the longer-term brickmaking and limekiln industries in the area, the road network spread northward from the settlement and supplemented the earlier tracks of the convict timber getters. Eventually they became the main north-south arterials of the new settlement.

The steep southwest-northeast trending coastline, central ridge, and often rocky banks of the Hobart Rivulet caused difficulties for north-south horse-drawn traffic, and even pedestrians. Stone steps were installed for pedestrians to negotiate the steep slopes from the foreshore to the town on Argyle Street and others were needed across Murray Street. It was not until the hill crests were cut down that horses could travel that way.

The earliest northwest-southeast connector road developed between the Sullivan's Cove camp and the New Town free settlers along the valley of the Domain Rivulet (Park Street Rivulet). This route is still followed by the Brooker Highway. The floor of the valley followed the approximate line of the faulted western contact of the Queen's Domain dolerite and formed a natural eastern boundary of northern urban expansion.

Anchorages and Land Reclamation

In 1804, a small, elongated oval-shaped dolerite island lay near the mouth of the Hobart Rivulet and was linked to the shore by a sandy spit that was dry during low tides. It was named Hunter Island and was developed as the first secure storage site for the stores offloaded from the ships anchored in the deep water nearby. Three days after landing at Sullivan's Cove, a convict party was at work constructing a new wharf on the island (Knopwood 1977, 44). The first major harbor works in Sullivan's Cove were the development of a formed jetty along the alignment of the natural sand spit to link the island to the mainland. A new Commissariat Store was built at the landward end of the spit and took over the function of storage. This early work led to successive phases of land reclamation along the foreshore of Sullivan Cove and the focusing of many industries around the store and first wharf areas. The first Government House and flagstaff, occupying the same approximate site as Collins' first tent, overlooked the wharf area that developed around Hunter Island.

Geohazards

Landslips and mass flows have shaped much of the western shoreline of the Derwent Estuary, particularly to the south of the study area, on the slopes of Mount Wellington, and on the slopes of Knocklofty. Occasionally, slope failures had catastrophic results, with large slips reaching the waters of the estuary. In recorded times, the most significant was the Glenorchy debris flow which killed several people and cut a path of destruction from the upper slopes of Mount Wellington to the shores of Prince of Wales Bay (e.g., Stevenson et al. 2016). Many scars remain on the mountainside today. And there were many more such catastrophic slips such as those that now form the Wrest Point Peninsula, where the unconsolidated Cenozoic mass-flow deposits caused engineering problems as a hotel was constructed (Stevenson 1969). Similar dolerite boulder deposits occur along the shoreline of Taroona (Cromer 2015) and the still active slopes near Taroona High School (Latinovic et al. 2001).

The debris fields and scree deposits act as subterranean water reservoirs that feed the many springs at the headwaters of rivulets (e.g., the Sandy Bay and New Town rivulets). The valleys formed by the drainage lines across the talus and boulder deposits provided excellent habitats for quality timber trees while the landslips and mass flows have shaped the shoreline, where they have also formed good agricultural soils. The boulder beds that provide such good agricultural soils and tree habitats are still forming (e.g., the Glenorchy debris flow, Mazengarb et al. 2007; Stevenson et al. 2016). The geology of debris flows and their slope stabilities still represent major planning considerations in the Greater Hobart area, particularly in the western suburbs.

In 2018, heavy overnight rainfall in the Hobart and Sandy Bay rivulet catchments caused flash flooding in the lower reaches of the water courses, particularly in central Hobart and Sandy Bay. The worst affected areas were city blocks bounded by Collins, Davey, Argyle Streets, and the Brooker Highway. The floodwaters overwhelmed the canalized and built over Collins Street drain and re-found the original course of the Hobart Rivulet to enter Sullivan's Cove at the approximate location of the original 1803 mouth at Hunter Island.

THE BUILT ENVIRONMENT

In 1944, Craig (1945, 99) observed that Hobart had retained a higher proportion of its early colonial character than any other Australian capital city. This is still the case and in 2009, Leaman (2009a, 49–60) was able to list over thirty-five buildings and landscape features within central Hobart that were present when Charles Darwin briefly visited in 1836. The urban character of central Hobart is still largely determined by this remarkable and unique collection of historic brick and sandstone buildings, which still form almost complete streetscapes recognizably those of the 1830s to 1860s. Such streetscapes exist nowhere else in a modern Australia capital city.

Colonial Hobart, like Sydney, was a sandstone city, largely built from materials that outcropped in the local area. But unlike its much larger older sister, Hobart does not rest on much of the rock from which it is built. While the earliest buildings sourced local stone, the later buildings that now form most of Hobart's grand, and mostly intact, colonial streetscapes often required higher-quality stone that came from further afield and not from the immediate settlement area. Some of the most important sandstone buildings are listed below.

- The Town Hall,
- The Old Hospital Building (old Lands Department),
- Old Treasury Building,
- Royal Society Meeting Room (Tasmanian Museum complex),
- Parliament House,
- Domain House—the original building of the University of Tasmania,
- Government House,
- Salamanca Place warehouses,
- St Georges Church Battery Point, and
- Royal Engineers House.

Surviving early buildings that are shown on maps of Hobart extending back to 1811 (e.g., Meehan 1811; Sprent 1846) are the original Commissariat Store building, which was later "twinned"with a newer Bond Store building, and then remodeled to become the Customs House. It is now the offices of the TMAG. The former governor secretary's cottage that was once within the grounds of the Old Government House survives and is now situated within the TMAG grounds.

Brickmaking Clay and the Hobart Brickfields

While stone was used for many of the Hobart's grandest colonial buildings, it has always been a largely brick-built town. Robert Knopwood noted that bricks began being made in October 1804, within a few months of Collins establishing the Sullivan's Cove camp. The first brickmaking was a piecemeal affair, utilizing local clay sources encountered in the early settlement area, and as where they were needed (Knopwood 1977, 62; Waight 2015, 3). The most historic and pivotal building in Hobart is the Commissariat Store, built on the foreshore from brick in 1810 (figure 2.2). This historic reference point has witnessed all subsequent development of the city.

By 1827, an area on the northern outskirts of the settlement, and adjacent to the main road from Hobart Town to New Town had become a focus of government brickmaking and became known as the Brickfields. This area is noted on all early maps of Hobart without any details of its layout or pattern of resource exploitation (figure 2.1). The area is underlain by mainly Permian carbonaceous siltstone and mudstone (Forsyth and Clarke 1999). In the modern era, all the brickmaking clay of the Hobart area was sourced from weathered Triassic mudstone. Permian beds in the Hobart area are usually too siliceous for use as a ceramic material (Leaman 1976, 78).

The site of the Brickfields was rapidly overtaken by urban development and was redeveloped by 1842 to become the location of a female factory, where female convicts holding probation passes awaited their work assignments in the colony. The hiring depot buildings eventually became an immigration depot and then a paupers' establishment before being demolished in the 1920s to make way for the North Hobart Oval, which was the headquarters and home of Australian Rules football in Tasmanian for over six decades.

Dolerite and Urban Character

Dolerite was rarely used as a building stone because it was extremely hard and difficult to work into regular shaped blocks. However, examples do exist, such as the base courses of Salamanca Place warehouses, which are otherwise built of sandstone, and one example of dolerite used for façade walls (see figure 8.2, panel 2I). A more usual use of dolerite was for construction fill, armor rock, and road base.

One of the subtle geological influences that dolerite has had on Hobart is the use of its weathered products as road base and gravel, particularly as red gravel. Dolerite gravel is the main component of most gravel roads (as top dressing and as road base), and most gravel driveways. The characteristic red-brown color of most Tasmanian gravel roads and tracks is the result of the use of weathered dolerite. Crushed dolerite is also used as blue metal of lesser quality.

Historic Quarries and their Current Landforms

Former quarries have been repurposed across the study area but particularly in the Domain. The quarries that supplied the high-quality sandstone used to construct the Government House were redeveloped as the site of the former Beaumaris Zoo and as ornamental lakes.

Other quarries that have been reused include the main council depot at the northern end of the Domain Hill, which occupies the site of a quarry shown on Krause's 1883 geology map. The line of buildings opposite the intersection of the Brooker Highway and Brisbane Street also occupies a former quarry recorded by Krause (1883). And another former dolerite quarry is now occupied by the Domain Aquatic Centre.

Sources of Lime

The first lime used in Hobart was made by burning shells from the extensive raised beds formed as strandlines around the estuary shoreline. As mentioned above, some of these deposits contain stone artifacts and charcoal layers and were probably also Aboriginal shell middens. The extensive shell deposits on the estuary shoreline were important raw materials in the early colony, even some resources that were remote from the main settlement. Robert Knopwood recounts how he had to assist two Ralph's Bay lime burners to return to Hobart after they were driven away by Aboriginals in June 1806 (Knopwood 1977, 157; Boyce 2019, 41). Charles Darwin later visited the Ralphs Bay shell deposits during his 1836 visit (Banks and Leaman 1999, 44–5).

Shell lime was not an inferior product and even after the discovery of limestone, it was still utilized. The limestone from an early quarry that was in operation by June 1816, and located just one mile from Hobart Town, was considered good for masons' work, but shell lime was better for plasterers (McPherson 2004, 119).

Permian limestone outcrops extensively just to the north of the study area, in modern Glenorchy. It was worked as an important lime source, which led to considerable infrastructure development on the northern fringes of the settlement, such a lime kilns, roads, and tracks. Of low quality, its most recent uses were as coarse aggregate and sometimes building stone in retaining walls (Leaman 1976, 70 and 77).

A spring-derived deposit of calcareous material was discovered by two convicts in 1813 and became an important source of lime (Banks and Leaman 1999, 45). The calcareous and fossil plant-bearing spring limestone of the Burnett Street deposit had probably been buried by an early mass-flow debris deposit from Knocklofty (Banks and Leaman 1999, 45). The site of the small quarry from which this material was won is now the John Doggett and West Hobart parks.

Lime Kiln Hill

Lime burning was established on the outer fringes of the early settlement, around a site that became known as Lime Kiln Hill. The hill became the location for several of the photographic panoramas taken of early Hobart. The lime kiln closed when there was no longer a place to dump waste (Marsden 1995, 30). However, the brickmaking industry persisted in this area until the end of the twentieth century, with quarries on Knocklofty and brickmaking furnaces in New Town only closing in the 1980s.

The New Town Coalfield

Several coal seams associated with carbonaceous mudstone and shale outcrop within the Permian strata immediately to the northwest of the study area (on the western edge of the New Town), on the lower northeastern slopes of Knocklofty. The geology of these mostly uneconomic occurrences was described in detail by Selwyn (1855), Thureau (1883), Krause (1883), Bacon (1986), and Bacon and Banks (1989). Examinations of these deposits led to the first geological map of Hobart (Krause 1883) and one of the first drilling exploration campaigns in the colony, at the Cascades, undertaken by Mr. Degraves of the Cascade Brewery. The exploration hole was drilled to search for extensions of the New Town coal seams beneath the western end of Hobart Town, where Krause (1883) had predicted they would occur. Drilling for deep coal seams near the Cascade Brewery followed the advice of the "eminent" New South Wales geologist C. S. Wilkinson. Later a private shaft was sunk in an old quarry near the brewery. None of this work was successful.

Locating Modern Infrastructure

Ever since the first log bridge was built across the steep banks of the rivulet to enable convict gangs to access better timber resources, the steep and deeply incised terrain of the Hobart Valley has affected the siting of important transport infrastructure. Fords and bridges also had to be built across the Maypole and New Town Rivulets and they formed fixed points that controlled the evolving northern road system (see figure 2.1).

When the railway into the town center was constructed in the 1870s, a coastal route was used that skirted the eastern foot of the Domain Hill and Macquarie Point. The city terminus was built on a section of land on the southwestern slopes of the Domain Hill. The only alternative level route would have been along the course of the Domain Rivulet. This may have been too difficult because it was still a flowing water course.

As motor vehicles became more common in the postwar era, the need for a major northward outlet route to the developing suburbs became necessary. The only alternative was a level route that followed the course of the still mostly undeveloped Domain Rivulet valley. Only when the rivulet was enclosed in drains could this corridor be redeveloped as Brooker Avenue and then the Brooker Highway. Subsequent design of the Southern Outlet, the major arterial outlet to the south of the city now partially follows the course of the upper reaches of the Sandy Bay Rivulet.

Plains and the Airports

The generally rugged and deeply dissected topography of the Hobart region has few lowlands and plains. One effect is that there is little broad-acre arable farmland. Consequently, most agriculture near Hobart involves grazing (Scanlon et al. 1990). The topography also makes it difficult to find locations for airfields. Hobart has two airports, the original Cambridge Aerodrome, which is the location for flying training, tourist light plane traffic and commercial flights, and the Hobart Airport further east (established in the 1950s). Both are located east of the estuary, on one of the few open and truly flat plains near Hobart. The flat land provided an ideal landing strip for aircraft and so has been the location of Hobart airports since the 1920s. These plains are largely formed from Quaternary marine sediments that were deposited as a sand bar at the mouth of the Coal River at Pittwater. There is a covering of windblown sand dunes that extends the length of Seven Mile Beach (Mathews and Donaldson 1994, 7).

Chapter 3

Ridgelines and Waterways

Brisbane, Queensland, 1825

In April or May 1825, the Moreton Bay Penal Settlement was moved from the Redcliffe Peninsula to a new inland site on the Brisbane River, but no one bothered to record the exact date (Steele 1987, 27). There were probably good reasons for this oversight. The coastal location on the shores of Moreton Bay lacked reliable water and agricultural land, and several alternative sites had been considered. Lieutenant Henry Miller of the 40th Regiment of Foot, Waterloo veteran and Commandant of the first settlement, chose the new location but left no record of why. He seems to have ignored, or misinterpreted, the suggestions of John Oxley, the surveyor general, and the governor, Thomas Brisbane, who had recommended a site further down the river, probably near modern Breakfast Creek (Statham 1989, 12). Ships larger than 100 tons could not cross the bar and enter the Brisbane River, so everything had to be man-handled and rowed up the river to the new site. The convict workforce available to do this was unfit, and probably unwilling. The work involved in managing the movement of people, food stocks, and building materials from one recently established settlement to another must have been considerable. It is perhaps because of the effort involved that Miller may have been too preoccupied to record the events and left no information about why he chose the new location. He was replaced as Commandant less than five months after the move and never returned to Brisbane. From this inauspicious beginning, the nucleus of a settlement was created that would take its name from the river and grow into the modern city of Brisbane.

The city of Brisbane, capital of the state of Queensland, lies in the subtropics, at 27.467°S, 153.027°E, approximately 15 km inland from the shores of Moreton Bay. Modern greater Brisbane is one of the largest urban corporations in the world by area, measuring 1,343 km^2, and now incorporates large

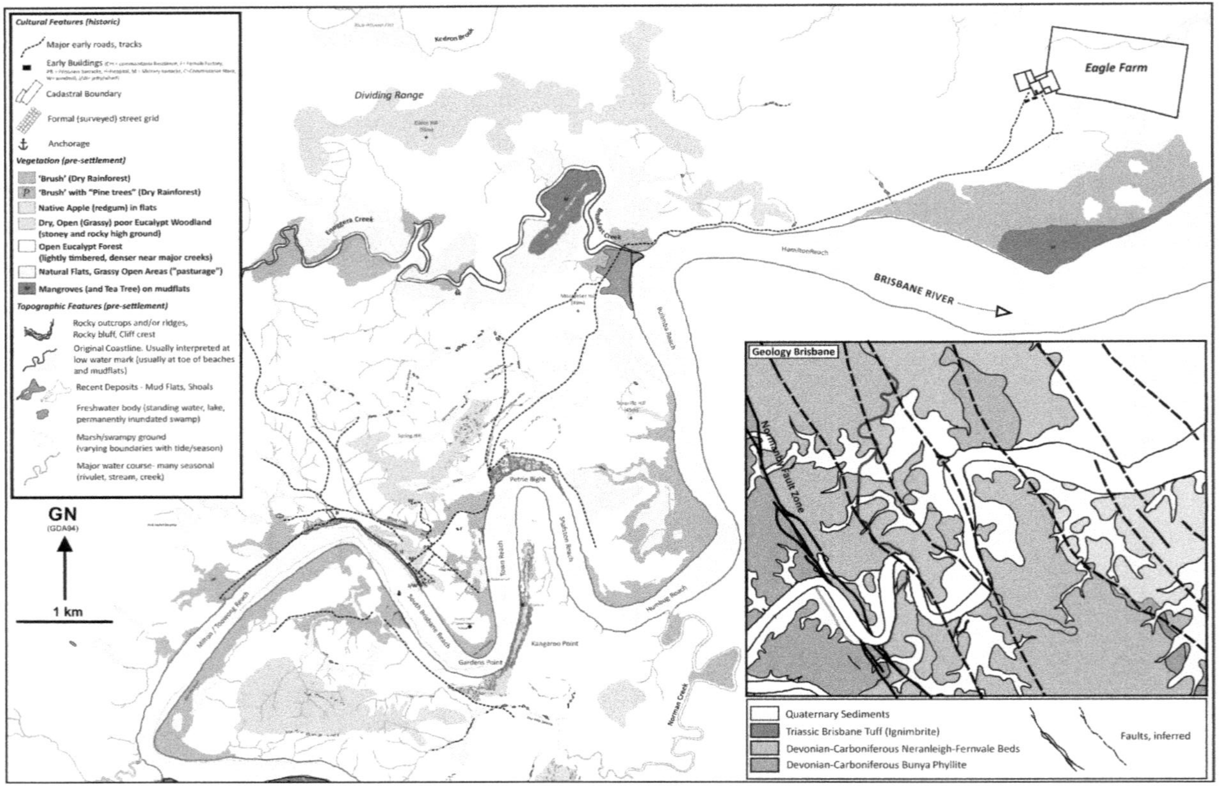

Figure 3.1 The Brisbane Region. Topographic reconstruction of the landscape surrounding the Moreton Bay Penal Settlement on the Brisbane Peninsula circa 1825 to 1844. The hinterland of the settlement was a landscape of curvilinear valleys and bounding ridges with variably marshy, low-lying valley floors, marshes, and wetlands. A general reconstruction of the original native vegetation circa 1825-1844 is also shown. This is largely based on the survey mapping by Henry Wade (1842, 1843, 1844). **Inset.** The geology of the Brisbane region. The geological units are simplified from Houston and Tucker (1965), with additions from Rands (1887), Grimes, Cranfield et al. (1986), Grubb (1989), and Bennett and Norbert (2014). Please refer to the extended caption at the rear of the book for additional information.

regions that were never part of the original 1825 settlement (see figures I.1 and I.2 for location).

The discussion in this chapter is focused on a 12 km^2 part of greater Brisbane that extends northward from modern Woolloongabba to York's Hollow and Fortitude Valley; and east from Milton and South Brisbane to Kangaroo Point and New Farm (figure 3.1). Within this small area, most early European settlement was focused on the Brisbane Peninsula, an informal geographic name adopted here to describe the V-shaped river pocket bordered by the South Brisbane and Town reaches of the Brisbane River and extending south from modern Wickham Terrace and Spring Hill to Gardens Point. This was the location chosen by Henry Miller for a permanent settlement in 1825, and it was the focus of European activity in the first two decades of settlement. The surrounding hinterland also became important, and particularly the region extending to Enoggera-Breakfast Creek and Eagle Farm to the northeast of the peninsula. The Brisbane Peninsula has remained a central focus of government and commercial activity since the 1820s and is now the CBD of modern Brisbane.

The first two decades of the growth of Brisbane are the historic focus of this chapter, extending from 1823 to approximately 1844. During this period, the Brisbane River was discovered and explored by Europeans, the Moreton Bay Penal Settlement was established and then subsequently opened to free settlement in 1842. During this formative period, early Brisbane developed in several distinct phases (information mostly from Steele 1975).

Phase 1. September 1823–May 1825: John Oxley explores Moreton Bay and chooses the Redcliffe Peninsula as the site for the first (possibly meant as temporary) penal settlement. Convicts and a military detachment are sent to build the new establishment and arrive in Moreton Bay on September 12, 1824.

Phase 2. April–May 1825: Settlement moved to a narrow pocket (peninsula) on the north bank of the Brisbane River. The construction of a new and permanent penal settlement (a military camp) began, initially comprised of tents and timber structures. Brickmaking, stone quarrying, and farming began shortly afterward.

Phase 3. Expansion into the hinterland: Limestone Station founded (1827) and then Eagle Farm (1829).

Phase 4. 1836–1842: Winding down of the penal settlement and final closure, free settlement, and Wade's land survey from 1839 to 1844 (Wade 1844).

Phase 5. 1859. Independence from New South Wales. Brisbane was established as the seat of colonial government and administration for the newly independent colony of Queensland. Government institutions initially utilized the mainly stone buildings of the former penal establishment. The modern street grid was established as an overprint of the old penal settlement layout. Through a slow process of demolition of the old buildings and the construction of buildings following the grid alignments, the new settlement was imposed, and the old settlement layout largely disappeared.

The focus of this chapter are phases 2–4.

EUROPEAN DISCOVERY AND EXPLORATION OF MORETON BAY AND THE BRISBANE RIVER

The Brisbane River had only recently been discovered by Europeans by 1824, but knowledge of the region had grown gradually over the preceding fifty-five years. In 1770, members of James Cook's *Endeavour* crew had noted the paler water as they sailed through the region and suggested that a river may be nearby. Cook seems to have doubted this but was unable to check the possibility because the ship "had the wind." He named Cape Morton [*sic*], and another feature as Morton Bay [*sic*] and sailed on (Steele 1972, 7–8). Matthew Flinders was the first to enter and explore Moreton Bay in April 1799, naming several features that included Red Cliff Point (named because the rocks were "a strongly impregnated iron stone," now Woody Point), Pumicestone Channel, and Point Skirmish (now South Point). Flinders recognized South Passage, and that Moreton Island was not part of the mainland (Steele 1972, 9–19). Flinders also noted a feature on his chart that was the mouth of the Brisbane River, but (apparently) made no comment about it (e.g., Figure 5 of Steele 1972).

Despite other brief visits to Moreton Bay, the Brisbane River was only discovered and named by John Oxley, NSW surveyor general, in 1823. His formal discovery followed information supplied to him by John Finnegan and Thomas Pamphlet, two castaway timber cutters who had been shipwrecked on Moreton Island the previous year. They had been welcomed and supported by the local Aboriginals and during their travels in the region, had followed the course of the Brisbane River upstream as far as Oxley Creek. They were eventually found by John Oxley at Point Skirmish in November 1823 (Steele 1972, 9–19). Oxley had been sent north by Governor Brisbane to seek out potential sites for further convict settlements. In December 1823, Oxley entered the Brisbane River, accompanied by Finnegan, and explored as far upriver as modern Goodna (Steele 1972, 87). The party stopped for a meal break near a steep bluff (formed from Brisbane Tuff, see below), which they climbed to see the coast and the surrounding country. Oxley decided during this reconnaissance trip that Moreton Bay would be a good location for the new convict settlement (Steele 1972, 87, 127).

In September 1824, Governor Brisbane ordered an advance party of convicts and army guards to Moreton Bay to establish an outpost. The military detachment and convicts were under the command of Lieutenant Henry Miller. Transport was provided by the brig *Amity,* under the command of John Oxley, who was also charged with choosing the site of the settlement and surveying the local region (Steele 1972, 127). On September 17, Oxley began exploring the river in two boats, accompanied by the botanist Allan Cunningham. There is now a memorial on Coronation Drive that marks the approximate place where John Oxley landed at the future site of Brisbane. This was the first of several exploratory expeditions of the river system over the next three years, which included those of Edmund Lockyer (in 1825), Patrick Logan (in 1827), and Allan Cunningham with Charles Fraser (in 1828) (Steele 1972). Edmund Lockyer and Charles Fraser, and the brig *Amity*, would also play significant roles in the subsequent settlement and early exploration of Western Australia (see chapter 4).

GEOLOGY OF THE 1825 BRISBANE SETTLEMENT

Three main rock units underlie the first settlement landscapes of central Brisbane. The basement is comprised of older metamorphic rocks; the Bunya Phyllite (western side) and the Neranleigh-Fernvale Beds (NFBs). To the east, the basement is unconformably overlain by a north-south trending belt of welded ignimbrite known as the Brisbane Tuff (figure 3.1 inset).

Basement Paleozoic Metamorphic Rocks

The basement rocks of the Brisbane region are comprised of low-grade, metamorphosed Paleozoic and late Paleozoic sediments and volcanics outcropping in discrete, north to northwest-trending structural blocks. In the study area, the NFBs and the Bunya Phyllite are the most important.

Around 315 million years ago, the sequence was severely deformed and metamorphosed by a subduction event and is now comprised of phyllites, argillites (cleaved shale), and greywacke (Wilmott 2012). The Bunya Phyllite was particularly intensely deformed, producing a mica-rich schist with common deformed white quartz veins. The NFBs occupy the central part of the study area beneath the modern CBD. They are mostly comprised of argillite, but include phyllite, greenstone (metamorphosed pillow basalt), chert, arenite, and other metasedimentary rocks (Bennett and Norbert 2014, 132–5). They were originally deposited as part of a thick sequence of medium- to fine-grained sediments formed as submarine fans around 370 million years ago (Devonian). The Bunya Phyllite underlies most of the area to the west of the South Brisbane Reach of the river, including modern South Brisbane and West End. It was formed from a finer-grained, muddy part of the same sedimentary sequence as the NFBs (Wilmott 2012). Shear bounded slivers and strips of argillite, greenstone and marble (altered greenstone) are incorporated into deformed Bunya Phyllite adjacent to the deformed boundary with the NFB on the western side of the Brisbane Peninsula, in the Queens Wharf area (Grubb 1989).

The basement metamorphics are strongly foliated, possessing a schistosity defined largely by micas, with some quartz veins. The recrystallization of minerals with a preferred orientation created a strongly favored grain in the basement. In central Brisbane, the foliations in the NFB strike slightly west of north. However, just to the west, the foliation trends are more curvilinear in the Bunya Phylite, and range from east-west to north-south, and define a broad antiform (Holcombe 1977).

A series of near-vertical shears and faulted zones are present throughout the basement metamorphics, which tend to parallel the dominant northwest-southeast trending foliation (Holcombe 1977; Grubb 1989; Bennett and Norbert 2014), but are mainly focused on the boundary between the Bunya Phyllite and NFB, where they form the Normanby Fault Zone (NFZ) to the immediate north of the settlement area (Bennett and Norbert 2014). The NFZ has multiple splays, particularly to the west, within the Bunya Phyllite. The NFZ probably extends southward, beneath the South Brisbane Reach of the river (Houston and Tucker 1965). The fault systems in the study area probably represent the southern end of the North Pine Fault System, which is part of a regionally significant series of north to northwest striking fault zones (Babaahmadi and Rosenbaum 2014).

It is more difficult for running water to erode across the grain of the rock than along it, so drainage lines have been influenced by the orientation of the metamorphic foliation. The weathering and erosion of the foliated but otherwise more homogeneous metamorphic basement rocks have created the low but hilly topography, and the more irregular drainage pattern of the city and the inner western and northwestern suburbs.

Shears and fault zones in the basement were also important influences on the landforms of the study area at all scales. Erosion and weathering were focused along zones of structurally weakened rock, preferentially etching them out as topographic lows. The course of the Brisbane River follows such a deformed zone at the boundary between the NFBs (underlying the Brisbane Peninsula) and the Bunya Phyllite (beneath modern South Brisbane and West End). River erosion has preferentially picked out the weaker boundary to create the course of the South Brisbane Reach. The course of the river follows this boundary until encountering the outcrop of the Brisbane Tuff at Kangaroo Point, where it is deflected sharply back to the north (see below).

Permo-Triassic Sediments and Volcanics

A sequence of Permian marine sediments unconformably overly the NFBs. They are comprised of mudstone, chert, and conglomerate, and associated volcanics comprised of rhyolite, andesite, and some agglomerate (Cranfield et al. 1976, 1). Marine deposition was terminated in the mid- to late Permian by the Hunter-Bowen Orogeny. A phase of gabbroic to granitic intrusion developed into the mid- to late Triassic, when the Enoggera Granite intruded the basement sequence. Where the Bunya Phyllite mantles the granite in the Mt Coot-tha area, west of the study area, the phyllite was strongly hornfelsed. Metalliferous vein occurrences were formed in the basement rocks from late-stage fluids associated with the intrusions. Some have been worked as mines, including silver-lead veins at Indooroopilly and a gold vein at Kingston (to the southwest) (Willmott 2012, 35). Permian to Late Triassic uplift disrupted the Paleozoic rocks, creating the D'Aguiler and Beenleigh Blocks and formed loci for subsequent sedimentation (Cranfield et al. 1976, 2).

Deposition of nonmarine sediments and volcanics in the Esk Trough began in the early (?) to middle Triassic. Sedimentation ceased at the end of the Middle Triassic when westerly directed compression tightly folded the rocks. In the late Triassic, uplift in the Brisbane region led to an extended period of erosion followed by localized continental sedimentation and the formation of the Ipswich and Tarong Basins, of which the coal measures and associated sedimentary rocks of the Ipswich Basin are the most economically significant. The Ipswich Basin contains basal volcanics, which include the Brisbane Tuff. The rocks of the Ipswich Basin were folded and extensively faulted late in the Triassic. Subsequent down-warping resulted in the formation of the Moreton Basin and Nambour Basin. Fluviatile deposition of the Bundamba Group characterized these basins from the late Triassic to the Middle Jurassic. Uplift and a long period of erosion was followed in the Early Cenozoic by localized continental sedimentation to form the Petrie Formation and the Oxley and Booval groups. The coastal physiography east of Brisbane was strongly influenced by late Cenozoic faulting and Pleistocene and Holocene sea-level fluctuations (Cranfield et al. 1976, Willmott 2012).

The recrystallized rocks of the aureole surrounding the Enoggera Granite created a resistant massif as the granite was unroofed by erosion and weathering and now form the high ground to the west of Brisbane. The elevated region is now a major influence on the regional drainage pattern.

The Brisbane Tuff

Approximately 25–30 percent of the eastern side of the study area is underlain by the Triassic Brisbane Tuff, which has been dated at 226 million years (Willmott 2012, 14). It outcrops as a linear belt trending north to north-east that varies in width from around 600 m to around 2 km. Originally interpreted to be an intrusive porphyry (Wilson 1856), the volcaniclastic origin was recognized by Rands (1887, 1–2) following his geological mapping of Brisbane, since when it has been known as the Brisbane Tuff (Richardson and Bryan 1933). The Brisbane Tuff is a welded ignimbrite, formed when superheated gas and lava was deposited as an ash fall, compacted, and then welded together by its own intense heat. The volcanic deposits are thought to have formed rapidly, from one or more close-spaced eruptions that ejected material from a large volcano somewhere to the north, though the vent has not been found (Willmott 2012, 15). The rock type was originally known as "porphyry" by stonemasons and the quarried stone continues to be referred to by this name.

Although defined as a single stratigraphic unit, the Brisbane Tuff is comprised of two distinct petrological types: a lower stratified type that is an ash fall deposit and an upper massive type that lacks tuffaceous characteristics (Richardson and Bryan 1933, 60). The rapidly cooling ash typically formed a high-strength, massive jointed rock that is layered. At the Royal Brisbane Hospital site, the jointing controlled the geotechnical characteristics of the excavations (Grubb 1989; Qualischefski 2001; Bennett and Norbert 2014).

The Brisbane Tuff lies at the base of the Ipswich Coal Measures and was deposited on an undulating land surface developed on the more ancient basement metamorphics of the NFBs and Bunya Phyllite. The falling ash buried an existing landscape (paleotopography) broken by a series of small lakes (Rands 1888; Briggs 1929, 147). The linear outcrop pattern of the Brisbane Tuff is the result of the ash infilling a valley system (Willmott 2012, 15). The unconformity at the base of the tuff is exposed in the old quarry face behind the Howard Smith Wharfs at Petrie Bight, and at the southern end of Kangaroo Point. It is marked by silty clays or weathered rock, river gravels, and blocks interpreted to be scree formed on the old basement land surface. Occasional fossilized tree trunks are also found (Willmott 2012, 15). The tuff immediately above the contact zone is often highly brecciated.

The tuff outcrops as several elongated ovoid-shaped subareas with north- to northwest-oriented long axes that create an en echelon pattern. The pattern suggests that an original northeast trend of the tuff belt has been dislocated by the NFZ and related basement faults and splays and is now comprised of a series of fault-bounded blocks. The dislocations in the tuff were important influences on the course of the Brisbane River and its major tributaries in the study area. The South Brisbane Reach, Town Reach, Shafston Reach, and the Bulimba Reach are strongly northwest-oriented stretches of the river that follow these structures. Kangaroo Point is underlain by one ovoid body of tuff and the topography of the peninsula (and Petrie Bight and Bowen Terrace on the north bank) strongly reflects the geometry of the underlying tuff.

Preferential weathering of the weaker strata underlying the ignimbrite has resulted in the development of vertical cliffs along the erosional shores of the outer bends of the river in central Brisbane, particularly on the southwestern side of Kangaroo Point and on the north shore, at Petrie Bight (see further discussion below).

Quaternary and Recent Cover

An extensive superficial cover of Quaternary and recent sediments formed during the erosion of the current landscape. The alluvial sediments were mainly deposited in low points of the landscape, such as wetlands, marshes and drainage lines, and as alluvial flats on the inside bends of the Brisbane River and tributaries. They are particularly extensive to the northeast of the 1825 settlement site, where the poorly drained deposits at Eagle Farm became important for early large-scale agriculture. Once adequately drained, the broad flats became racecourses and the location of Brisbane's main airports (see further discussion below).

THE BRISBANE LANDSCAPE, 1825

The Moreton Bay Penal Settlement was established on a "pocket" (peninsula) on the north bank of the winding, southeast to eastward flowing Brisbane River. The river is 340 km long, stretching from its headwaters near Mt Stanley before entering the sea at Moreton Bay, a shallow embayment of the Pacific Ocean that is bordered by extensive offshore sand islands. Most of the upper reaches of the river occupy a low, wide valley that is informally referred to as the Brisbane Valley. The river is joined by the Bremer River approximately 21 km southwest of Brisbane, at modern Ipswich, before traversing the greater Brisbane metropolitan area eastward to the sea. River dredging, rock ledge removal, and port developments have significantly changed the flows and salinity levels of the lower reaches of the river (figure 3.2).

The sinuous and meandering form of the lower Brisbane River results from the interaction between the river and the underlying geology, creating both erosional and depositional shores. Erosional shores lie on the outside of bends and are occupied by cliffs and bluffs, usually formed by the undermining of resistant rock types. In the study area, the most significant erosional shores are on the bends of the river, at the boundaries of the Brisbane Tuff. Most importantly, on the southwestern shore of Kangaroo Point and on the northern

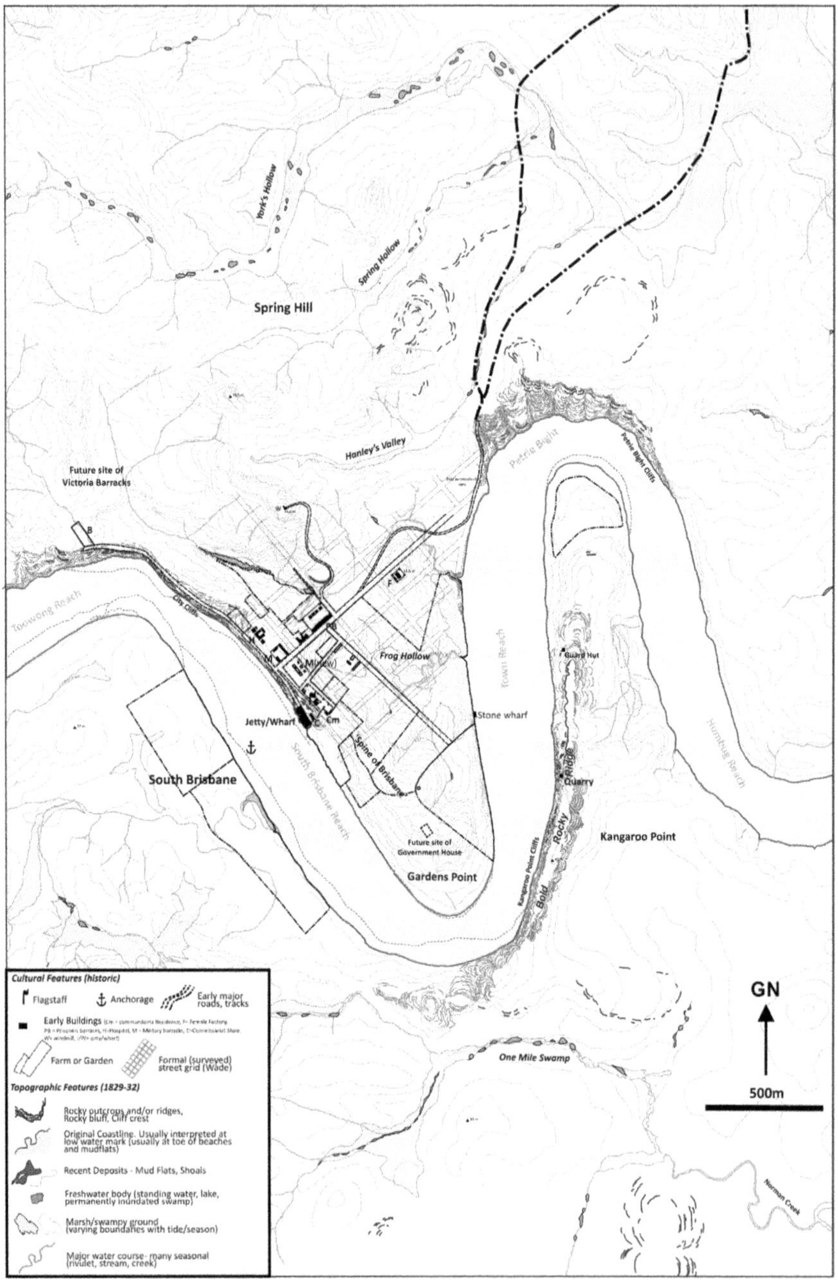

Figure 3.2 The Brisbane Settlement. Topographic reconstruction of the Brisbane Peninsula and the Moreton Bay Penal Settlement, circa 1826-1830. The topographic contours are at 5 m intervals and are derived from the 3D terrain reconstruction model.

Figure 3.2 (Continued)
The 1842-44 street grid of Henry Wade is also shown. The feature labelled 'Jetty/Wharf' indicates the likely first landing place chosen by Henry Miller. This is the location where a rocky shoreline first appeared on the peninsula and was developed into Queens Wharf. The first road of the European settlement must have been a track that followed the rise of the land from the landing place to the north, up the gentlest incline of the steep bank, to the low linear ridge where the camp was formed. The first farms of the penal settlement were established on areas of well-watered but low-lying Quaternary sediments on the inside bends of the river. Please refer to the extended caption at the rear of the book for additional information.

shore of Petrie Bight. Depositional shores are located on the low-lying inside bends, such as the tips and immediate downstream shores of Gardens Point and Kangaroo Point. These places were low and marshy, and usually occupied by natural levees, with wetlands formed behind them that captured the small creeks channeling runoff from the dryer and rockier hills behind. The low-lying and wetter environments behind the natural levees were often occupied by rainforest remnants, usually below the 6 m contour (figures 3.1 and 3.2).

Several straight reaches of river between the bends are individually named. Those near the study area include the Toowong Reach, South Brisbane Reach, Town Reach, Shafston Reach, and Humbug Reach. Quarries Reach is further downstream and quarries on the southern bank were an important early source of sandstone for the construction of several key historic buildings. In addition to the river reaches, several other geographic features of the river are named, including Petrie Bight, Gardens Point (the southern tip of Brisbane Peninsula), Kangaroo Point, and the West End Peninsula.

The South Brisbane Reach and the Southern Shore

The historic heart of modern Brisbane is the South Brisbane Reach of the river. This 1.8 km stretch of the river lies between the Toowong Reach, where it makes a sharp turn to the south, and Gardens Point. This section of the Brisbane River was the place where the first landings were made by European explorers, such as Oxley and Lockyer, and the east bank was chosen as the best landing place by Henry Miller in 1825 (see the section on Miller's first landing place, below). From that time onward, it became the focus of activity in the new penal settlement, particularly after a wharf was constructed at the landing place.

This stretch of the river became the first port area and commerce and settlement spread across the river to South Brisbane, where large trees near the shoreline provided some of the early mooring points for larger ships. The first bridge across the Brisbane River, named Victoria Bridge, spanned the South Brisbane Reach to link North and South Brisbane, which both became important commercial centers after free settlement began (Johnston and Gregory

1989, 247). From the 1850s, early Queensland colonial government institutions were focused on the north shore, and it remains the center of modern state government administration.

The shoreline of South Brisbane is a depositional shore, where river-borne sediments accumulated on the inside bend of the river. Where the river turned sharply southward from the Toowong Reach into the South Brisbane Reach, there was a small sandy beach on the riverbank that was a favorite swimming place for Aboriginal people. The river was shallow here because of the river-deposited sediments, and so was a good place for the local people to swim across to the north shore. During flood events, the south bank could be impacted by the backwash of heavy flows encountering the erosional shore of the city cliffs and the low-lying shore could be inundated.

In 1825, the bank was fringed by low-lying and marshy alluvial sediments with wetlands and was occupied by a narrow strip of rainforest ("brush") with a few Hoop Pines (*Araucaria cunninghamii*) standing tall above the other trees. The river's edge would have been thickly overhung by the forest, and the water level would have varied with the tide. There may have been debris from previous floods hanging high in some of the trees.

Southward along the bank, the rainforest stand became wider and there was a small winding creek entering the river near modern Riverside Green. It may have only been a trickle in the undergrowth and hard to see. There was a small mudflat at the mouth. This little creek drained a network of swamps starting beneath the site of the modern Cultural Centre but extending inland. In a few years after the settlement's foundation, the forest would have been gone, and there would be crops growing here.

West of the alluvial soils of the river's edge, the land became dry, rockier, and harsher, and was covered in open eucalypt forest with stands of wattles. Highgate Hill and West End were low rolling, but stony hills. Modern Grey Street and the railway line follow a ridge of higher ground that once separated the low-lying riverside rainforest from the inland swamps. The higher ground to the west was much drier and occupied by open eucalypt forest with a grassy ground cover. Higher hills further inland were dry and stony, with rocky outcrops forming low banks and covered in scattered eucalypts and grass (figure 3.2).

The Brisbane Peninsula

The Brisbane Peninsula was created by multiple geological processes that reflect the entire history of the region. The outcrop pattern and character of the underlying rocks influenced the form of the landscape through their variable resistance to weathering and erosion. Those that were more resistant to weathering and abrasion stand out as ridges, while the more easily weathered and eroded form depressions or valleys. Many drainage lines are also focused along faults and geological boundaries because the rocks have been structurally weakened and are more susceptible to weathering and erosion (most notably, the South Brisbane Reach, see figure 3.2 inset).

The rocks have been weathered and sculpted by the present subtropical climate and relatively high rainfall, and by the conditions that existed during earlier climatic periods. These processes formed the modern landscape of rolling hills and valleys, and the often-intricate, dendritic drainage pattern. Of particular importance was the variable resistance to erosion of the NFBs and Bunya Phyllite and the deformed geological boundary between them.

The orientation of the metamorphic foliation within these rocks influenced the action of the erosion and down-cutting action of the river in the South Brisbane Reach (the western side of the peninsula) and dominant northwest striking metamorphic "grain" of the foliation and bedding in the basement rocks influenced the orientation of ridgelines. Of most historic importance was the "spine" of the Brisbane Peninsula, where the shallow easterly dip of bedding and foliation resisted erosion by the easterly river flow and created bluffs that faced west, forming the cliffs of the western shore. This important spine of the peninsula was partly formed because the strong metamorphic foliation of the NFBs trends approximately north to northwest and the down-cutting river erosion and weathering has tended to follow that grain.

Where the terrain is underlain by NFB, the hills are more symmetrical and the drainage pattern more radial in pattern. The imprint of the ancient radial and arcuate drainage system and erosional surface of the metamorphics is best preserved in the study area by the alignments of the street network of Spring Hill and Spring Hollow. Inner Brisbane has a series of "spines"; long ridges that formed the places where many of the settlements' major arterial roads developed (e.g., Dornoch Terrace, Wickham Terrace, Gregory Terrace, and George Street and Queen Street in the CBD).

There is a strongly arcuate, generally northeast-southwest trend on the ridgelines formed on the NFB to the north of the Brisbane Peninsula, the most significant of which is the ridgeline now defined by Wickham Terrace. This imposes a strong northeast trend on the marshy valleys and drainage lines of this part of the settlement area. In 1827, a windmill was built on the high point of this ridge to catch enough wind to grind maize. Unfortunately, this was a failure and for a time, the grinding was partly achieved by using a human-powered treadmill. Later, the tower was used as a signal station and time ball tower for communications with the ships moored in the South Brisbane Reach (from which the windmill tower was highly visible at the time). The building was subsequently used as an observatory (Steele 1987, 67–68).

The highest ground in the study area is also underlain by NFB, including Montpelier Hill, Teneriffe Hill, the Spring Hill ridgeline, and much of Highgate Hill. Elevated positions at the crests, or on the upper slopes of several hills formed on the NFB, are the sites of modern hospitals and churches, including the Royal Brisbane Hospital, which is located at the tuff contact, and the Mater Hospital which occupies a high point south of the river formed by basement metamorphics.

The drainage pattern and topography developed on the younger Brisbane Tuff is significantly different to that developed on the basement rocks. Where the landscape is underlain by tuff, the topography is generally smoother and is a lower elevation. The tuff forms a north-south trending series of flats along the eastern side of the study area. In Fortitude Valley, Woolloongabba and the southern end of Kangaroo Point, the landscape is gentler and less hilly than that of the basement metamorphics.

The topography formed on the Brisbane Tuff (and the underlying NFB) is subdued and plain-like and has a distinctly different drainage pattern to that developed on the more hummocky Bunya Phyllite to the west. The tuff belt is deeply dissected by several sinuous, northeast flowing drainage lines, including Spring Hollow, York's Hollow, and the Brisbane River. These drainage lines have few minor tributaries while the drainage developed on the basement metamorphics is strongly dendritic (see further discussion below).

The boundaries of the tuff have been an important influence on the course of the river, causing the sharp "U" turns around the Brisbane Peninsula and Kangaroo Point. The Brisbane Tuff has had a major impact on the landscapes of central Brisbane. The welded ignimbrite is a very strong and resistant rock; however the basal unconformity is very weak. The weaker unconformity was preferentially eroded by the river on the faster flowing outside bends, which undercut the overlying tuff in a process known as sapping. The undermining process created the Kangaroo Point Cliffs at the southern end of Town Reach, where the river is deflected to the north. The cliffs of Petrie Bight were created at the northern end, where the river was deflected back to the south. The cliffs formed by it on the western side of Kangaroo Point, and at the Howard Smith Wharfs of Petrie Bight, remain prominent landscape features of the river, despite extensive quarrying.

The outcrop boundaries of the tuff also cause the sharp changes in orientation of other drainage lines near Brisbane. The east to northeasterly drainage of York's Hollow, Spring Hollow and Fortitude Valley all coalesce and swing sharply to the southeast where they encounter the tuff. The lowest section of Enoggera-Breakfast Creek forms a remarkable hairpin bend where it crosses the belt of tuff just before it enters the Brisbane River (figure 3.1).

Wetlands and the Drainage System

The largest drainage lines of the study are all tributaries of the Brisbane River. They generally have an arcuate geometry and a northeast flow direction that crosscuts the regional northwest strike of the foliation in the basement metamorphics and the north to northeast strike of the Brisbane Tuff. Most of the valleys had marshy and swampy floors, with occasional chains of ponds and rivulets. The northeast flow direction is very strongly impressed on all the main drainage lines and valleys, despite the pervasive regional foliation. The same arcuate southeast to east to northeast trajectory persists in drainage lines south of the river, including the Woolloongabba Chain of Ponds (One Mile Swamp), and in many of the meandering bends of Norman Creek (figure 3.1).

The meandering and flood-prone Breakfast-Enoggera Creek is the largest tributary of the Brisbane River in the study area. The flood-prone character of creek would probably have been obvious to Henry Miller in 1825, and it may have deterred him from establishing the new penal settlement anywhere nearby, despite his apparent instructions from the governor. The flood plains of the creek are now mostly used for industrial purposes.

Five other small drainage features in the study area profoundly influenced land use and the placement of transport infrastructure throughout the modern history of Brisbane. From south to north they were Frog Hollow (drained by a small unnamed creek), Wheat Creek, Hanley's Valley (the southwestern extension of Fortitude Valley), Spring Hollow, and York's Hollow. Three of these features were little more than marshy valleys containing chains of ponds, but they were difficult to drain in the early days. The valleys were difficult to develop because of the poor drainage and were often reserved for water supplies and recreational grounds (e.g., York's Hollow and Spring Hollow). Gradually, as drainage works improved, some were built on, such as Hanley's Valley, Fortitude Valley, and Spring Hollow, while others remained undeveloped, such as York's Hollow (figure 3.2).

A sufficient source of clean freshwater was an important driver for the early settlement and led to the reservation of lands in Spring Hollow and York's Hollow, where spring-fed swamps and alluvial valleys could be developed into water supplies. The locations and orientations of modern transport infrastructure, particularly railways and inner-city motorways, were then able to utilize these reserved lands to access the central city. The main transport corridors entering the Brisbane CBD now follow the alignments of these swampy, low-lying valleys that were reserved for water supplies (or were simply too flood prone to be built on). The railway servicing the line to Fortitude Valley, later extended by tunnel to Roma Street Station, was also built on such a swampy valley floor.

Wheat Creek

Wheat Creek was a minor but well-incised water course near the northern base of the Brisbane Peninsula. It had a large catchment, which drained the broad low valley between modern Petrie Terrace, Wickham Terrace, and Leichardt Street. Wheat Creek followed the same arcuate east to northeast drainage trend as all the other major drainage lines of the study area. It was so strongly controlled by this trend that it traversed the almost 1.3 km width of the peninsula before debouching into Town Reach near Eagle Street and yet came within 140 m of the river on the western shore, near Roma Street Station. The creek was the first freshwater supply for the penal settlement, and immediately afterward, and was developed into a small reservoir. Wooden pipes channeled water to distribution points near the military barracks and other important buildings of the establishment.

The broad valley at the head of Wheat Creek became a focus of railway infrastructure, such as railway goods yards, and the site of the main city railway station at Roma Street. It also became the site of a cricket ground, and the jail, police barracks, and a military barracks. The railway yards have since been converted into the extensive Roma Street Parklands.

Wheat Creek is now buried beneath modern development and forms part of the subterranean drainage network of the city. The site of the reservoir is now beneath the law courts precinct, and the location of one of the natural ponds lies beneath the Brisbane City Hall and King George Square. Only the name of Creek Street marks the location of part of its former course at surface, where the course of the creek swung sharply to the southeast before entering the river.

Frog Hollow

Small patches of low-lying swampy ground could severely impact early urban development. One such place became known as Frog Hollow. Frog Hollow was a swampy low point in the landscape that originally lay within the boundary of the government farm. It was subsequently overtaken by urban development and came to lie near the intersection of Albert and Margaret streets, in the area bounded by Mary and Margaret streets. Frog Hollow lay at the head of the small serpentine creek, where two small rivulets draining the northeastern side of the spine of the peninsula joined. The creek then flowed through the convict farm before entering the river on the eastern side of the peninsula. Another small creek to the north drained the central spine and east-west ridge now occupied by Queen Street. This creek fed into the hollow near the modern intersection of Albert and Charlotte streets, but the channel faded away and it did not join with the serpentine creek (e.g., Dixon 1840; Figure 49 in Johnston and Gregory 1989).

Low-lying and swampy areas such as Frog Hollow were unhealthy places in urbanizing Brisbane, tending to harbor disease and mosquitoes, and they had poor sanitation. They were also flood prone, as shown in the 1864 photograph of the intersection of Charlotte and Albert Streets (Q-Album 2020).

They were therefore less desirable residential locations so became places where poorer people lived, and industrial activities occurred (figure 3.2).

Spring Hollow/Spring Hill

The inner suburb of Spring Hill was once known as Spring Hollow, named for the spring-fed, low-lying and swampy ground and ponds that originally occupied the valley floor. This valley became part of Brisbane's second water supply, before becoming too polluted, after which it was replaced by a pipeline from Enoggera Dam.

Higher ground was favored for residential development by wealthier members of society and the name of the suburb was changed to Spring Hill when they began to build on the ridgelines, in locations such as Wickham Terrace. The marshy and low-lying ground of the hollows was less desirable real estate, but once drainage was improved, it became a working-class suburb, with a notorious reputation as a red-light district (figure 3.2).

York's Hollow

York's Hollow, to the north of Spring Hollow, was a favorite camping area for the Aboriginal people and named after a local leader known as "The Duke of York." The long marshy valley was dotted with waterholes and was reserved as a water supply for the growing settlement. Land in the valley was later utilized for recreational purposes, including Victoria Park (the largest parklands in central Brisbane), a rifle range, and the Exhibition (Ekka) Grounds.

Later, when railways developed, the reserved land provided a rail corridor into the center of the growing town and became a focus of modern transport infrastructure. Subsequently, this same corridor became a focus of motorway development. The area also became the original home of the Queensland Museum and has been the site of the Royal Brisbane Hospital since the 1860s (figure 3.2). Brisbane Grammar School, one of the oldest and most prestigious private schools in Australia, has occupied a site on the drainage divide between York's Hollow, to the north, and the Wheat Creek catchment to the south since the 1860s.

Fortitude Valley

Fortitude Valley (originally known as Bell's Valley) is a curved valley that extends approximately 1 km to the northeast of the Brisbane Peninsula but extended into the modern city area where it formed an isolated hollow that was known as Hanley's Valley. Fortitude Valley was one of the locations of the first successful settler farming and originally developed as a separate satellite village to Brisbane. It subsequently evolved into an important commercial center (figure 3.1).

The lowest points of the valley were also marshy, with a series of ponds linked by a rivulet, and so remained undeveloped. The unused land was later adapted for use as a railway corridor, which is why the rail lines in Fortitude Valley still occasionally flood. The rail line was linked to Central Railway Station via a tunnel to allow direct rail traffic from Roma Street to the north. A tunnel was required to cross the spur of high ground between the main valley and the shallower Hanley's Valley to the south.

1825 Vegetation and Environment

Contemporary maps of the Moreton Bay Penal Settlement show that by the mid-1830s most of the native vegetation was cleared from the Brisbane Peninsula (e.g., Barney 1839) and no records exist from the time of its establishment. However, John Oxley (in 1823–1824) and Edmund Lockyer (in 1825) recorded general descriptions of the vegetation types bordering the banks of the Brisbane River. Both were military personnel who had long familiarity with Australian environments, and they made notes as the explored upriver. Oxley noted the vegetation and landscape at each survey station he stopped at along the river and was accompanied by the botanist Allan Cunningham on his 1824 journey (Steele 1983, 107–114), while Lockyer also recorded the broad vegetation types along the riverbanks (Steele 1983, 182–200). Both are a prime source of information about the pre-European environment of the Brisbane Peninsula and surrounds (figure 3.1).

The land surveyor Henry Wade provides the most important record of the pre-European environment of inner Brisbane region as it was being opened to free settlement (Wade 1844). He went much further than the earlier observers when he mapped the distribution of vegetation communities, water courses, wetlands, ponds, and mangroves from Eagle Farm to the east, to modern West End.

A reconstruction of the vegetation based on these early descriptions (particularly that of Wade 1844) shows that geology and topography had a profound influence on the distribution of the original native vegetation (figure 3.1). There was an ecologically rich and diverse mosaic of dry rolling hills, winding creeks, wetlands, and marshy valleys, with rainforest remnants in the low-lying wetter areas (particularly on the inside bends of the river) (Webster 2016).

Three main vegetation types were present; sparse dry open grassy eucalypt woodland on stony high ground; open undulating eucalypt forest, and dry rainforest (referred to as "brush" in early records) (figure 3.1). Higher hills further inland were dry and stony, with rocky outcrops forming low banks, and covered in scattered eucalypts and grass. There were large stands of red gums (described as "apple trees") and natural pastures in the flood-prone flats on river and creek bends, and particularly along the course of Enoggera-Breakfast Creek.

The density of large trees (>30 cm diameter) in the open eucalypt forest was sparser on the Brisbane Tuff (80 per ha) than on the Brisbane Metamorphics (110 per ha) despite similar species being present. Bloodwood (*Corymbia intermedia*), Mahogany (*Eucalyptus acmenoides*), and Stringybark (*E. carnea*) were more common on the tuff while Gum (*E. propinqua / E. major*) and Ironbark (*E. crebra*) were more frequent on the Brisbane Metamorphics (Specht and Specht 2007, 12–13).

Mangroves did not extend any further upstream than Hamilton Reach in 1825. Harbor developments, river channel dredging, and sand and gravel extraction have made the upstream river more tidal and saline. The mangroves on the sand and mudflats of Gardens Point have colonized the sites since 1825.

The Aboriginal Landscape

The inhabitants of the Brisbane area spoke the Turbal language. This language group extended south to the Logan River, and northward to the Pine River, and inland to Moggill. The language was also understood by inhabitants of Fraser, Bribie, Stradbroke, and Moreton islands. A different dialect was spoken at Redcliffe (Steele 1984, 121). Five clans were identified within the language region; the "Duke of York's" clan, occupying the Brisbane metropolitan area on the north side of the river; the North Pine or Petrie clan; the Coorpooroo clan on the south bank of the Brisbane River; the Chepara clan of Eight Mile Plains; and the Yerongpan clan of Oxley Creek (Steele 1984, 121).

There is widespread archaeological evidence for the long Aboriginal occupation of the rich environments of the Brisbane region. Middens are documented on the south side of the Brisbane River, near the mouth of Norman Creek, and multiple artifact sites have been identified in West End, and in York's Hollow to the northeast of the city (Kerkhove 2015). The Aboriginal sites identified by Kerkhove (2015) and Steele (1984) were plotted onto the pre-European landscape reconstruction to provide an environmental context for the otherwise isolated locations. It suggests that the local people camped close to water sources but on slightly elevated places nearby. There were also several fighting and ceremonial grounds in the area, showing that the population was large, and that there was a rich and active cultural life.

There are many Aboriginal natural feature names recorded in the 1825 settlement area (Steele 1984; Kerkhove 2015), including,

- Mianjin: Gardens Point (spike place or tulip wood)
- Barrambin: Yorks Hollow
- Mahreel: Spring Hill (stepmother)

- Walan: Royal Brisbane Hospital
- Kurilpa: South Brisbane, West End. The bend of the river on the South Brisbane Reach, was a favored crossing places, possibly because the sediments deposited on the inside bends of the river created shallows.
- Beenung-urrung: Highgate Hill
- Baneraba: Toowong
- Binkinba: New Farm (place of the land tortoise)
- Buyuba: Newmarket

Historical records show that the local Aboriginal people had an immediate impact on the British attempts to establish their settlements. It is possibly due to the Aboriginal resistance to the first settlement at Redcliffe that it was decided to be unsuitable for permanent settlement. And it is possible that it was because the Brisbane Peninsula was a more defensible landscape that Lieutenant Miller chose it as the final position for the penal establishment.

LANDSCAPE, GEOLOGY, AND THE EARLY DEVELOPMENT OF BRISBANE

Modern Brisbane began as a small and remote convict outpost of New South Wales. It was established to move hardened and troublesome convicts as far as possible from Sydney, where they would have no reasonable chance of escaping. Port Macquarie had previously served this purpose but had recently been opened to free settlers. Brisbane was not meant to be a desirable place to go, or to be sent in 1824, and it was (apparently) never planned to be a major town.

Choosing the Site of Brisbane

Henry Miller managed the establishment of the first settlement at Red Cliff Point, at the site chosen by John Oxley on the shore of Moreton Bay, but then eventually received a vague instruction from Governor Brisbane to abandon it and move to a site "27 miles up the Brisbane River." Miller chose the new site, possibly assisted by John Gray, a pilot from Sydney (Steele 1972, 129; 1987, 27, 36). While no specific reasons were recorded, historians since have suggested the following reasons for Miller's choice of the site (e.g., Steele 1987):

- There was a reliable freshwater supply (Wheat Creek).
- The high ground was above the flood line, which provided a safe place for the main buildings of the settlement.

- The low-lying land was fertile and suitable for farming.
- The peninsula provided a good defensive position (because the Aboriginals were troublesome at the first camp).

Subsequent Commandants, such as Captain Bishop, agreed with Miller's choice and argued against relocating the settlement yet again (Steele 1972, 129).

The site that Miller chose was a narrow pocket of high ground bordered on three sides by the Brisbane River. The land of the peninsula rose from low-lying, marshy valleys and flood-prone flats on the riverside, to sinuous ridges and rounded hills reaching a height of 85 m. The eastern shore was generally low-lying and marshy, where several small creeks and marshes entered the river but rose steadily toward the cliffs at Petrie Bight and modern Bowen Terrace. The western shore was generally higher and rockier and rose to low cliffs to the north. The peninsula has two long, low ridges of higher ground formed on the metamorphic rocks. The main ridge trends approximately north-south and forms a prominent spine of the peninsula. This north-south ridge became the site of the most important buildings of the penal settlement. The second and more subtle ridge extends east-northeast, from the main ridge to a small knoll, which became the site of the Female Factory. The 1825 settlement developed south and eastward along these ridgelines from the first landing place on the western shore. A permanent water source was provided by a small easterly flowing freshwater creek that traversed the northern neck of the peninsula, which became known as Wheat Creek. Beyond Wheat Creek the land rose steeply to a series of sinuous ridges with deep marshy valleys between them.

There were many advantages to building along the high ground. It was well ventilated and above the flood level, and the nearby less healthy marshy and swampy ground. The ridgelines were also more open, being originally occupied by open eucalypt forest that was easier to clear for campsites and the first structures. The nearby better watered lower slopes and marshy flats were good for farming.

Old Government House and the first Government Domain were eventually placed on a prominent position at the southern tip of the north-south ridge. When Government House outgrew the vice-regal role, a new property was acquired, and Old Government House became the first home of the University of Queensland. When the university subsequently moved to St Lucia, the site was turned over to the precursor of the Queensland University of Technology (QUT), which developed around Old Government House. QUT now occupies much of the southern part of the spine of high ground that forms Gardens Point. The sharp topographic contrast between the high ridgetop, moderate to steep eastern slopes, and the low-lying land adjacent to the riverbanks is still well preserved here (see figure 8.1, 3C). Main Drive separates the ridgeline campus from the formerly marshy lands (and convict farmlands) of the City Botanic Gardens.

The "City Cliffs" and Miller's First Landing Place

The landing place chosen by Miller on the Brisbane Peninsula was probably a meta-sandstone outcrop at the waterline of the western shore. This place provided the first secure footing at the shoreline and developed into the first jetty, eventually becoming Queens Wharf (figure 8.1, 3B). The low ridge above this riverside outcrop became the administrative center of the new settlement and the nucleus of the modern Brisbane CBD. A stone Commissariat Store built just behind the jetty in 1829 remains one of the few surviving reference points from this first phase of development of modern Brisbane (figure 3.2).

The eastern bank of the river here was quite different in character. It is an erosional shore where the river was actively cutting into the outcropping NFB at the neck of the Brisbane Peninsula and creating low rocky cliffs. The first landing site was probably located at the place where the low-lying mudflats and rainforest of the southern end of the peninsula (now Gardens Point) gave way to a low rocky shore and drier forest. From this point, the shoreline begins to rise gently to the north and becomes rocky, and eventually forms the "city cliffs" (now heavily modified and mostly obscured by the Riverside Expressway and riverbank stabilization works). This important change to the shoreline was an ideal place to land, offload stores, and to transport them up a relatively gentle slope to the high ground of the settlement. Any further north and the rocky banks would have been too steep to achieve this.

This landing place was probably selected by Lt Miller because it offered the best place to efficiently offload stores. It then became a focus of the activities of the burgeoning settlement. This is the case in earlier British colonies, such as Sydney Cove (NSW) and Sullivan's Cove (Tasmania), and later, in Melbourne (Victoria). Once proven to be useful, the first landing sites are cleared, made safe, and temporary structures are built for securing stores and to relieve the labor of trans-shipping heavy goods. Activity of the new settlement moves out from this spot. And the rocky shore may have provided a foundation for the first jetty—Kings Wharf—later, Queens Wharf. In 1829, the stone-built Commissariat Store fixed this site in the city's geography.

The first road of the European settlement must have been a track from the landing place, up the gentlest incline, following the rise of the land to the north, to the low ridge where the camp was formed (figure 8.1, 3A). This route was dictated by the crest of the steep and rocky riverbank. This road appears on every early extant map of the settlement and is the only road to survive with an alignment established from the first camp. Now, Queens Wharf Road is the last significant remnant of the original street pattern of the Moreton Bay Penal Settlement.

The convict settlement wharfs and stores became focused on the easily accessed western side of the Brisbane Peninsula. Ships that could travel upriver therefore moored in the South Brisbane Reach of the river, where they could be moored to a large tree on the south shore. From this simple beginning, South Brisbane grew into one of the first locations for more permanent moorings and eventually vied with North Brisbane to be the commercial center of the new town. However, the 1893 floods devastated the low-lying area and ended that. The higher ground of North Brisbane made it a more secure place for business.

Kangaroo Point Cliffs and the Petrie Bight Cliffs

The Kangaroo Point Cliffs are vestiges of a bold natural ridge and bluff of Brisbane Tuff that once extended to the waterline of the southern side of the river (figure 3.2). They were formed as an erosional shore on an outside bend in the river. The original ridge has been significantly modified by quarrying since at least 1829, when Cunningham recorded a quarry there, with a stone wharf (the first jetty built on Kangaroo Point) to transport stone across the river to the main settlement (in Steele 1987, 117).

Settlement anchorages were originally located in the South Brisbane Reach of the river, with ships mooring to trees on the opposite shore. Better port facilities were soon required once free settlement began, and the first focus of modern port development was at Petrie Bight, along the Eagle Street frontage. This location is still the site of the historic Customs House building, and the Howard Smith Wharfs, the last surviving pre–World War II port infrastructure. It is an erosional shore on an outside river bend, and steep cliffs are cut into the resistant Brisbane Tuff. The Petrie Bight Cliffs have also been extensively quarried, to provide space for port infrastructure and as a stone source (the unconformity at the base of the tuff is exposed). Harbor facilities spread further downriver from the Petrie Bight, into New Farm and beyond and are now focused on the river mouth, where extensive reclamation of the mangrove mudflats has taken place.

Quaternary Sediments and the First Farming

The first farming in the convict settlement was on the low-lying but well-watered and fertile marshy areas on the eastern side Gardens Point, at the tip of the Brisbane Peninsula (now the City Botanic Gardens). Fields were also established on the tip of Kangaroo Point, and on the south bank, at South Brisbane. Wheat fields were also sown along the banks of small freshwater stream that was dammed to create the first important freshwater supply for the settlement, giving Wheat Creek its name (figure 3.2). All these locations are located on the Quaternary and recent alluvial sediments deposited on the inside bends of the river. These small patches of sediments played a very important part in the early self-sufficiency of the settlement.

The modern City Botanical Gardens evolved from land originally dedicated to the convict farms on Gardens Point. It was reserved for recreation after the penal settlement closed and has never been developed, partly because the land is too low-lying and flood prone. The higher ground to the west, beyond the line of modern George Street, was also a part of the early convict farm but was reserved as a Domain surrounding the new Government House. It is above the flood level and when Government House was relocated, the building and grounds were eventually developed into the first home of the University of Queensland. The City Botanic Gardens were abandoned as a location for important scientific plantings after several floods inundated the site. Significant parts of the collection were lost in the 1974 floods and the decision was made to create a new botanic garden at Mt Coot-tha.

Larger and more extensive farming was developed in the well-watered lands of Eagle Farm to the northeast of the main settlement in 1829. This marshy land provided the broader expanses of flat, fertile, and well-watered land needed for large-scale farming. Eagle Farm was the largest and most successful farming venture near Brisbane in the convict era (Steele 1987, 113). The site later became the location of a new convict Female Factory, to place it away from the main settlement. A track developed from the main settlement to the farm and developed into one of the first significant arterial roads. A small part of the alignment of this track remains as Eagle Street.

The first burial grounds were also located in the alluvial soil areas at the fringe of the settlement, possibly because of the easier digging at these locations. The first formal burial ground was on Eagle Terrace, but a much more extensive denominational burial ground was created to the northwest of town, on the site of modern Lang Park Stadium.

Landscape and Infrastructure Placement

Prominent hills across the inner metropolitan area have been chosen as the sites for important government buildings, early defense facilities, hospitals, and churches. One of the most prominent sites in early Brisbane Town was the high point at the tip of the peninsula (Gardens Point), where Government House was placed to overlook the river, the main port, and its approaches. The Royal Brisbane, Mater and Queensland Children's hospitals have all been located on hills, probably because these sites are well ventilated.

Victoria Barracks and the former police barracks once occupied the most prominent hilltop site on the northwestern edge of the growing town, on high ground above the Wheat Creek catchment. While originally reserved for a jail (1858 map), by 1890 Victoria Barracks had been built on the adjacent block. This site once commanded a panoramic view of the South Brisbane Reach and the whole town area. The hill has been extensively modified by railway and motorway earthworks and the vistas that it once enjoyed are lost in the modern city. The former Boggo Road Jail was built on a similar prominent hilltop site in Dutton Park, on the south side of the river.

The Road Network

The route up the gently rising shoreline from Miller's 1825 landing place to the high ground where the camp was formed must have followed the slope of the bank from the riverside outcrop of NFBs to the crest of the central ridge. The route was dictated by the crest of the often steep and rocky riverbank. It is now Queens Wharf Road, one of the few surviving road alignments of the Moreton Bay Penal Settlement and the oldest street in Brisbane. North Quay (now Coronation Drive) continued this alignment northward as it followed the city cliffs of the western shore of the peninsula.

Queen Street, the main east-west axis of Brisbane, and the most important commercial street, evolved from the important track that developed to the first Female Factory (now the site of the Brisbane GPO), which was built at some distance away from the main settlement, on a separate knoll.

Ridgelines were the most important locations for other early roadways as the settlement developed beyond the Brisbane Peninsula, and they are the defining landscape element of the original inner Brisbane road network. Many of the original main arterial roads follow the sinuous ridgelines formed by the NFBs and Bunya Phyllite. Key examples are Petrie Terrace, Wickham Terrace, Leichardt Street, and Gregory Terrace. Other important examples include Main Street, which follows the eastern flank of the central spine of Kangaroo Point (in this case formed by Brisbane Tuff), and Grey Street, Gladstone Road, and Dornoch Terrace in South Brisbane, Highgate Hill, and West End.

Many important early roads probably developed along ridgelines because the locations provided more level routes through the undulating terrain, but also because the elevated locations kept important transport corridors above potential flooding. Residential areas were also developed along some ridgelines, to access cool breezes and to be above the more disease-prone lower-lying land. In such ways, the landscape contributed to social stratification in the developing city. This process continues to some extent, with the wealthy seeking out coastal or riverside properties, while the less well-off live in the outer suburbs of modern Brisbane.

Many of the former villages dotted around the modern suburbs (and now redeveloped as shopping centers) are located on elevated sites, such as Toowong and Indooroopilly. The roads that originally linked them at least partly followed ridgelines. The main roads of the pre-motorway road system are generally named after the places they went to: Moggill Road, Annerley Road, Milton Road, or Ipswich Road. These roads all tend to follow the ridgelines. The first arterial roads were the tracks to Eagle Farm (Eagle Street is a final remnant) and to Limestone Station (now Coronation Drive). The latter followed the north bank of the river in the settlement area. Then there are the "terraces" high above the riverbanks and parallel to the waterways: Wickham Terrace, Dornoch Terrace, Petrie Terrace, Gregory Terrace, Highland Terrace.

Modern motorways now dissect this old road system, leaving the earlier street pattern as islands in small, disjointed pockets between modern developments. The result is that modern motorways have little relationship to the landscapes they traverse, and high-density development corridors along the alignments reduce or remove any vistas from them.

The ridgelines and hill crests have remained important throughout all subsequent development phases of Brisbane. The use of the highest landscape features in the central area of the first settlement set in place the pattern of the street grid. The central ridgeline determined the north-south axis, and the location of the Queen Street "knoll" established the east-west axis. When Henry Wade surveyed the city street grid in 1842, this existing layout strongly influenced the orientations of the new layout. This road network later spread to the new wharf areas on Petrie Bight. Ultimately, the same west to east development spread along the alignment of Fortitude Valley as the commercial district grew.

A Seat of Government

The early colonial government of newly independent Queensland utilized many of the buildings of the former penal establishment. These were gradually replaced over subsequent decades as the colony prospered, with new buildings being constructed on the road alignments surveyed by Wade. Eventually, the old roads mostly disappeared.

The "Victorians" knew something about building with style; building to impress and building with grandeur. When the proud new colony became independent from NSW, it used the many elevated sites it inherited from the old penal settlement to the best advantage. They built to impress arrivals to the city from the main anchorage in the South Brisbane Reach, the most important port area at the time. Buildings constructed in this phase include Old Government House, on a prominent site at the tip of Gardens Point, Parliament House, the Lands Department Building, the Treasury Building, and the Government Printing Office.

The modern city is less river focused, and the port facilities have moved progressively further down river, so the impressive government buildings have become steadily more obscured by new development. The Victorian vision has been largely lost from the river, but the central spine of the Brisbane Peninsula continues to be the focus of government institutions along George Street and Main Drive.

Wetlands, Sporting Facilities, and Transport Infrastructure

Low-lying and potentially flood-prone ground could not be easily drained and used for residential development, so marshy areas were often used for sport and recreational purposes. Several important facilities were placed in such areas, including the City Botanic Gardens, the Brisbane Cricket Ground (built on One Mile Swamp in Woolloongabba), and the Acclimatization Society Grounds developed in York's Hollow (which became the Brisbane Exhibition Grounds).

Not all sporting facilities and parks were created on poorly drained sites. Several have been developed on former burial grounds. Of note is Lang Park and the small grassed area at the intersection of Eagle Terrace and Roma Street, which was Brisbane's first burial ground.

Previously reserved land provided undeveloped corridors for transport infrastructure to reach the inner city once railways (and later motorways) became important. The modern railway and motorway systems developed along such low-lying land corridors reflects this old southeast, east-northeast arcuate trend of the former wetland drainage systems of York's Hollow and Fortitude Valley almost perfectly. Subsequence development of the Inner-City Bypass and various interchanges also followed the alignment of this ancient drainage system.

As Brisbane grew, the relatively rare expanses of broad flat land formed by Quaternary sediments in the Eagle Farm area became increasingly important. As land drainage continued, horse racing took advantage of the broad-acre land that was freed up by the drainage works around Eagle Farm in the 1860s, and the area is now home to the most important racing facilities in the state. From the 1920s, the flats became the location of the Eagle Farm Airport, and an important wartime air base. Postwar, the facility grew into the first Brisbane International Airport. When the needs of a major capital city outgrew the Eagle Farm location, Brisbane International Airport was constructed further to the east, on partially reclaimed coastal salt marsh, wetlands, and mangrove mudflats on the shores of Moreton Bay. The old Eagle Farm site was redeveloped into a modern industrial estate, but the land drainage canals remain prominent features of the area.

Other low-lying and flood-prone riverside land, swampy valleys, and marshlands were also developed as industrial estates and transport hubs (e.g., railways and rail yards), of which the course of the meandering Enoggera-Breakfast Creek is a key example. Another historic example is the site of the former South Brisbane coal wharfs on the southern shore of the river at Woolloongabba, directly opposite Gardens Point (and now beneath the Captain Cook Bridge). The broad acreages of marshy land on which Eagle Farm was established remains a modern center of transport (airports and wharfs), racetracks, and industries requiring large floor areas.

River Flooding

The Brisbane River and Bremer River systems have a long flood history. River flooding and the health issues associated with polluted and disease-carrying urban wetlands are ongoing themes of colonial Brisbane history. Serious floods have occurred in the Brisbane Valley on multiple occasions since European settlement: including 1864, 1936, 1974, and 2011. The damage caused in South Brisbane by the devastating 1893 floods ended its bid for commercial primacy, and afterward, most business activity moved to "North Brisbane," and particularly to the Eagle Street wharf precinct.

Flood mitigation works were undertaken in the river several times during attempts to improve flow rates. These works included the removal of the tips of both Gardens Point and Kangaroo Point to widen the river at perceived bottlenecks. Many of the meanders were removed from the flood-prone Enoggera-Breakfast Creek and several of the lower reaches have been straightened and canalized. Despite all these works, floods still occur. This could be some vindication of Henry Miller's choice not to form his settlement on Enoggera Creek.

THE BUILT ENVIRONMENT

Brisbane retains several grand stone-built public buildings that remain historic anchor points in the ever-changing cityscape. Three stone sources were the most important in their construction, Brisbane Tuff, sandstone from local Permo-Triassic sequences, and the more distant Helidon Sandstone quarries near Toowoomba.

Brisbane Tuff is the most characteristic and enduring building stone used in the early colony and remains the most unique and defining character element of the historic Brisbane built environment. Tuff was mainly sourced from the first quarries on the Kangaroo Point Cliffs (figure 8.1, 3E). It was used throughout the penal settlement for major buildings (e.g., the Prisoners Barracks and Commissariat Store) and continued to be used for foundations, retaining walls and kerbing into the twentieth century. The most historically significant building is the Commissariat Store, constructed from Brisbane Tuff sourced from Kangaroo Point (figure 8.1, 3D). This building has been a landmark in the city since 1829. Many other buildings of the convict settlement were built from this stone but have not survived. The most significant (and largest) extant building constructed from the tuff is St Johns Anglican Cathedral (von Gnielinski and Siemon 2012) (figure 8.1, 3F).

Many minor features are also made of locally quarried tuff. Most historic retaining walls and curbstones are made from Kangaroo Point tuff, but there are many smaller quarries across the inner urban suburbs that were once also local sources of stone (figure 8.1, 3D).

Sandstone sourced from quarries in the Permo-Triassic sequences near Brisbane (chiefly the Raceview Formation of the Woogaroo Subgroup) was important for some early buildings, such as Parliament House and Old Government House (now QUT, Willmott 2012, 36–37). Other local sandstone sources were used to construct buildings, such as the GPO (figure 8.1, 3H) (von Gnielinski and Siemon 2012; Willmott 2012, 36–37). Many of the later landmark public buildings were constructed from sandstone sourced from the Helidon quarries. Important buildings that used this stone include the ANZAC memorial (figure 8.1, 3I) and Brisbane City Hall.

There is a fourth type of stone that receives little recognition for the role it plays in determining the urban character of Brisbane, despite its widespread use. This is the hornfelsed blue-gray Bunya Phyllite sourced from the Brisbane City quarries near Mt Coot-tha. This blue-gray foliated stone, with common deformed quartz veins, is usually used as irregular, rough stone blocks. There are many characteristic retaining walls and smaller feature walls built from this stone throughout inner Brisbane. Many others have been built from stone salvaged from cuttings and roadworks.

Historic Quarries and Their Current Landforms

Many former suburban quarries from which tuff was once extracted occur throughout the inner suburbs, and particularly in Spring Hill. Most have been redeveloped as parks or incorporated into building complexes. The most historically significant quarry in the study area is the Kangaroo Point Cliffs. Once a major source of tuff for building works around the settlement and city, these former quarries began in the earliest days of the colony and did not cease being worked until the 1970s. They were originally cliffs that outcropped at the southern end of the peninsula (where the cliff bases can still be seen at the shoreline) but have since been extensively modified by quarrying.

Brickmaking Clay and Lime Sources

The first bricks made in Brisbane used clay extracted from the banks of Wheat Creek, near the original reservoir. This source may have been discovered as the reservoir wall was being constructed. Allan Cunningham recorded the site of the "old brick kiln" near the first burial ground (at the corner of modern Eagle Terrace and Roma Street) just to the northwest of the Wheat Creek reservoir during his survey of the settlement site in September 1829 (in Steele 1987, 117).

The importance of both stone and brick construction created a demand for lime as the settlement became more established. However, there were no sources of lime for mortar near the 1825 settlement. When Captain Patrick Logan recognized low hummocky hills of limestone on the north bank of the Bremer River while exploring the region in 1827, it was a significant discovery (Cunningham 1828, in Steele 1972, 281). Logan named the low hummocky hills the Limestone Hills and the deposits there were developed as a small convict outpost called Limestone Station. The deposits lay approximately 38 km west of Brisbane (by land) and 80 km by river, which was the main method of transporting the lime (Fraser, in Steele 1972, 234–235). The magnesian limestone probably formed during the Cenozoic, by the precipitation in shallow lakes of carbonates produced from the weathering of the surrounding basalts (Trezise 1981, 53–54).

Limestone Station became the first inland settlement developed beyond Brisbane and, being at the head of the navigable river, developed into a service center for the hinterland as agriculture spread across the district. In 1843, Limestone Station was renamed Ipswich, after the town in England. Coal was discovered in the area shortly afterward and Ipswich developed into the first significant coal-mining district in Queensland. The site of the original limestone quarry is now modern Queens Park.

Timber Structures

A discussion of the built environment of colonial and modern Brisbane would not be complete without some mention of buildings built from timber. In many ways, Brisbane is a timber city. The subtropical climate led to the development of dwellings that were more purpose-built to adapt to the climate than those of more southern settlements. This led to perhaps Australia's most unique architectural style, the high-set "Queenslander." This style of timber building, with open or partly enclosed verandahs, was adapted to all dwelling types, from workers' cottages to ornate mansions. This city is an exception as an Australian capital city because so much of the unique historic built character of its suburbs was determined by using timber, rather than stone, during the colonial era and beyond.

Chapter 4

Limestone and Lakes

Perth, Western Australia, 1829

The Swan River Estuary became the first mainland Australian landscape to be significantly explored by Europeans in 1696. European knowledge of the Australian west coast had begun much earlier however, when the Dutch began to seek shorter and faster trading routes from Europe to the East Indies (Indonesia) via the high latitude westerly winds of the Indian Ocean. Several of these earlier encounters were unexpected, and ended disastrously (Brearley 2005, 77). It was difficult to accurately calculate longitude and mariners occasionally came upon the Australian coast before they had commenced their turn north. Dutch ships such as the *Batavia* were lost when they unexpectedly hit the low-lying limestone islands and coral reefs of the western Australian coast (Brearley 2005, 78). The Dutch named one hazardous group of islands Abrolhos—"watch out" (figure I.2).

In 1696 William de Vlamingh, commanding the ships *Gellvinck, Nijptangh,* and *Weseltje*, discovered Rottnest Island and the Swan River Estuary while searching for the wreckage of one of the lost ships, the *Ridderschap van Holland.* During this brief exploratory visit, he discovered and named several key features of the drowned limestone and dune landscape of the west coast, most notably the Swan River and Rottnest Island. One hundred and thirty-three years later, the Swan River Estuary would become the focus of the British colonization of the west coast of Australia and the site of the city of Perth.

GEOGRAPHIC SETTING AND STUDY AREA BOUNDARIES

The modern city of Perth is located on the southwestern coast of Australia at 31.95°S, 115.85°E (Perth General Post Office, Google Earth). It is located approximately midway across the Swan Coastal Plain (SCP), between the Darling Ranges in the east and the Indian Ocean to the west.

The focus of this chapter is a broad southeast projecting promontory of the north bank of the Swan River Estuary that marked the position in the westward flowing Swan River where estuarine influence began. This small promontory had steep southern banks and low-lying wetlands and mudflats at the southeastern point and was a moderately wooded landscape of low linear, sandy ridges and hills interspersed with seasonal and permanent rush-rimmed lakes and wetlands. The land rose westward from sea level at the shoreline to a prominent 80 m high limestone bluff (Mount Eliza) (figure 4.1). This promontory was chosen by James Stirling in 1829 as the location of the administrative center of the Swan River Colony, which he named Perth (after a provincial Scottish city). The promontory is now occupied by the CBD of modern Perth and lies at the center of an urban region that has expanded to occupy most of the central part of the coastal plain. The small promontory remains the heart of government, commercial and cultural activity in the modern State of Western Australia.

Stirling labeled the entire promontory as Fraser's Point on his first 1827 published chart of the Swan River, in the same manner as he labeled Point Walter and Point Belches (Stirling 1829a; Watson 1923, Map 1). The name seems to have since been specifically applied to the original finger-like peninsula at the southeastern tip of the promontory, which was contiguous with the shoals and mudflats of Heirisson's Islands. It is uncertain from Stirling's annotations whether he originally intended to apply the name to the entire promontory, but it will be used in the following discussion to refer the landscape between Mt Eliza and Heirisson's Islands.

The influence of the small administrative center spread rapidly into the surrounding hinterland, and particularly along the lake system to the north and northwest of Point Fraser. They were the most significant historic landscapes occupied by European settlers in the first decades of the colony and they are important to the following discussion.

The northern boundary of the study area follows the southern shoreline of Lake Monger, one of the two main survivors of the 1829 Point Fraser Lake System, which drained into the Swan River at Clause's and Walter's brooks. The western boundary is defined by the foot of the Mt Eliza Escarpment (modern King's Park). The eastern boundary is the Maylands Peninsula and Burswood Peninsula, including parts of Victoria Park and Belmont. The southern boundary is defined by the fringes of the southern shore of Perth Water, and includes Mill Point (Point Belches) and the former coastal wetlands of South Perth. A total area of approximately 40 km^2 (8 × 5 km) is discussed. Most of the earliest suburban growth of Perth took place in this area.

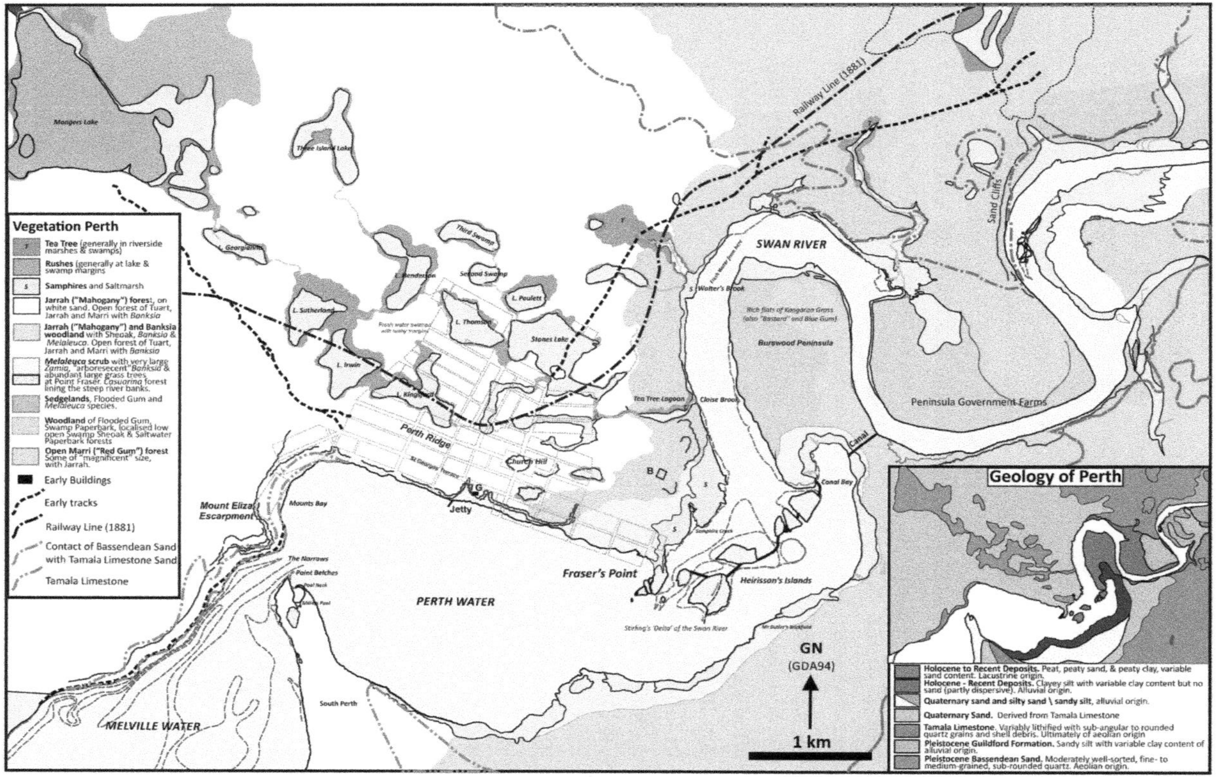

Figure 4.1 The Perth Region. Reconstruction of the landscape of the central Perth region at the time of British settlement, circa 1827-1831. The map view is centered on Frasers Point and the Perth Ridge, where the terrain was dominated by a linked system of groundwater-fed lakes, swamps, and marshes formed behind sand ridges (particularly the Perth Ridge). The reconstruction of the main vegetation types of central Perth, c. 1827-31 is based on the contemporary written descriptions of Charles Fraser (1827), and observations by surveyors on land survey charts in the early 1830s. The contemporary information is augmented by modern vegetation reconstructions. **Inset**. The geology of central Perth (simplified after Gozzard 1986). Please refer to the extended caption at the rear of the book for additional information.

Historic Time Frame

The historic period discussed encompasses the first two decades of British settlement of the Swan River Colony, from 1829 and 1850. During this brief period, most of the key decisions were made about the locations for towns, and land was surveyed and allocated in the newly settled areas. Land grant surveys made at this time set patterns in place that now strongly influence the modern urban geography of Greater Perth.

The focus of the discussion is on the development of Perth; however, the Swan River Colony was unusual in Australian history in that three almost contemporaneous settlements were founded in 1829. Fremantle (which predates Perth by several weeks) was created at the mouth of the estuary, to be the port. Guildford was established inland, to service the agricultural hinterland from the head of the navigable river. The land transport links between the three settlements were key elements of the early evolution of the modern urban road (and rail) system and the spread of suburban development.

The geology and landforms at the mouth of the Swan River Estuary had a profound influence on the foundational years of the colony. This requires some discussion of the early development of Fremantle, but space precludes a discussion of Guildford (see figure I.2 for locations).

There were five phases in the development of Perth of which the first two are the focus of this chapter.

Phase 1, 1829: Temporary camps and selection of town sites: Temporary encampments were created to house the first settler arrivals while the mainland was explored for settlement sites and land surveys undertaken. The first camp of the *Parmelia* crew members, and settler families was on Garden Island. Another was created on the shore at Bathers Bay, Fremantle, where the settlers then stayed while grant allotment surveys were completed. During their residence at Bathers Bay,

Stirling formed his temporary government and survey camp at Perth and from there parties completed the town and rural land grant surveys.

Phase 2, 1829–1834:. Town and rural surveys and initial settler influx: Following the survey of the main townsites, the surrounding region was surveyed, dividing the SCP into land grants. The geometry of these blocks influenced the subsequent orientations of rural property boundaries and the placement of access and boundary roads and is now imprinted on the urban layout of modern Perth.

Phase 3, 1834–1850: Growth of the free settlement and expansion of the colony, and labor shortages: Settlement spread across Western Australia and multiple new towns were established. The transport networks of Perth evolved from early tracks.

Phase 4, 1850–1868: Convict transportation period: At the request of the colony, convicts were sent to Western Australia from 1850. The arrival of convicts led to an expansion of public works, and the construction of several important building, including the Perth Town Hall, Fremantle Jail, and the Round House.

Phase 5, 1892 onward: The Gold Rush Boom: Gold was discovered 550 km to the east of Perth in 1892, leading to an influx of people and wealth into the colony. Many of the most significant limestone and sandstone buildings surviving in Perth were constructed at this time, including Parliament House (1904) and the Old Perth Fire Station (1901) (refer to figure 8.2, 1G and 1K).

EUROPEAN DISCOVERY AND EXPLORATION OF THE SWAN RIVER

After initially mooring near Rottnest Island ("Rat's nest") in January 1696, de Vlamingh explored the mainland by foot after probably landing near Swanbourne Beach (Brearley 2005, 78). Another party traveled up the newly discovered river after dragging their boats across the limestone bar at the estuary mouth. Over the next few days, they explored the estuary, probably reaching as far as modern East Perth. De Vlamingh noted the estuarine and brackish nature of the water and the sandy and heavily wooded (prickly and thorny) landscape, possibly describing the Crawley-Dalkeith foreshore. He was impressed by aspects of the beauty of the landscape and the abundance of birdlife, naming the river after the abundant black swans, but he was also troubled by the flies (Brearley 2005, 77–79).

The French explored the Swan River region during Baudin's 1801 expedition. Joseph Bailly made the first geological observations of the Swan River as he accompanied Midshipman Heirisson's exploratory boat party upriver, beyond the site of Perth (Peron 2012). They made many accurate observations of the coastal geology of the Swan River and estuary and the rolling but mostly flat topography of the forested country along the banks. Like de Vlamingh's boat's crew, they had great difficulty passing the shoals of the Heirisson Islands on both the upriver and downriver journeys. Despite the difficulties of the journey, Heirisson produced the first chart of the Swan River, estuary, and the landscape of the riverbanks, which included the future site of Perth (Heirisson 1801).

British Exploration

The British began to arrive on December 25, 1826, initially to counter perceived French interest in the southwest. Major Edmund Lockyer arrived in

King George Sound (the site of modern Albany) with a detachment of soldiers and convicts from New South Wales to establish an outpost (Nayton 2011) (see figure I.2 for location). He was sent there by the NSW governor to establish a penal settlement for recidivist convicts (Statham 1989, 123–124). Later, Major Lockyer would be one of the first explorers of the Brisbane River (see chapter 3).

Stirling's Reconnaissance of 1827

On March 5, 1827, James Stirling's first exploratory expedition to the Swan River arrived off Rottnest Island and anchored just outside Thomson Bay (Statham-Drew 2003, 69). After negotiating the rocky entrance, Stirling's party explored the river. They initially encountered smooth sailing and progressed upstream through Melville Water but were grounded at Point Belches, subsequently named after the lieutenant that accompanied Stirling (now "the Narrows"). They could not find a channel through the shoals, so the seaman dragged the boats along the banks for much of the length of Perth Water, a distance of around 3.2 km (Statham-Drew 2003, 72). While the boats were being ported, Stirling climbed the nearby high hill, which he named Mt Eliza in honor of Governor Darling's wife. Garling sketched the view from there, showing Point Belches and Mill Point (now South Perth). From Mt Eliza, they observed the entrance to the Canning River and realizing that it was not an arm of the sea as described by the earlier French explorers, named it the Canning River after the British prime minister. They probably camped at Point Fraser and their need of freshwater led to the discovery of Clause's Lagoon (Claisebrook) about a mile to the northeast (Statham-Drew 2003, 74), which they named after the surgeon (Statham-Drew 2003, 74). They moved the camp there and had an encounter with three Aboriginal men who made angry gestures at them, which they believed showed they wanted them to leave (Fraser, in Hay 1906, 16).

Accompanying Stirling was the botanist Charles Fraser. Fraser described the landscape, soils, and particularly the vegetation in several key locations along the lower reaches of the Swan River Estuary (figure 4.1), up to the Heirisson's Islands (Hay 1906). He observed flowing springs near the waterline and recognized that Mt Eliza was probably the eastern limit of the limestone (Hay 1906). Like Stirling, Fraser recognized that the estuary ended at the Heirisson Islands and that the Swan was a true river upstream of this point.

The river became more navigable above Heirisson's Islands as it wound inland and gently swung more northward across the coastal plain, following the trend of the changes in the underlying geology. As they traveled further, the river became progressively narrower and more difficult to navigate due to fallen trees and eventually, on March 13th, the cutter was holed by a tree.

This happened just before they encountered the confluence with a significant creek entering the river from the west, which Stirling named the Ellen Brook, after his wife. The ever-narrowing Swan River turned east above this point, becoming rocky and full of rapids. Stirling decided that the river was too obstructed to continue and turned back (Statham-Drew 2003, 76–77). This is the point where the Swan River leaves the coastal plain, crosses the Guildford Formation and then passes over the fault escarpment eastward onto the Archean rocks forming the Darling Range.

They began the return journey on March 15 and made good time because they were better-prepared for some of the navigation hazards they encountered. They camped at Point Garling—possibly the modern Burswood Peninsula (Statham-Drew 2003, 78) and the next day were at Point Fraser, where they more easily navigated the shoals. Stirling sent Lieutenant Belches to explore the Canning River, who was able to report back that it was a freshwater river like the Swan (Statham-Drew 2003, 78).

THE GEOLOGY OF THE PERTH BASIN

The Swan River Colony was established on the southwestern coast of the Australian continent, in a part of the landmass underlain by the Perth Basin. This elongate and tectonically complex north to northwest trending sedimentary basin extends for 1,300 km along the southwestern continental margin and measures over 172,300 km^2. It has both an onshore and a more extensive offshore component. The Perth Basin developed as a half-graben rift valley that began forming in the early Cretaceous, during the breakup of Gondwana and the separation of Australia and India. The Perth Basin is bordered to the east by the steeply dipping Darling Fault, which has over 10,000 m of displacement but has been largely inactive since the late Cretaceous. To the east of the fault is the Archean Yilgarn Craton, a stable block of igneous and metamorphic rocks that occupies approximately 25 percent of the geographic area of modern Western Australia. To the southwest lies the Precambrian Leeuwin Block. The Northampton Block and Carnarvon Basin lie at the northern end. The former western margin probably now lies beneath the Himalayas. The rift valley became the focus of marine deposition and infill by a sequence of Permian to Cretaceous, mainly continental, sediments that reaches a maximum thickness of 15 km (Commander 2003, 7; Geoscience Australia 2019). The Mesozoic strata are presumed to unconformably overlie Archean basement.

The sedimentary rocks of the Perth Basin rarely outcrop because they are usually unconformably mantled by a complex sequence of Cenozoic sediments, informally named the superficial formations (Allen 1976). There is then a thin covering veneer of the variably weathered and redistributed components of the Cenozoic sediments.

Deformation during the early Cretaceous caused uplift and erosion followed by marine deposition that ended in the late Cretaceous. The Cretaceous rocks were gently folded at regional scale, creating the Swan Syncline, the Pinjar Anticline, and the Yanchep Syncline (Davidson and Yu 2008, 17). The center of the Swan Syncline was eroded to form a deep valley cut by an ancestral Avon River in the Perth area (Commander 2003, 7). This deep valley was subsequently drowned and filled (Beard 1999, 163; Davidson and Yu 2008, 83).

A series of north-northwest-trending curvilinear lineaments propagate outward from the Archean craton and traverse the basin in the Perth area. They appear to influence trends in the basin geology (possible structural offsets), particularly the boundary between the Bassendean Sand and the Tamala Limestone, and the orientation of river reaches (e.g., the Canning River).

In the time interval between the deformation of the Mesozoic sediments and the onset of the marine transgressions leading to the deposition of the superficial formations, a deep channel was cut into the Cretaceous strata by a precursor to the Avon River. This buried channel corresponds approximately to the axis of the Swan Syncline in places, and in the subsurface, to the course of the modern Swan River. It was infilled by the Paleocene to Eocene Kings Park Formation composed of 469 m of gray calcareous, glauconitic siltstone, and shale beneath Kings Park. The Kings Park formation is then unconformably overlain by the superficial formations (Playford et al. 1976; Commander 2003, 11; Davidson and Yu 2008, 18).

The Superficial Formations

A 5–90 m thick sequence of largely horizontal Pliocene to Quaternary marine, alluvial and aeolian sand, limestone, silt, and clay was deposited during eustatic sea-level changes above a relatively planar, west-dipping unconformity formed during Pliocene erosion. Known informally as the superficial formations (Allen 1976), they are particularly important to this discussion because they underly Point Fraser (figure 4.1 inset).

Sedimentation was shallow marine to continental onshore (but there was little deposition) but was dominated by deep marine clastics and carbonates offshore (Commander 2003, 7; Davidson and Yu 2008, 18). The stratigraphy of the superficial formations is laterally variable, complex, and difficult to date, with further revision likely (Commander 2003, 8). They comprise mainly calcareous marine sands and aeolianites or coastal limestone (Tamala Limestone and Safety Bay Sand) near the coast, and variable fine and medium sand with minor silt and limestone (mainly Bassendean Sand) inland. To the east, the sediments interdigitate with the generally brown silty to slightly sandy clay and clayey sand at the base of the Gingin and Darling Scarps known as the Guildford Clay (Davidson and Yu 2008, 18, 21), or the Guildford Formation (e.g., Gozzard 2007; Hillman et al. 2003). The clay was also deposited along river valleys where it can be difficult to distinguish from river alluvial deposits (Commander 2003, 11).

The most important strata of the superficial formations in central Perth are the Tamala Limestone, Bassendean Sand, and the Guildford Clay (figure 4.1 inset). The Quaternary Tamala Limestone is an aeolianite that was deposited between Shark Bay (north) to near Albany (south). It is formed from coarse- to medium-grained calcarenite, windblown skeletal fragments, and quartz sand. Large-scale cross-bedding is characteristic and soil horizons and calcified tree roots are common (refer to figure 8.2, 1A). It was deposited as several linear belts of coastal dunes in the Perth Basin during the Pleistocene (late Pleistocene in the Perth region), with the greatest development in the central region. The most extensive belt is the Spearwood Dune System.

The aeolianite was subsequently lithified by the dissolution of calcium carbonate by percolating groundwater and its re-deposition at lower levels in the sandy formations to form limestone (Playford et al. 1976). The Tamala Limestone has been broken down by carbonate dissolution through most of central Perth area and is mantled by a veneer of sand derived from it by leaching. An 8m thick aeolian sand unit named the Spearwood Sand forms a ridge in the CBD, with the highest points between Hay Street and Murray Street (figure 4.2) (Hudson-Smith and Grinceri 2008, 45). It is comprised of an upper yellow poorly graded sand with some silt and a lower white to light brown poorly graded sand, which extends to the west and north of the CBD and eastward to East Perth. The southern boundary is to the north of the William Street Esplanade intersection. The Bassendean Sand underlies the eastern margin of the study area. The Guildford Clay, which lies at depth beneath the city, is important geotechnically for deep excavations, and in the modern eastern suburbs where it lies at or near the surface.

The landform expression of the superficial formations are three north-trending dune systems that lie approximately parallel to the coast. They decrease in age to the west, from the oldest Bassendean Dune System (the surface expression of the Bassendean Sand) against the Pinjarra Plain; the Spearwood Dune System, which underlies the study area and represents the surface expression of the Tamala Limestone; and the Quindalup Dune System bordering the modern coastline, which began to form 10,000 years ago (the end of the last ice age) and is still forming in some coastal sites (Gozzard 2007, 17). The landforms associated with the superficial formations are discussed in detail below.

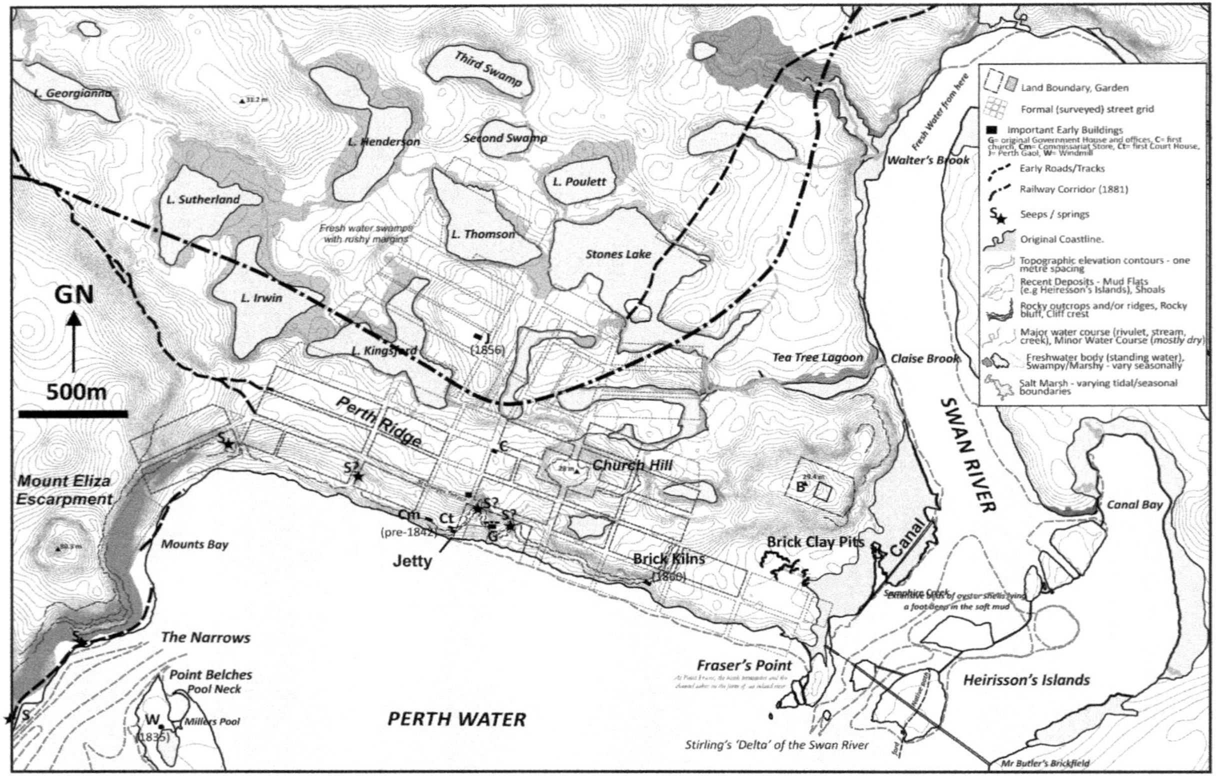

Figure 4.2 The Perth Settlement. Reconstruction of the landscape of the Perth Settlement c. 1830. This plan shows the location of the main buildings of the camp as recorded by Arrowsmith (1839). The first settlement was a temporary camp on the crest of the steep riverbank, where a gully provided easy access to the high ground of the Perth Ridge. The landing place at the base of the gully became the focus of river traffic to the camp and a jetty was developed, which became a major transport hub. The lake system that dominated the landscape north of the Perth Ridge was the major influence on the location of land available for town block subdivision, and strongly influenced their geometry when the streets were surveyed between 1830 and 1832. Key sources for this map include Roe, Arrowsmith and written descriptions by Charles Fraser, James Stirling, and others. The topographic contours are at 1 m intervals and are derived from the 3D terrain reconstruction. Please refer to the extended caption at the rear of the book for additional information.

Alluvium and Colluvium

The Swan and Canning rivers have deposited an extensive sheet of alluvium across the SCP, which extends as far west as Point Belches (the Narrows) (Gozzard 1986). Alluvial sediments have been identified beneath the central city area where they consist of complexly interlayered sand, silt, and clays underlying the aeolian dune sands lying above the Kings Park Shale (Formation) (Andrews 1971, 49).

The Swan River is flanked by recent floodplain clay and loam deposits, and terraces that reach as far downriver as Point Fraser and Heirisson's Islands, with the Burswood and Maylands peninsulas being particularly important (Gray et al. 1951; Davidson 1995, 6).

Colluvial deposits accumulated along the base of the Darling Escarpment, and formed largely from the erosion of the Archaean rocks to the east of the Darling Fault. They are composed of sandy and gravelly silt with pebbles of granite, dolerite, and laterite (Gozzard 1986).

Peat and organic-rich lake sediments accumulated within the low points of the Fraser Point topography, which usually represented dune swales of the Spearwood Dune System (Commander 2003, 11). The organic-rich sediments were the most fertile soils of the SCP and were important as sites for early agriculture, particularly market gardens, in the 1830s and onward.

The Swan Coastal Plain

The SCP is bounded to the east by the Darling Scarp and Gingin Scarp, which mark the eastern extent of Cenozoic marine erosion. It is 23 km wide south of the Swan River and has an average elevation of 25 m. North of the river it is 34 km wide and averages approximately 75 m in elevation. Davidson (1995, 5–6) recognizes five distinct landforms that approximately parallel the coast. From east to west, they are:

- Ridge Hill Shelf: Colluvial slopes forming the foothills of the Darling and Dandaragan Plateaus and which are dissected remnants of a sand-covered, wave-cut platform.
- Pinjarra Plain: A piedmont and valley flat alluvial plain comprised of clayey alluvium originating from the Darling and Dandaragan Plateaus.
- Bassendean Dune System: A gently undulating aeolian sand plain that was originally comprised of mainly calcareous sand, quartz sand, and minor fine-grained heavy mineral concentrations. Most of the carbonate has been leached, resulting in quartz sand as the main constituent. The dunes probably formed at the coast during the higher sea levels associated with interglacials. They generally have higher topographic relief north of Perth than to the south.

- Spearwood Dune System: A weakly calcareous aeolian remnant formed from the leaching of the underlying limestone. The exposed leached upper surface of the Tamala Limestone is coincident with the Spearwood Dune System.
- Quindalup Dune System. The most westerly and immediately coastal landform of the SCP is a windblown deposit of lime- and beach-sand that forms ridges and dunes that runs parallel to the modern coastline.

Several rivers cross the plain from east to west, including the Murray, Serpentine, Canning, and Swan rivers. Each of these major rivers begins its westerly course in the high ground of the Darling Escarpment but is then deflected to more northerly or southerly courses as it traverses the north-south trending landforms and geological units of the SCP. The Swan and Canning rivers are the most important to this study.

The Swan Valley

The course of the Swan River forms a valley with three distinct parts: the Lower Swan, from Fremantle to Heirisson's Islands, which is an estuarine drowned river valley; the Middle Swan, between Heirisson's Islands and Guildford, which is a wide meandering valley lined by dunes of the Spearwood and Bassendean Dune Systems; and the Upper Swan, from Guildford to the Darling Range escarpment, which follows a mature course that has undergone several rejuvenations characterized by a succession of erosion terraces (Gozzard 2007, 94–95).

Geological boundaries have a profound influence on the course of the Swan River as it traverses the SCP from east to west, particularly in the study area. The boundary between the Bassendean Sand and Tamala Limestone (and its leached derivatives) occurs between Point Fraser, the Burswood and the Maylands peninsulas (Gozzard 1986). The course of the river is sharply deflected here, by as much as 80–90 degrees, with river reaches lying parallel to the trend of the boundary. The course of the Canning River swings sharply northward at this same boundary just prior to entering the Swan River Estuary at Point Heathcote.

The Swan-Canning River Estuary

In 1827, James Stirling considered the Swan River to end at the entrance to Perth Water (then considered a part of Melville Water) at Heirisson's Islands, with the estuary extending to the rocky headlands of its mouth at Fremantle. He considered Heirisson's Islands to mark the end of the inland river and the start of the estuary and the marine influence, an observation also made by de Vlamingh and Heirisson in their earlier explorations.

This is correct because the river only became fully fresh at a point just to the north of this position, where Walter's Brook joined the river. Stirling noted this point as an "abundance of freshwater" on his 1827 chart (Stirling 1829b). Removal of the limestone bar at Fremantle heads during harbor works in the 1890s, and continuous dredging of navigation channels has since made the estuary saltier in summer when previously it was brackish (Brearley 2005, 77–79).

The Swan River Estuary was originally a more complex and varied waterway than it is today. Apart from the many narrow points and shoals already mentioned, it once possessed limestone cliff-lined shores along the southern shoreline of Blackwall Reach (which survive) and on the northern shore of Rocky Bay. Of most significant was the limestone reef that obstructed the mouth of the estuary (this is discussed in detail below). There were also many low sandy (and occasionally rocky) islands, islets, shoals and sandbanks, including Heirisson's Islands; at the Narrows between Point Belches and Eliza Point, and there were sand-spits such as Point Walter, the Karrakatta Bank in Freshwater Bay, and the Rocky Bay shoals. An extensive sand bank and shoal known as Ferry Point once existed on the south bank, near the mouth of the estuary at Fremantle.

The estuary widened significantly west of Point Fraser where it became known as Perth Water . Downstream of the narrow point at Point Belches, the estuary further increased in width to form Melville Water and Freshwater Bay (figure 4.1).

THE PERTH LANDSCAPE IN 1829

Perth was founded in a landscape developed on the Bassendean Dune System, and by the sand deposits and limestones formed from the leaching and redeposition of carbonates of the Spearwood Dune System. The re-deposition of carbonate at depth within the sand deposits formed the Tamala Limestone. Subsequent erosion exposed the deep limestones as resistant topographic high points such as the Mount Eliza Escarpment (figure 8.1, 1A) and Buckland Hill. Further leaching of the exposed limestone formed the veneer of Tamala Sand that is encountered across the surface area of Point Fraser. To the northeast of the main study area, the Bassendean Sand predominates north of the Burswood Peninsula. It is the small-scale, localized landforms and geology of these features of the Perth Basin that had greatest impact onPerth urban development.

The riverine landscape developed on the partially inundated SCP played a key role in how the European explorers moved about the country, and how and where they chose to establish the first settlements and exploit the landscape. Quaternary sediments supplied the greater part of the building materials from which the developing settlements were built, and ground water was an important natural resource (as it still is). The urban character and landscapes of Perth have been more influenced by Quaternary geology and geological processes than any other Australian city.

Reasons for the Choice of Point Fraser

It was largely because of Charles Fraser's positive report about the agricultural potential of the Swan Valley, made after the 1827 reconnaissance of the river, that the Swan River was chosen as the site for settlement. However, Perth is located where it is because of the landforms and geology of the Swan River Estuary. Multiple settlements were required to service the settler population because no single site provided all the necessary features required for a single large population center (deepwater anchorages, fertile soils, and adequate freshwater).

Several versions of Stirling's first charts of the region are prominently annotated as "Abundance of Freshwater" on or near Point Fraser (e.g., Stirling 1829b; Watson 1923, Map 1) and it was there that he camped during his exploratory visit, and located Clause's Brook. So, the freshwater availability was probably an important point in its favor. The site would have also given protection from any seaward bombardment during any possible future attack (Statham-Drew 2003, 139).

Stirling was aware that Sir George Murray preferred an inland settlement location at Point Heathcote, where the Canning River entered the Swan River Estuary (Statham-Drew 2003, 139) but believed that the site he chose would allow better communication between the port and the important inland agricultural settlements exploiting the rich alluvial soils of the river valley. This was an important factor because at the time, most heavy transport was by water and Point Fraser was the last place on the river that larger boats could navigate upriver because of the shoals around Heirisson's Islands (Statham-Drew 2003, 139–40).

There were several key pinch points between Point Fraser and the mouth of the estuary created by alluvial sandbanks, spits, shoals and mudflats, which made the river unsuitable for the passage of anything other than small vessels. The shoals of Heirisson's Islands made navigation upstream of the islands even more difficult. So, Fremantle was established on the coast, Perth at a well-watered point inland as an administrative center, and Guildford was established approximately 17 km further upstream, at the head of the navigable river, to service the agricultural hinterland. The modern geography of Greater Perth is a result of the characteristics of the Swan River and the estuary, which made these multiple settlements necessary.

Some writers mention the beauty of the Point Fraser and the river landscapes (e.g., Friend 1830, 20–21) and there are other occasional hints that the site of Perth may have been chosen partly because of the beautiful setting it would provide for an administrative capital.

Landscape Features of the Perth Settlement Site

Point Fraser lies at the eastern fringe of the outcropping limestone and at the transition from landforms dominated by limestone outcrop to landforms more strongly influenced by residual and superficial geology, and by the boundary between the Tamala Limestone (Spearwood Dune System) and the Bassendean Sand (the Bassendean Dune System) (see figure 4.1 inset).

Erosion by water and wind deflation, and sand and carbonate re-deposition within the older dune systems, has created many subtle landscape features within central Perth that are important influences on the layout of the modern city. The most important of these factors was where landscapes have interacted with groundwater and its varying seasonal variations to produce the Point Fraser Lake System. Two distinct linear topographic trends are observable in the reconstructed 1829 topography (figure 4.2). There is a prominent northwest-southeast trend that approximates the mapped boundary of the Bassendean Sand and the Tamala Limestone. This geological boundary coincides with one of the most fundamental landscape boundaries of the study area. To the south of the boundary, smaller-scale landscape lineaments with a northerly trend are developed on leached Tamala Limestone (Gozzard 1986) and probably represent dunes. It is the small north-trending landforms that were most important to the location and geometry of the central Perth lakes, and subsequently to the development of the layout of the roads and surveyed blocks of the settlement.

The Perth Ridge

In 1830, the town of Perth was described as "situated on an eminence" (Friend 1830, 21) that rose to a height of 20 m through most of the heart of the settlement and would have towered above the landing place. This southeast-trending ridgeline was mainly formed by the Tamala Sand, windblown and remobilized material derived from the breakdown of Tamala Limestone further to the west (Gozzard 2007, 67). It will be referred to here as the Perth Ridge (figure 4.2).

The Perth Ridge formed a central spine of the settlement that extended for approximately 2 km eastward from the foothills of Mt Eliza. The southern flank of the ridgeline was deeply embayed by south-draining gullies extending to the shoreline in the zone between the modern Supreme Court and

Stirling Gardens, and Victoria Avenue. These gullies would play an important part in the subsequent growth of Perth.

A slightly northern offset of the main ridgeline continued eastward for a further 550 m where it reached a maximum height of 28 m, forming a prominent knoll that became known as Church Hill. The highest place on Point Fraser was a second prominent hill located 1 km east of Church Hill. It was offset from the main ridge and largely detached from it. Despite the prominence of this feature, it does not seem to have been named.

The topography of the Perth Ridge was initially used to advantage for the placement of important government buildings, particularly Government House and the Domain, which were placed on a prominent riverside site, on a high point of land overlooking the principal landing place.

As John Roe surveyed the settlement, he laid out a square around Church Hill, which he originally conceived as a focus for the administrative district of the colony. It would later be named Victoria Square. This vision was soon lost, and the focus of government moved west, and commerce became centered around Hay Street. East Perth and Victoria Square (and Wellington Square) did not become the administrative center of the settlement. The hill site was offered to the Anglican Church, but it declined because it was considered too isolated. They preferred to worship in the more fashionable St Georges Terrace (Bolton 1988, 146). In the 1840s, as Catholic clergy arrived in the colony, the square became the location of the first convent and St Mary's Catholic Cathedral. In the 1850s the Colonial Hospital was also constructed in this east end area, possibly because the elevated site offered good ventilation and cooling breezes. Victoria Square is still the site of the Cathedral and the Royal Perth Hospital.

The prominent hill at the eastern end of Point Fraser was earmarked for use as a Government Domain and the site of a new Government House but this never happened. Ultimately, the prominent hill became the main cemetery for colonial Perth.

The Steep Southern Shoreline of Point Fraser

The southern shoreline of Point Fraser followed the base of Perth Ridge and the land rose steeply upslope from the shoreline, forming low bluffs. Land reclamation of the foreshore has significantly modified the shoreline, but the cliff-like slopes remain.

The topography of the original coastline is still discernible in the city center. Traveling northward along the alignments of Mill Street and William Street from Mounts Bay Road and the Esplanade traverses the steep slope of the original river terrace for which St Georges Terrace was named. The way the ground falls steeply away from St Georges Terrace toward the river in the Government House, Supreme Court (which once backed directly onto the shoreline), and the Perth City Council building precinct is also a remnant of the original steep shoreline. It is well preserved in this area because these historic government buildings were placed early in the town's history. Recent buildings in the area, such as the Perth Concert Hall, exploit the steep slope of the bank by having subfloor levels in their southern wings below the current street level. The recent re-excavation of reclaimed land to create Elizabeth Quay has returned a small part of the shoreline to something approaching its original position.

St Georges Terrace—the First Elite Suburb

St Georges Terrace once marked the southern edge of the Perth settlement and followed the crest of a river terrace that sloped steeply down to the riverbank from the southern slopes of the Perth Ridge. Low gullies in the steep shoreline allowed easier access to the prominent ridge behind and probably dictated where Stirling landed and established his first tent camp in 1829 (see section below).

Pressure for land on St Georges Terrace grew rapidly because of the access it offered to the cooler afternoon westerly breezes and the prestige its location attracted (Bolton 1988, 146). It lies on the southern upper slope of the high ground of the central Perth Ridge, facing the river. Official policy required that buildings on the terrace must be worth at least £200. It became the best street in Perth, and the western extension up to Mt Eliza, which became Mounts Street, housed Western Australia's elite families until the mid-twentieth century (Bolton 1989, 146).

Hay Street, set one block back from St Georges Terrace, developed into the major commercial thoroughfare because it followed the crest of the Perth Ridge and was not subject to the flooding associated with the lakes that lay near Murray Street and Wellington Street (Bolton 1989, 144). Land situated among the lakes and wetlands north of Wellington Street was cheaper because of the higher risk of flooding.

The Mt Eliza Escarpment

The western end of the settlement rose from the Perth Ridge to the foothills of a limestone escarpment that reaches a height of 80 m. Stirling climbed to the top of this high ground from the river in 1827 and named it Mt Eliza. Mt Eliza is a tableland formed by Tamala Limestone of the Spearwood Dune System, which formed from the leaching of calcareous sands and re-deposition of calcium carbonate at deeper levels within the dune sands. The escarpment is mainly an erosional feature, with cliff faces formed as the Swan River receded. The tableland diminishes to the west and southwest to merge with the lower country of modern Subiaco and Nedlands. Perched water tables associated with sandy loams sheets are important water sources for vegetation (Botanic Gardens and Parks Authority 2019). At the western end of the city grid, Mount Street follows the rising ground to the top of tableland and was an important residential area for the Perth elite for much of the late nineteenth and twentieth centuries.

Mt Eliza is the location of modern Kings Park, the Botanical Gardens, and the Perth War Memorial. It is also the site of important water supply infrastructure. It has also been a site of significant limestone quarrying to supply materials for Perth public buildings.

Heirisson's Islands and the Shoals

The earliest narratives of European exploration of the Swan River come together at the low-lying islands and shoals of Heirisson's Islands. This landscape pinch point had long been used as a crossing place by the Aboriginals but became a pivotal location in European history. The Dutch ended their river explorations there after the boat party was unable to negotiate its way past the shoals in January 1696 (Brearley 2005). In 1801, the French exploration party also grounded on the shoals and only passed them with great difficulty, naming them Heirisson's Islands (Peron 2012). James Stirling also stranded there and spent considerable time negotiating this pinch point in the river on his 1827 upriver journey. The presence of the shoals contributed to James Stirling's decision to create two separate inland settlements on the Swan River and were a major factor in the choice of Point Fraser as the site for Perth.

The islands and shoals have been extensively re-shaped and recontoured using dredging spoil since 1829, creating the modern form of Heirisson Island. This important place remains one of the main river crossings of the modern city (The Causeway).

Burswood Peninsula (Island)

Burswood Island was created in 1830 when a canal was cut through the narrow southern base of the Burswood Peninsula, an elongate, horseshoe-shaped bend (peninsula) in the river east of Point Fraser (figure 4.1), to shorten the river journey. The canal and diversion dam had been abandoned and filled well before 1906 and the peninsula reinstated but the name persisted. Slag, dredge spoil, and power station ash were used to reclaim low-lying land on Burswood Island, beginning in the late 1930s and the natural lagoons on the western side of the peninsula, adjacent to Heirisson's Islands were filled in (Brearley 2005, 88).

Another canal was cut through the mudflats at the eastern fringe of the settlement, in East Perth, to improve the passage of river traffic past the shoals and flats of the Heirisson's Islands but was unsuccessful (Hay 1906, footnote 51, 14).

Buckland Hill and the Limestone Cliffs

Limestone cliffs and low bluffs are a feature of the lower reaches of the Swan River Estuary, giving their name to features such as Rocky Bay and Blackwall Reach. Limestone outcrops also form Rous Head and Arthur Head, and they formed the limestone bar that formerly blocked the mouth of the estuary. Many of the riverside limestone outcrops, particularly those of Rocky Bay, were extensively quarried for industrial purposes and the original landforms are lost. The limestone riverside bluffs have survived on the south bank of the river further upstream, at Blackwall Reach, named after similar features in the River Thames near London. This locality is a popular swimming place.

The cliffs on the north shore of Rocky Bay are contiguous with limestone outcrops extending to Buckland Hill, one of the most prominent topographic features between Fremantle and Perth. Buckland Hill was the highest of a group of hills knows as the Seven Sisters. Fraser (in Hay 1906, 9) seemed to imply that the high ground in this area was being considered in an early proposal as a possible location for the primary settlement of the Swan River Colony. The idea was rejected by James Stirling once he had re-examined the region in 1829.

Buckland Hill became a focus of intensive limestone quarrying to supply the nearby Mt Lyell superphosphate works, which reduced its height by at least 15 m and leveled many of the surrounding hills. Subsequently, it became the site of a reservoir, trig station, and military gun emplacement (Rocky Bay Heritage Trail 1988). Once the limestone resource was exhausted, the now flattened surrounding landscape became the site of industrial development.

Wetlands, Lakes, and the Drainage System

The landscape and urban development of the Perth CBD has been strongly influenced by the many wetlands and lakes that were originally present in 1829. Wetlands and lakes formed in the interdunal swales of the Bassendean Dune System, in the inter-barrier depressions between the Spearwood Dune and Bassendean Dune Systems, and within the Spearwood Dune System (Davidson 1995, 6). The influence of the landforms and geology has produced a strong northwest-southeast linearity in the geometry and distribution of wetlands of the SCP in the Perth region, where they are focused on key geological boundaries. One exception is the zone immediately to the northeast of the Perth settlement site, where the lake systems are deflected to the southeast by the high ground of the limestone underlying the Mt Eliza tablelands, and drain through the central Perth area (figures 4.1 and 4.2).

The largest lakes, such as Lake Monger and Herdsman Lake lie on the same trend as the deepest parts of the Swan River Estuary, including Perth Water and Melville Water, and the deep hole at Rocky Bay. The course of the Swan River may have exploited topographic features that originated as lakes and wetlands adjacent to its course.

The water level and surface area of the wetlands in the study area varied seasonally in extent, reflecting the level of groundwater in superficial aquifers. Survey plans compiled at different times record such rises and falls and show the varying extent of the lakes and swamps. In February 1830, Mary Friend visited the Perth camp and described the three lakes in the settlement she walked past on the way back to the boat as "rather marshes than lakes" (Friend 1829, 22). See figure 4.2.

The Point Fraser Lake and Wetland System

Stirling chose the site for his administrative settlement where there was abundant freshwater in a semiarid region. Most of the first water supply came from the freshwater wetlands and lakes nearby, several springs and rainwater tanks, or directly from the Swan River (particularly from the eastern edge of the settlement where the water became fresh). Water was also drawn from wells which, as the city grew after the 1890s gold rushes, became a source of diseases such as typhoid (Engineers Australia 2012, 1).

The lake system occupied much of the space to the north of the Perth Ridge and formed an interconnected network of wetlands lakes and low-lying areas that may have flowed to some extent during the wetter times of the year. The lake landforms were created through the interactions of the shallow groundwater with the surface topography and geology. The outlet of the flowing water to the Swan River was via Clause's Brook and Walter's Brook and through the springs along the southern foreshore of Point Fraser (e.g., near Spring Street).

There were originally fourteen named lakes or lagoons forming the Point Fraser Lake System, including Three Island Lake (Smith's Lake), Lake Georgiana, Lake Henderson, Lake Sutherland, Lake Irwin, Lake Kingsford, Lake Thomson, Stones Lake, Second Swamp, Third Swamp, Lake Poulett, Tea Tree Lagoon (Clause's Brook), and Walter's Brook and lagoons. Further to the north, Herdsman Lake and Lake Monger were part of the same system and are the only significant survivors of the system (figure 4.1). Several other unnamed smaller wetlands and lakes were also present.

Many lakes were developed as market gardens in the 1840s because the organic-rich lacustrine sediments accumulated in them was a rich substrate for horticulture. Despite their agricultural value, the wetlands quickly limited the northward development of the town. With continued urban expansion, they became valuable inner-city residential and commercial land. The water level in the lakes varied seasonally and as more native vegetation was cleared, the water table rose, resulting in flooding problems, particularly during very wet winters (Brearley 2005, 85; Bolton 1989, 152). Early drainage works in the 1850s did not prevent further floods and related health issues, and a major drainage program was required to connect the lakes to Clause's Brook. The low-lying land that became available was less desirable than the elevated sites with river aspects so developed into working-class suburbs. The first Perth Jail was built in this less desirable area in 1855 (Bolton 1989, 152). The old jail was repurposed as a Geological Museum in 1891 and became the nucleus of the modern Perth Cultural Precinct.

Most lakes were drained or filled by the mid-nineteenth century. Often, because they became polluted and unhealthy. Some, such as Third Swamp, were formed into European-style parks (Hyde Park), while many others became sports grounds and reserves. A good example is Three Island Lake, which survives as a sports ground that retains a small remnant of the original lake in the Smith's Lake Reserve. The largest lakes of Point Fraser Lake System remain to the northwest, including Lake Monger and Herdsman Lake. They have become desirable places to live and are now mostly surrounded by fashionable housing.

Samphire Creek and the Point Fraser Salt Marshes

Extensive salt marshes occupied the flats at the southeastern tip of Point Fraser, opposite Heirisson's Islands and a shallow and brackish creek drained them. At some point prior to the production of Arrowsmith's (1833) map, only the small, curved tip of the peninsula, and its associated islets and shoals had retained the name Point Fraser. Immediately to the north was a small inlet, which drained the salt marshes and flats in this area. This was named "Samphire Creek" by September 1838 (Lands Department 1838). This creek formed one of several outlets for the surface drainage and groundwater discharge from Point Fraser.

The low-lying salt marshes and wetlands became the focus of sporting events, including horse racing and cricket matches. The areas were eventually leased to sporting clubs and remain the location of Western Australia's most important sporting infrastructure, including the WACA Ground and Gloucester Park Raceway. Low shorelines of the river further to the east are now the sites of most of the state's most important sporting venues, including Belmont Park Race Course and the Perth Stadium (on the Burswood Peninsula), and the Ascot Racecourse beyond.

Tea Tree Lagoon, Clause's Brook, and Walter's Brook

The Perth wetland system debouched into the Swan River immediately upstream of Point Fraser, through a narrow, linear, tea-tree–lined lagoon and stream that Stirling named Clause's Brook (modern Claisebrook). The brook was less than 5 m wide and passed through thick undergrowth where it entered the river (figure 4.2).

Immediately upstream there was a second east-flowing stream that Stirling named Walter's Brook in 1827. It entered the Swan River near the north end of Burswood Peninsula, at the approximate position where the Swan River became fully fresh in 1827 (Stirling 1829). This creek was the exit point of much of the drainage from the wetlands on the northern fringes of the study area, being fed by two small lagoons immediately behind the shoreline.

Groundwater

The location, abundance, and quality of Perth's groundwater have been a defining influence on urban development throughout its history. The French explorers collected water from features that were probably Aboriginal wells developed on springs as they traveled upriver (Peron 2012). Stirling and Fraser made similar observations of springs and wells during their journey upriver in 1827. Fraser also observed how easy it was to dig into the sand in an otherwise "barren" place on Point Fraser and find abundant fresh water that, with a little more digging, would become plentiful (Charles Fraser, in Hay 1906).

Wells and some springs remained the primary source of potable water for the population of Perth until the first reticulated water supply scheme was built in the 1880s. The only other sources available were rainwater and springs (Engineers Australia 2012). The first publicly funded water distribution scheme piped water by gravity feed from Victoria Dam in the Darling Range to a service reservoir excavated by hand on the summit of Mt Eliza. From there, it was distributed through the settlement by a reticulation network. The scheme was opened on October 1, 1891. While the original service reservoir has been demolished, it was replaced by larger reservoirs that are still important infrastructure elements of the modern city water supply (Engineers Australia 2012, 3).

Both superficial and deeper aquifers remain important freshwater sources for modern Perth. Groundwater provides up to 50 percent of the scheme water requirement and up to 80 percent of the total water requirement of the

Perth Region (Davidson and Yu 2008). The water is sourced from multiple freshwater aquifers in the Leederville Formation and the Quindalup Dunes. The Gnangara Mound Groundwater System, north of the metropolitan area, is one of the most important sources (Leyland 2012).

Springs

Springs were common along the southern shores of the estuary between the heads and the Canning River, particularly around Point Heathcote, where a garden was established to exploit them (e.g., Stirling 1829b, and in Watson 1923, Map 1). Springs were also a common feature of the southern shoreline of Point Fraser and particularly further westward along the foot of Mt Eliza at Mounts Bay. They were probably outlets for the groundwaters and wetlands at the western end of the settlement. Clause's Brook and Walter's Brook provided the outlets at the eastern end of Point Fraser.

The site of the first brewery in Perth was chosen in 1834 because the Spring Street site contained a natural spring. Further west along the shoreline, another spring was later developed as the Kennedy Fountain and was an important public water source for the early town. A third spring issuing from the base of Mt Eliza was a source of freshwater for the former Swan Brewery site and a pump house remains at the location.

1829 Vegetation and Environment

The foundation of the Swan River Colony was unusual compared to the beginnings of contemporary Australian settlements because professional scientists and artists were present to record the landscape and natural history prior to the commencement of organized settlement. The result is a largely unique record of Swan River landscapes surrounding the Perth site prior to the rapid environmental modifications that followed British settlement. These records have been used to create figure 4.1.

When Stirling was sent to the Swan River to assess the region's suitability for settlement, he had NSW Colonial Botanist Charles Fraser with him. Charles Fraser made the first systematic botanical observations in the Perth area. Places where he made botanical observations include Point Fraser (named for him by Stirling), Clause's Brook, and Heirisson's Islands (Hay 1906). There were also written descriptions of the pre-European vegetation and local environment from the earlier Dutch and particularly the French exploration but those by Fraser and Stirling in 1827 are particularly detailed.

Uniquely, there are accurate paintings of Perth landscapes from 1827 because Stirling also had Frederick Garling with him. Garling was a highly regarded, self-taught watercolor marine and landscape artist who was to record many of the landscapes seen during the 1827 expedition up the Swan River (Statham-Drew 2003, 66). Of note to this study are Garling's views of Mt Eliza (now Kings Park), Point Belches (Mill Point and the Narrows), and of Heirisson's Islands viewed from "Fraser's Point."

The low hills behind the banks and flats of Point Fraser were described as very "barren," except for the western Australian red gums (known as Marri), and *Banksia* that could reach over 20 m and had girths of over 1 m (Charles Fraser, in Hay 1906, 15). And as the survey of town allotments along (the as yet unnamed) St Georges Terrace was completed in October 1829, the Perth Ridge, along which it ran, was described as "adorned with lofty trees and a variety of lovely flowers" (Martens 2011, 22). Occasional patches of kangaroo grass were seen on the banks (Charles Fraser, in Hay 1906, 16).

The tip of Point Fraser marked the end of the steep bank and of the estuarine influence of Perth Water. To the east, the channel of the Swan narrowed to become a true "beautiful" inland river. There were broad, fertile alluvial flats and levels located here that included extensive salt marshes (probably Samphire Creek). A scarlet-flowered species of *Meleleuca* "of great elegance" formed thickets with two similar species on the flats, and there was a variety of other smaller flowering plants. There was also evidence that the flats were occasionally inundated, in the form of drift timber strand lines 2 m above the water level. Large flocks of cockatoos flew overhead (Charles Fraser, in Hay 1906, 14–16).

The narrower outlet of the river channel was largely filled by extensive mudflats and shoals around which the river bifurcated. They were covered in extensive beds of oyster shells lying 30 cm deep in the soft mud and these lacerated the explorer's feet. Later, they would be used to make paths in Perth (Charles Fraser, in Hay 1906, 14). The alluvial deposits reached sufficient heights above water level to form low vegetated islands, with *Meleleuca* and *Casuarina* rimming their margins and interiors covered with salt-tolerant succulents (Charles Fraser, in Hay 1906, 37–8; Garling painted the islands in 1827). A mile further upstream, on the western bank, the party saw a small freshwater creek draining an extensive lagoon (Claisebrook) surrounded by abundant *Melaleuca, Calytris*, and *Acacia,* with *Zamia and Xanthorrhoea.* Gigantic red gums occupied the flats (Charles Fraser, in Hay 1906, 16).

Additional information about the original vegetation was recorded by surveyors along survey transects and water bodies as they laid out land grants in the districts surrounding the main settlements. Roe's (1829) "eye sketch" of the settlement noted that jarrah forest ("mahogany") occupied the western end of the Perth Ridge between Lake Kingsford and the river, and then extended westward from modern Barrack Street, following the alignment of Monger's Road along the northern flank of the Mt Eliza Escarpment to the southern shores of Lakes Monger and Herdsman. The annotation on Arrowsmith's

1833 chart of Perth town allotments notes the "rushy margins" of freshwater swamps north of Lake Kingsford (e.g., Martens 2011, 23).

The Fremantle Landscape in 1829

There are more records of the appearance and layout of the first camp at Fremantle than there are of the Perth camp. This is because the settlers were first landed at Fremantle, in a temporary encampment and more people wrote about it, and even painted it. Perth was only established when the region had been explored and assessed. Once Stirling had decided that multiple settlements were required, he selected the site for Perth as an administrative hub, and a base for the surveying parties. No civilians were present to create artworks, and the personnel were too busy to create detailed plans of the camp layout.

The sterility of the soils in the coastal environment deterred Stirling from creating a single port and administrative settlement at Fremantle (or immediately inland at Rocky Bay and Buckland Hill). The coastal environment of Fremantle is rocky, sandy, windswept, and harsh. The soils are sandy and only thinly mantle the Tamala Limestone, which forms all the prominent ridges and headlands at the mouth of the river, and to the north and south along the coast. Freshwater was limited, and groundwater was the only reliable source. The result was that the site of Fremantle could not also serve as the location of a major administrative and agricultural hub for the colony.

The Limestone Bar

The most significant geological feature to influence British settlement in 1829 was the limestone bar at the mouth of the Swan River Estuary, between the prominent coastal limestone ridges of Rouse's Head to the north and Arthur's Head to the south. Like Heirisson's Islands, this natural pinch point in the river played a significant role in the early exploration of the Swan River, and strongly influenced the way that the region was settled and developed by the British.

This bar was a low-lying limestone reef that prevented any vessel drawing more than 1.2 m (larger than a longboat) from entering the river estuary. It was also difficult for even smaller boats to enter (e.g., de Vlamingh; and Heirisson's description of the 1801 exploration quoted by Peron 2012). In heavy weather the sea broke over the bar from point to point preventing any vessels from entering (e.g., Charles Fremantle, in Dowson 2011, 24), a fact that Mary Friend illustrated in 1829 (Friend 1830). Friend (1830, 14) described the part of Fremantle around the camp as "beautifully undulating and thinly wooded" but noted that the soil was only white sand and that the glare it created was hard on the eyes. Upstream of the bar was a broad, low sand spit that became known as Ferry Point. The associated shoals became dry at very low tides and impeded almost 80% of the estuary, leaving only a narrow navigable channel for small vessels close to the north bank (figure 4.2).

The inability of large vessels to pass the bar and Ferry Point resulted in the development of a jetty-based port infrastructure extending outward from the south head. Western Australia's original port lay outside the river mouth and used long jetties until the bar was removed in the 1890s. This work required a ton of dynamite a week to construct the modern self-scouring inner harbor designed by C.Y.O'Connor (Dowson 2011, 24).

The limestone bar at the mouth of the river prevented larger vessels navigating upriver and led to the need for twin settlements, a port, and an inland agricultural service center. This in turn led to the development of the colony's earliest transport networks, including jetties, then roads and later, the Fremantle railway line, which transported materials from the ocean wharfs to Perth and beyond. This also controlled the direction of urban expansion, which was to the west along the rail corridors, creating some of Perth's oldest suburbs. Later, the Midland Line to the east would have a similar influence. Beach culture, which arrived later, leading to seaside developments and day-trip facilities.

The Aboriginal Landscape

The SCP was well watered and relatively resource-rich in 1829, so resident Aboriginal family groups only needed to move small distances to exploit seasonal resources. They did occasionally make longer journeys to participate in larger social and ceremonial gatherings. The populations were generally permanent and highest around the Swan and Murray rivers, with around five major communities spread from north to south. The groups of the plain shared a common language and material culture with those of the greater southwest, the *Nyungar* or *Bibbulman.* The Darling Range represented a boundary that defined them as a distinct subgroup (Gibbs 2011, 4–5).

Ecological transitions were important boundaries of Aboriginal cultural groups and their divisions, and group names were based on landform features. The Darling Range formed the eastern boundary of the *Beeda-kala*, *Wadarndee*, or coastal *Bibbulmun* and there was only limited penetration inland along valleys beyond the lower foothills. The *beelagur* (river people) followed the course of the rivers inland, beyond the Darling Range and the *boyun-gur* (stone or hill people) lived over the range (Gibbs 2011, 5).

Aboriginal Natural Feature Names

Most of the recorded information about Aboriginal life, culture, and interaction with the landscapes and geology of the Swan River region was made by Europeans. There are almost no surviving firsthand observations made by Aboriginals. The most specific are those collected and recorded by Daisy Bates in the late nineteenth century (e.g., Bates 1992). Hughes-Hallett (2010, 42) compiled many of the place names of Perth and the following list is based on her compilation. Additions have been made from Bates (2018).

- Byerup: Near the rise of Kings Park
- Yogarin: Kings Park place of Zambia plants and its nuts place for catching kangaroos
- Dyeedyallup Nyoongar: Near the Esplanade campsite located near Burt Way
- Kooraree: Government House Fanny Balbuk's grandmother is buried in the grounds
- Gumap: Foreshore between William and Barrack Sts—fishing area—(now reclaimed to create the Esplanade)
- Goodinup: Perth City—west end—favorite camping place of Yellagonga
- Beeabboolup (Beeabboollup): Somewhere near the original Pt Fraser—place of many fish
- Buneenboro: Perth Water (between Narrows and Causeway)
- Booneenboro: the north bank of Perth Water
- Matagarup: Heirisson Island(s) place where the river is only leg deep
- Kakarooomup (or Kakkanoomup): one of the Heirisson Islands where Balbuk's mother was born
- Boodjar-Gorden: Heirisson Island
- Danjanberu: Lake Henderson-place of the shallow lake OR Three Island Lake—where the zamias or byers grow
- Goongarnula: Lake Henderson- place of the stinking spring with moss (also referred to as yarreenup)
- Boodjamooling: Hyde Park area the earth ground where it looks like a nose pegging takes place, referring to the initiation of young men
- Gabbee darbarl: very big river
- Gabbee kowangoolup: Melville Water
- Goodroo: Eliza Bay
- Wandaraguttagurrup Nyoongar: meeting place in Kings Park where there were holes toward the ocean
- Gabbikalga: Kings Park—near the Statue of Queen Victoria once was an old hollow tree in which rainwater collected
- Jin'jeejerudup: Mt Lawley

- Joo'albub: Lake Jualbup (formerly Dyson's Swamp)
- Wagoorjup: the mouth of the Canning River

Camp Sites

One of the most historically significant family groups encountered by Swan River settlers in 1829 was that of Yalgonga, whose family group comprised his two wives, their daughters, and their husbands. Yalgonga's favorite *kala* (hearth) was at a spring at the base of Mount Eliza, near the shoreline of modern Mounts Bay. This spring was one of several, including the original source of the Kennedy Fountain, that rose along the shoreline from the mount, to Spring Street (named for them) and eastward. It may have been the presence of the eastern springs that attracted Stirling to the Domain site. This group occupied the Perth Ridge between the lakes to the north and the shoreline. In 1833, Lyon (in Hallam 1991, 41–42) recorded that this ridgeline was known as *Byerbrup* and was important to Yelagonga because it overlooked and commanded the flats, and one of the most important river fords across the shoals of Point Fraser and Heirisson's Islands. After the establishment of the British settlement, Yalgonga moved his main camp to Lake Monger (Lyon 1833, in Hallam 1991, 42). The lake provided turtle, mudfish, gilgies, and wildfowl, and camps were located on rising ground to its west and south (O'Connor et al. 1985, 30).

River Crossings

Three places were used as river-crossing places by the local people; the limestone bar at the mouth of the estuary, the Heirisson's Islands (where the Aboriginal track and crossing place is marked on the earliest cadastral survey plans, see figure 4.2), and Point Walter. Noel Morrison mentioned that the crossing at Point Walter, southwest of the settlement area, was an important crossing point used by women and children (Hughes-Hallett 2010, 30).

Stone Tool Technology

People have been using the stone resources of the Perth area for over 6,000–7,000 years, and prior to the most recent sea-level rise (Dortch 1975, 15). The limestone and dune sand systems of the Swan River Estuary region are poor sources of raw materials for cutting tool manufacture. The result is that most of the important raw materials used had to be imported over long distances.

Rock flakes found in the Perth Basin consist of bryozoan chert, non-fossiliferous chert, Quartzite, dolerite, and granite feldspar. They usually range between 1 and 5 cm. Most artifact accumulations result from the erosional deflation of dunes that overlook wetlands and lakes. Dolerite grindstones and chopping tools are also present, as are tools made from broken glass and china, showing that site usage and tool manufacturing continued after European settlement in 1829.

All dolerite, granite, quartzite, and non-foliaceous chert originated from the Precambrian terrains to the east, with the nearest possible source of some artifacts being over 60 km away (Glover and Cockbain 1971, 545). The distance between source and find spot shows that lengthy and distant trading networks existed and that such networks persisted for thousands of years (Glover and Cockbain 1971, 545).

Artifacts made from a distinctive form of Eocene bryozoan chert are common in the Devil's Lair site near Cape Leeuwin, and in sites throughout the Spearwood Dune System, and in time-equivalent contexts north of Perth. The sites have been dated to Holocene or late Pleistocene based on stratigraphic context and typology. Bryozoan chert artifacts were excavated during rescue archeological investigations in the ancient Spearwood Dune System in Mosman Park (5 km north of Fremantle) and charcoal from the site was dated (McArthur and Bettanay 1960, in Dortch 1975, 15). The primary source of the chert was intersected in Eocene limestone during offshore drilling but has no mainland equivalent (Glover and Cockbain 1971, in Dortch 1975, 15).

Ochre and Pigments

Fanny Balbek inherited wilgee-garup, an ochre ("wilgee") pit from her father, but Bates (2018, 8) does not record the location. The ochre had to be burnt to produce the red pigment. White paint was also a common body adornment, and several early photographs show men and boys wearing intricate body patterns. The source of the white pigment is not recorded (Bates 2018).

LANDSCAPE, GEOLOGY, AND THE EARLY DEVELOPMENT OF PERTH

Three almost concurrent settlements were established in the first days of the Swan River Colony in 1829: Fremantle (May 2, 1829), at the mouth of the Swan River Estuary, and Perth (August 12, 1829), as an administrative and commercial service center for the agricultural hinterland. Guildford was founded shortly afterward on a U-shaped bend at the head of the navigable part of the river, where it serviced the best agricultural land of the Swan River Valley and acted as a river port (Bolten 1989, 143). The colony developed rapidly, with the three foundation settlements established and surveyed within a year and the surrounding land grants issued within the next three years.

Stirling realized that he could not establish an inland settlement with its own port because of the poor navigability of the Swan River Estuary, so the colony would need a port settlement (Fremantle). The lack of fertile soils in the coastal location meant that the port area could never support significant agriculture. Ships were prevented from entering the estuary and proceeding upstream, so they had to anchor offshore. This problem was initially solved by offloading stores onto the beach, and then by the construction of jetties at Bathers Bay, which became the first port of Fremantle.

Perth was created as an administrative center located midway between the port and the agricultural hinterland. On July 27, 1829, Stirling issued a proclamation stating that on August 12, King George IV's birthday, a stone would be laid at the new inland settlement to be named Perth and that from that date forward all government business would be conducted there (Statham-Drew 2003, 142). In the event, no stone was available at the place, so Stirling formalized the choice of the settlement with a ceremony involving the felling of a tree. The first blow was struck by Mrs. Helen Dance, wife of the *Sulphur*'s captain, who was the only woman present (the others were all unwell or minding children) (Martens 2011, 20). Several tents were set aside as temporary government offices, and over the succeeding months, they formed the base for the twelve civil officials who began the administration, exploration, charting, and surveying of the site of Perth (Martens 2011, 21) (see figure 4.2).

The Garden Island and Fremantle Camps

When Stirling first arrived at the Swan River, he established a temporary base camp on an offshore island lying 14 km to the southwest of the Swan River. Stirling's ship, the *Parmelia*, had grounded and been damaged while entering Cockburn Sound and the weather had remained poor afterward. It seems that Stirling elected to take the safer option of establishing the first camp on the more sheltered site at Garden Island rather than risking the transfer of passengers and goods to the mainland before it had been adequately explored (Statham-Drew 2003, 134). This first camp was comprised of wooden huts and tents and included a temporary Government House on Cliff Point, at Sulphur Bay. The camp was used as a base to explore the river for suitable permanent settlement sites.

Prior to Stirling's arrival, Captain Fremantle of HMS *Challenger* had established a small redoubt on the mainland at Arthur's Head to guard the mouth of the estuary, raised the union flag, and toasted the King (Statham-Drew 2003, 137). The first private settlers arrived on August 14 and were landed at a small bay near Arthur's Head, where they set up camp near the beach (Statham-Drew 2003, 144–145). Mary Anne Friend, who visited this camp in 1830, described it as looking like a county fair, with many white and multicolored tents. She preferred the Fremantle camp to Perth, despite its white sand and glare, because she disliked the black sand, mosquitoes, flies, and fleas of the latter. She did, however, find the banks of the river beautiful and the site of the Perth camp picturesque (Friend 1830, 20).

The Perth Camp, 1829–1834

Stirling led a party to a landing place near the foot of Mt Eliza, near the current Supreme Court, and adjacent to what would become the Government Domain. There were two deep embayments (almost gullies) in the steep north bank of the river at this site in 1829, where runnels that were possibly spring-fed, drained the higher ground of the Perth Ridge and Church Hill (figure 4.2). The broader expanse of flatter ground on the shore and the gentler slopes of the deep gullies probably gave easier access to the crest of the ridge and these features may be the reason why this place was chosen as the landing location.

The land was cleared on a patch of higher ground and tents erected (Martens 2011, 21). The first Government House (temporary) was built on a small promontory of higher ground on the bank, between the two gullies and the first government garden in the Domain was planted on the flat ground at the foot of the bank, near the water's edge. The garden may have benefited from a spring-fed runnel occupying the gully.

A modest government house was built on the Domain in 1829, with walls and roof covered in weatherboards. Despite the simplicity of the structure, it became the social center of the small colonial elite, and a venue for a ball in 1831. It became so severely damaged by termites by 1832 that it was uninhabitable, and Stirling had to travel from his residence at Guildford (Martens 2011, 24–26).

As Perth developed and St Georges Terrace became a major thoroughfare, a bridge had to be built across the head of the main western gully because it was so deeply incised into the steep bank. The temporary camp came to an end when the first permanent Government House was built in 1834–1837. The landing place became the nucleus of the new settlement, and the location is still the administrative and legal heart of Western Australia, as home to Government House and the Supreme Court.

The first Perth jetty was built at or near Stirling's original landing place near the incised gullies and was in place by 1838 (e.g., Lands Department 1838). A town jetty opened in 1842 and became the transshipment point for goods and people being brought upriver from Fremantle to the inland settlements (Bolton 1989, 15). The location has remained the principal landing place of Perth since 1829 and has been progressively developed as foreshore land was reclaimed. It still serves as the most important landing place in the modern city.

Land Grant Surveys and Urban Expansion

John Roe, the surveyor general, had the Perth streetscape and land allotments planned and mapped out within two weeks of the reading of the proclamation and felling of the tree (Martens 2011, 21). The street grid was laid out to parallel the central east-west long axis of the Perth Ridge. Later allotments to the north of Lake Kingsford and Lake Thomson (now Northbridge) would have a more northwesterly alignment because they were positioned to avoid the wetlands in this sector (figure 4.2). Roads were cleared, crops planted, and the first buildings were constructed. By 1832, all the principal streets and land allotments had been surveyed and named. The blocks along St Georges Terrace (originally named Front Street) immediately became the most sought-after addresses in Perth (Martens 2011, 21–22; 25).

The development of Perth continued into the 1840s, with progressive land clearances and wetland drainage, until the passing of the Drainage Act. The works resulting from this Act removed most of the significant areas of wetland in the Perth area as an aid to public health and to make the land available for further northward urban expansion.

Land Grant Surveys

Access to water (and river transport) was of prime importance, so the original rural land grants were designed to provide water access frontage. On August 28, 1829, Stirling stated in regulations that all grants would be in sections of one-quarter of a mile and that no grant would have more than one-quarter of its exterior boundary on a river frontage (Statham-Drew 2003, 146). The result was that Roe laid out many elongate rectangular blocks (the so-called ribbon grants) that were oriented normal to water courses, particularly the twists and turns of the Swan River. The geometry of these blocks influenced the subsequent orientations of rural property boundaries upriver of Point Fraser and the placement of access and boundary roads. The remnants of this pattern remain embedded in the orientation of the road networks of the Perth suburbs that developed from the urban subdivisions of these blocks. In this way, the geometries of water courses of the SCP have played a significant part in the layout of Perth suburbs.

Other landscape elements also influenced the size and orientation of surveyed blocks, including the shape of lakes, the locations of more fertile soil (e.g., alluvial or lacustrine soils), and the locations and steepness of hills. The richest and best-watered agricultural soils were often the lake deposits and so they are frequently associated with smaller allotments.

First Farming

The growing of crops began almost immediately. The Governors Garden at the foot of the gullies became a focal point of agricultural experimentation with imported crops, and a nursery, which was made possible by its proximity to the river, and probably the freshwater from the spring-fed runnel. The garden soon became an important source of vegetables for the settlement (Martens 2011, 24). A garden plot is shown on John Roe's (1829) survey plan, adjacent to the river on the flats of East Perth, at the far eastern end of the settlement. And on the same plan, Roe annotated the broad southern point of the Maylands Peninsula as the "Peninsula Government Farms." All these early gardens and farms were planted on the alluvial banks of the river and were mainly focused above point where the river became fresh water.

Several other nearby locations became important for agriculture, and particularly the growing of vegetables. After the discovery of freshwater at Point Heathcote, near the confluence of the Canning River, another garden was established there (Statham-Drew 2003, 79). The wetlands north of Perth were taken up for market gardening in the 1840s and early 1850s (Bolton 1989, 152). Lakes further to the north, and south of the river, would remain a focus of horticulture up until recent time, as would the shores of South Perth, which became a focus of Chinese market gardening.

Development of the Road Network

Most heavy transport in the new colony was water-borne, but shortly after the new settlements were established in 1829, track networks began to radiate outward from the administrative center at Perth. They linked it to Fremantle, Guildford, and properties developing along the lake system to the northwest (Lake Monger). Most of these early track networks have evolved into the major arterial roads of modern Perth, including the Stirling and Canning highways. By 1836, the first trail had been blazed to Albany (Bolton 1989, 151).

One of the earliest tracks forming a western exit route from the surveyed settlement in 1838 is labeled "Road to the lake near Mongers" (Lands Department 1838). The track was a continuation of St Georges Terrace and Hays Street beyond Milligan Street, which was then the western-most surveyed street of the settlement. Two tracks coalesced and then skirted the northeastern base of the Mt Eliza Limestone escarpment and the southern shores of Lakes Irwin, Sutherland, and Georgiana. This direct route alignment has subsequently been overprinted by modern gridded streets and has not survived.

The first Perth street grid was a simple rectangle focused on four main east-west streets; St George Terrace, Hay Street, Murray Street, and Wellington Street. The alignment of the long axis of the grid conforms to the topography of the Perth Ridge, and fills the available space south of the main lakes and north of the steep 1829 shoreline. The same alignment does not persist in the streets to the north. When they were laid out shortly after the main grid, they had to be rotated approximately 5 degrees to maximize the space available for rectangular land allotments while avoiding the undrained lakes and wetlands that covered much of this area in 1830 (figure 4.2). Later, the main Perth railway line to Forrest Place station was constructed along part of the under-developed lake alignment to access the inner city. Perth Railway Station was built on swampy land formerly occupied by Lake Kingsford and the alignments of the station building and platforms exactly parallel the former northwest trend of this elongate lake. The modern Perth Arena occupies the center of the former diamond-shaped Lake Irwin, originally left vacant as railway land.

Roads laid out to avoid the sodden ground in the early nineteenth century, before the wetlands were drained, were later filled in, but the street pattern could not be altered. The 5-degree change in the east-west street alignment north of the railway lines was done to create the most regular allotments between the linear wetlands of this area. The major transport corridors that radiate outward from the CBD follow the alignments of the original lake system, particularly the freeway and railway corridors, often exploiting the sites of former, and still existing lakes (e.g., the eastern shore of Lake Monger).

Bridging the river was also a necessary part of the evolving transport infrastructure. The narrow points in the Swan River and estuary that hampered the earliest European exploration are now the locations of the key crossing points of the modern city—Heirissons Island and The Causeway link the city to Victoria Park, and The Narrows Bridge links Mill Point (Point Belches) to the city. The first causeway was constructed across the Heirisson's Islands in 1840 and a bridge was built across a narrow point near the mouth of the Canning River in 1843 (Bolton 1989, 151). The Narrows Bridge, linking South Perth to the north shore, via Mill Point, was opened in November 1959 (Crowley 1962, 101). These crossing points became key anchor points in the development of the arterial road network.

Most of the lakes have gone but their former presence is ghosted by the road and rail alignments. Motorways now follow the alignments of former lake system to the northwest of the city. The transport corridors along which arterial motorways such as the Mitchell Freeway exit the city to the north and northwest follow the line of some of the most significant of the former lakes and wetlands.

Geohazards

River flooding following heavy rain has been a constant hazard in Perth, with low areas of the riverbanks occasionally being inundated. In the 1926 floods, large areas of the Esplanade and South Perth were particularly affected (Brearley 2005, 86). The risk of flooding also deterred other development in

South Perth. However, the fertile alluvial sediments deposited along the river shores and in the adjacent wetlands by previous floods was the reason that the South Perth foreshore remained a center of Chinese market gardening well into the twentieth century (Crowley 1962).

Upriver navigation in the Swan River still requires caution. Obstructions still hamper navigation along the course of the estuary and the shoals require constant dredging to keep channels clear (Brearley 2005, 87–88). Much of the dredging spoil was dumped at Heirisson's Islands to create the modern landform from several smaller and lower-lying islands and shoals.

Groundwater Contamination

Perth has always been highly dependent on groundwater supplies and contamination has been an ongoing problem. Contamination of the groundwater supplies of the developing city became a major issue in the 1880s (even before the intense growth of the 1890s goldrush era). The town was largely dependent on wells and rainwater tanks for domestic supplies and contamination from cesspits had become a serious health problem. A Commission was formed in 1885 to investigate the water supply and sanitation needs of the town, which led to the recognition that the water courses of the Darling Range provided the only realistic source of supply (Engineers Australia 2012; Select Committee 1885).

Subsequently, the large groundwater resources of the SCP have been developed and exploited (see discussion above). The management of potential groundwater aquifer contamination remains one of the most important issues of the modern city, which depends on groundwater for around 50 percent of its scheme water requirements to support an ever-growing population.

THE BUILT ENVIRONMENT

Two types of building stone have been used extensively in Perth (and Fremantle) since the earliest days of the Swan River Colony. The locally sourced Tamala Limestone is the most unique, and was once abundant in outcrops throughout the modern urban area (particularly at Buckland Hill and the riverside outcrops of Rocky Bay). The other is the sandstone sourced from quarries near Donnybrook.

The Influence of the Tamala Limestone on Urban Character

No other rock defines the urban landscape and built environment of post-1829 Perth than the Tamala Limestone. This rock has been used as dimension stone, rubble, aggregate, and a source of lime since the first days of the colony and remains the most significant defining character element of Perth today. Limestone has been used for building at every scale, from foundations and rubble walls to the freestone walls of major public and private buildings (see figure 8.2 1C–1K). This beautiful pale yellow to creamy-white cross-bedded rock is the building material used for most of the surviving early colonial buildings, and for the most beautiful later buildings of the pre-Federation era. Limestone use was particularly prevalent in Fremantle, which was closer to the quarry sources and lacked the larger and more opulent administrative and commercial buildings of central Perth where sandstone was a preferred material.

The rugged limestone terrain of Fremantle prevented it from becoming a major port for over sixty years. The limestone that formed Arthur's Head and Rous' Head also formed a bar across the mouth of the estuary, which meant that ships had to anchor offshore and offload their cargos onto long ocean jetties. Only when the limestone bar was finally removed could the inner harbor be developed and safe, sheltered berthing become available.

The built environment and landscape of Fremantle have been more strongly influenced by the geometry and landforms of the outcropping limestone than Perth. The geography of the town is determined by the limestone ridge outcrops and many of the important historic buildings are built from it. Several important Fremantle landmarks are hewn directly from the rock, including the ridge at the end of High Street, capped by the Roundhouse, and the Whalers Tunnel cut through it to provide easier access to the settlement from the beach. The inner Fremantle beaches are backed by limestone cliffs and Arthurs Head and Rous' Head are limestone outcrops that form the mouth of the Swan River Estuary. To the north, the War Memorial caps Monument Hill, a prominent limestone pinnacle. There are tunnels (water storage cisterns) excavated into the limestone beneath the old Fremantle Jail. There is the low and formerly more prominent Buckland Hill, capped by wartime anti-aircraft and anti-shipping gun emplacements, which still represents a prominent landmark despite the extensive quarrying. And then there are the cliff-lined shores of the lower reaches of the Swan River Estuary. Further to the north, at City Beach, the weathered limestone bluffs (former faces) of a repurposed quarry form the backdrop to the Quarry Amphitheatre on Reabold Hill.

While the built environment and landscapes of inner Perth have not been so obviously influenced by the limestone as Fremantle, it underlies the city's preeminent landmark, the Mt Eliza Escarpment and Kings Park, the backdrop to the modern city. Quarries at the foot of the escarpment once supplied raw materials for some of the earliest Perth public buildings. Freshwater springs emanating from its base were once important water sources for both the Aboriginal people and the later European settlers.

Landmark Limestone Buildings

Despite the building and demolition boom that began in the 1960s, Perth still retains several important examples of historic limestone building, particularly at the eastern end of the CBD. They include the Perth Mint (constructed from Rottnest Island limestone), St Mary's Catholic Cathedral, and Fire Station No 1 (See figure 8.2, 1G and 1J). Further west in the city are the Old Perth Boys School and the old Technical School on St Georges Terrace (limestone base with sandstone detailing), and the arches and trim of St Georges Anglican Cathedral. Other key limestone buildings include the old Perth Jail (now part of the cultural precinct, see figure 8.2, 1I), the Redemptionist Monastery in North Perth and Winthrop Hall, the landmark building of the University of Western Australia.

Most buildings in Fremantle have a proportion of limestone used in their construction, from the many humble workers cottages and rendered rubble walls, to major public buildings, such as the Roundhouse (Western Australia's oldest public building) and the former Fremantle Jail. Other landmark Fremantle buildings include the Western Australian Maritime Museum, the Fremantle Boy's School, the Fremantle Museum and Arts Centre, the German Consulate Building (5 Mouat Street), and St Patrick's Basilica. The Basilica is built from both Tamala Limestone, which forms the spires, flying buttresses and window dressings, and Donnybrook Sandstone, from which the main entrance is built (Fetherston 2007).

Minor Buildings and Features

Limestone also forms the foundations for many red-brick Federation and interwar era houses. And it was also used extensively for walls, fences, and garden features throughout Perth suburbs. Weathering, carbonate dissolution, and removal of the sandier material by wind and water etches the stone to produce a deeply textured surface (see figure 8.2, 1D–1E). Mosses and lichens grow on the limestone front walls and footings of the old houses as the stone ages to a khaki or honey-colored patina. This is the characteristic heritage look of much of Fremantle and of the older inner Perth suburbs. This is also the color that defines the character of Greater Perth and which subsequent generations of architects, builders, and home renovators have often sought to emulate. Modern feature bricks, pavers, and pillar blocks in many homes also seek to emulate the color and texture of the limestone. And large blocks of limestone have been used as fill and armor rock in the Fremantle Harbor groynes. In many ways, limestone also defines the character of Western Australia's coastal infrastructure.

Donnybrook Sandstone

High-quality freestone was discovered in the Cretaceous sedimentary sequences outcropping near the small farming and mining settlement of

Donnybrook sometime during the short gold-mining boom of the late 1890s (Frost 2001, 55–76). The stone is a fine-grained sedimentary rock formed in shallow waters, as shown by ripple marks and vertebrate trackways (Frost 2001, 93–94). This fine freestone has been used to construct many landmark buildings, particularly in Perth.

The first use of the stone was in the construction of the Supreme Court building (Frost 2001, 94), which led to the choice of the stone for the construction of the State Parliament building. Quarrying in the Donnybrook area was given impetus by the establishment by the government of two quarries, and other quarrying leases were taken up by various building contractors. The use of the stone in the high-status Parliament building created a demand for the stone, resulting in its use in several other buildings, including the new General Post Office and the Commonwealth Bank building (Frost 2001, 94–95).

Examples of other important buildings for which Donnybrook Sandstone were used include the old Claremont Teachers College, key elements of the museum and cultural precinct (including the administration and old court building, and the decorative elements of Hackett Hall), the Royal Insurance Building, Newspaper House, the former Chief Secretary's Office (57 Murray Street), and the font of St Georges Anglican Cathedral.

Brickmaking Clay

Brickmaking began using clay found in East Perth (figure 4.2), which was used to create the bricks for buildings such as St Georges Anglican Cathedral. The clay pits were rapidly exhausted and were eventually rehabilitated to create Queens Gardens, one of the only formal "English" style parks in Perth.

More extensive clay deposits were found further east in the Swan Valley, in the extensive floodplain alluvial clays and silts bordering the river. These alluvial deposits became the primary source of clay for Perth brick manufacturing. The first large-scale brickmaking industry developed to the immediate east of Perth, sourcing clay from the Maylands Peninsula but then moved progressively further upriver to Midland and beyond, where it is still a major industry that continues to exploit the alluvial clay deposits.

Chapter 5

On the Edge of the Delta

Melbourne, Victoria, 1835

The final stages of the breakup of Gondwana and the opening of the Southern Ocean triggered widespread volcanic activity that fundamentally changed the landscapes of southeastern Australia. Grasslands and open woodlands developed on the poorly drained and youthful volcanic terrain that was formed as lava flows slowly inundated a more ancient topography. In the early nineteenth century, the native grasslands would attract the largest unauthorized mass settlement of Europeans in Australian history and lead to the foundation of the city of Melbourne.

GEOGRAPHIC LOCATION AND SETTING

Melbourne is in southeastern Australia at Latitude −37.81°, Longitude 144.96° (the Melbourne GPO, Google Earth), at the head of Port Phillip Bay, one of the largest inland bays in the world. It was established on the north bank of the lower reaches of the Yarra River in 1835, at a site about 10 km upstream from the river mouth.

The primary focus of this chapter is the site of the 1835 settlement and the first town site laid out there in 1837. It was a gently rolling, moderately hilly and wooded landscape dominated by two hills, a linear western hill that reached approximately 35 m high and an eastern hill reaching to over 45 m. The hills were separated by a deeply incised south to southeast flowing seasonal creek. The southern border was formed by the Yarra River. This small area, measuring only 2.5 × 2 km, is the most historic part of central Melbourne and is now the modern CBD (figure 5.1). It remains the focus of Melbourne's cultural, sporting, political, and financial life, and still contains its most valuable real estate. It is also the ultimate focus of the modern

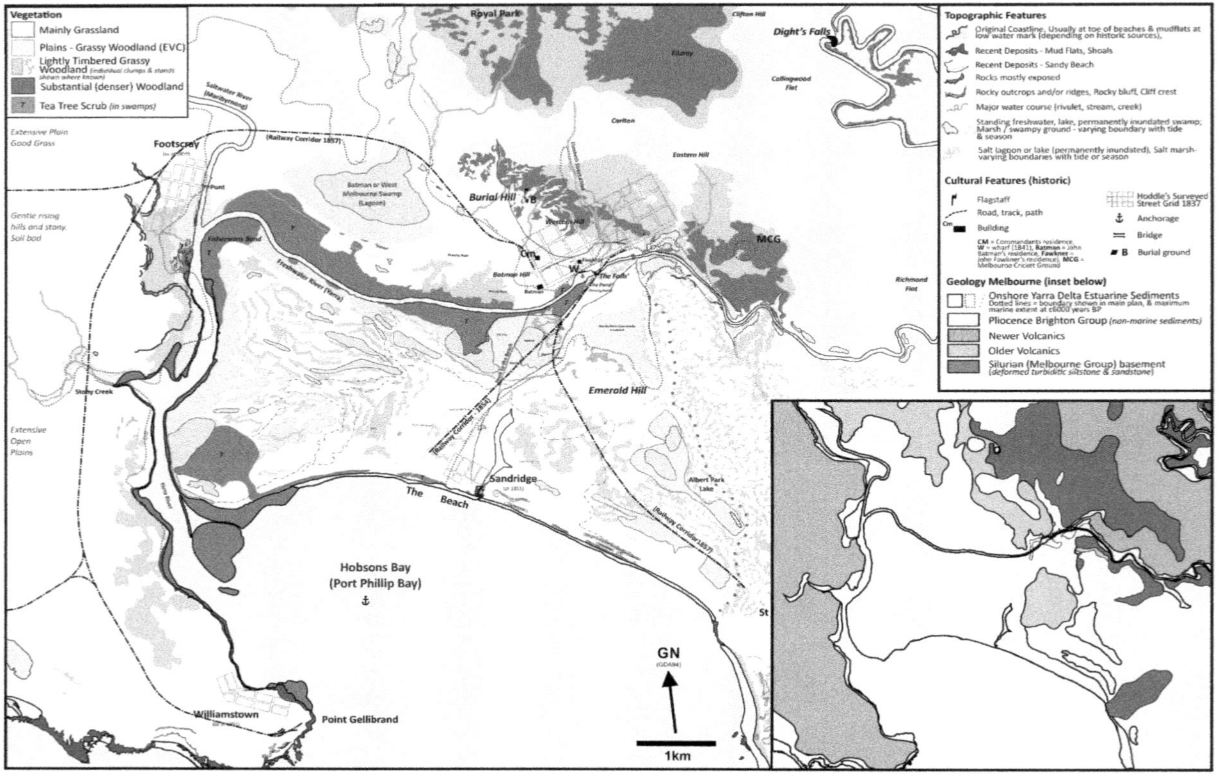

Figure 5.1 The Melbourne Region. The regional landscape context of the 1835 settlement at Melbourne, based on a compilation of historic map information (mainly Hoddle 1835 and 1837; Selwyn 1860, and Cox et al. 1866) and 3D topographic reconstructions. The distribution of native vegetation is a non-specific reconstruction of the density of woodland, mostly interpreted from the survey maps of Robert Hoddle, and various other early cadastral land survey maps. Where no vegetation distribution information was recorded by contemporary observers, some conjectural infill interpretations are made (dotted lines). **Inset.** Simplified geology of the Melbourne settlement site, based on Neilson (1989), with additions from Vandenberg (1975), Selwyn (1860), and Neilson (1989).

transport infrastructure, which includes railways, motorways, tramways, bridges, and tunnels.

The impact of the first settlement spread rapidly into the landscapes of the surrounding region and by the 1860s the first Melbourne suburbs had spread to Yarra Bend in the east (including Dights Falls and Studley Park), and Royal Park to the north. The salt marshes and wetlands of the West Melbourne Swamp and Fisherman's Bend impeded westward development, and the beach-bordered shoreline of Hobsons Bay defined the southern extent of early urban growth. This 80 km^2 area encompasses most of the mid-nineteenth century extent of the urban area and now encompasses the modern municipality of the City of Melbourne. These now inner-city quarters are also important to the following discussion (figure 5.1).

By the 1860s, several satellite villages had become well established in the surrounding hinterland. Emerald Hill, Sandridge, Footscray, and Williamstown (among others) were all important centers and are also discussed below.

HISTORIC TIMEFRAME DISCUSSED

The first twenty-five years of Melbourne were the most important in its urban development. There were three important phases to this formative period, which saw the settlement evolve from a small village into a planned town with a surveyed street grid. The village was adapted to its landscape setting, and with a layout and landscape utilization controlled by the natural environment, but the formalized town came to dominate and then massively alter its landscape. The focus of the following chapter is on the first two phases.

Phase 1, 1835–1837, The Western Hill: Batman's and Fawkner's village: An unplanned settler community forming a scattered and unplanned village that was focused on the Western Hill, and near the riverbank at The Falls and the Turning Basin.

Phase 2, 1837–1850, The Latrobe Era: Under the guidance of the first superintendent, and then lieutenant governor of Victoria, Charles Latrobe, Melbourne grew into a planned town, with a regularized street grid imposed on the natural topography of the site (known as the Hoddle grid). The first government land sales took place, and the first satellite villages and suburbs were founded. Most importantly, large acreages of public lands were reserved in a fringing ring around the town for recreational and sporting facilities.

Phase 3, 1850–1880, Gold Rush to "Marvelous Melbourne": The significant land-use patterns of the Latrobe Era were well in place by the time gold was discovered in the early 1850s, but the small town had to rapidly adapt to the wealth and population boom of the gold rush. This was a period of massive urban expansion and prosperity that saw the construction of railways, major harbor redevelopment, and the modification of the course of the Yarra River. Most significant wetlands were drained or reclaimed, and there was an administrative shift to the eastern fringe of the town, when Parliament House and the Treasury Building were constructed on the slopes of the Eastern Hill. This remarkable period of wealth and urban expansion is only briefly discussed below, but reference is made to the many exceptional contributions made to the city's architectural heritage during this period.

EUROPEAN DISCOVERY AND EXPLORATION

In early 1798, Matthew Flinders and George Bass were tasked with confirming reports from sealers working in the Furneaux Island Group that a strait existed between Van Diemen's Land and the mainland (Owen 2010, 15). They entered the strait from the eastern end and by December 11 had passed through, confirming that it did indeed exist (Flannery 2006, 26). European knowledge of Bass Strait, Port Phillip Bay and the Yarra River grew rapidly in the next few years, particularly between 1802 and 1803, when expeditions led by John Murray and then Charles Grimes further explored the region.

In January 1802, Lieutenant John Murray, commanding the *Lady Nelson*, was steering westward as he surveyed the southern coastline of New Holland but was finding this difficult because of the large volume of smoke obscuring the low shore, which he attributed to "numerous native fires" (Flannery 2007, 25–26). Murray was continuing the earlier surveying work of James Grant, who had discovered Westernport in March 1801 (Murray 2013, 22). At 3:00 p.m. on January 5, 1802, he observed a headland, and "an opening in the land" that he recognized as the possible entrance to a significant harbor. He chose not to enter because of the complexity of the entrance, which had an island and reefs behind the entrance, a rapidly shelving sea floor and heavy seas that broke short. Rocks were also observed ahead. He abandoned the attempt and steered to deeper and safer waters. After many observations at the masthead, Murray decided that the reef did not stretch across the entire entrance and saw a "fine sheet of smooth water of great extent" beyond. He intended to come back (Murray, in Flannery 2007, 26).

On February 14, 1802, Murray returned to explore the entrance. After careful observations of the entrance, by noon, he had decided on the conditions at the entrance. He saw that although the entrance was shallow, it was wide enough to "work" any sailing vessel in. He also realized that much of the breaking water he had taken for a reef on the previous observations was the "ripple effect" produced by the strong outward tidal flow and the onshore wind (Murray, in Flannery 2007, 27) that is now known as "The Rip." Sometime after noon the *Lady Nelson* had been "carried in" in 6–14 fathoms (11–125 m) of water and was under full sail up the harbor (Murray 2013, 50). Murray named the new harbor Port King after the NSW governor (Murray, in Flannery 2007, 28), but King later changed this to Port Phillip, to honor his predecessor, Arthur Phillip (Flannery 2007, 25).

Murray explored along the channel that followed the southern and eastern shores of the bay and noted that the bush here was more open and drier, with large mature trees, and unlike the "brush" of Western Port (Flannery 2007, 27). He named the highest visible point Arthurs Seat, after a mountain near Edinburgh, Scotland. Further north, the shore was a low sandy beach. Exploring inland from this place, he noted the widely spaced trees and the lack of significant undergrowth (Flannery 2007, 28). There were also native huts and several acres had been burnt off. Although finding no water, he felt that there must be an abundant supply nearby because of the number of Aboriginal huts and fires he saw (Flannery 2007, 28). After three weeks of exploring the southern part of the bay, Murray left. He never reached the northern coast, so never saw the landscapes of modern Melbourne.

The grasslands of the northern shores of Port Phillip Bay were first seen by British explorer Matthew Flinders, and the crew of his ship *Investigator* in April 1802. Flinders had also discovered Port Phillip Bay, ten weeks after Murray and explored the northern shores while completing the first accurate survey of the harbor.

Seeking out high points to gain a better idea of the extent of the bay, Flinders undertook a traverse to the top of a "bluff mount" he later learnt had been named Arthurs Seat by Murray. And on May 3 he walked across the western volcanic plains and climbed to the high point of some distant hills (the You Yangs), which he had named Station Peak. During the walk he noted that the soil was shallow and clayey and only supported "small-bladed grass," and that water appeared to frequently lodge in places. No significant trees grew until they reached within 1 or 2 miles (1.6–3.2 km) of the foot of the hills (Flinders, in Flannery 2006, 110). Even from both these elevated positions, the extent of the harbor was too great to observe the Yarra River (Flannery 2007, 34, 36). Flinders also noted the extensive evidence of the Aboriginal presence, and the well-timbered nature of the country along the eastern shore.

The Yarra River was finally discovered on February 2, 1803, by a small party of surveyors exploring the northern part of the bay on foot. Following Murray's report, Governor King sent the NSW deputy surveyor general, Charles Grimes in HMS *Cumberland* to accurately record the new discoveries of King Island, Western Port, and Port Phillip. A record of the journey was kept by botanist, James Flemming, who accompanied Grimes (Flannery 2007, 37).

The *Cumberland* probably anchored somewhere near the Carrum Carrum Swamp, which the explorers encountered after traversing about 2.5 km across a region of light sandy soil. The wetlands were dry but obviously inundated during wet seasons. Further north, there was a large salt lagoon, probably modern Albert Park Lake, and then the river. The ground was swampy on one side and high on the other and the birdlife was abundant (Flemming 1802 unpaginated). The following day Grimes and Flemming, with five seamen, explored "the great river" by boat following the "left hand stream" (the Maribyrnong) northward past the Yarra confluence. Climbing a high hill, they saw the large lagoon that was later named the West Melbourne Swamp. There were no trees for many miles, at least as far as the You Yangs and only a few "straggling oaks" by the river. Traveling as far as an Aboriginal stone fish trap, they found the river to still be salt (Flemming 1802, unpaginated).

On February 4th, they explored east along the second branch of the river, then landed to climb a hill "between the two rivers" where there was "fine grass, fit to mow; not a tree on it." The soil was a "rich black earth." This was probably the Batman's Hill of 1835 Melbourne. Two miles further on, the land was better, with larger timber, and they saw Aboriginal huts and a canoe. Over the next four days they explored the river valley beyond the "little hill," ultimately reaching as far as Dights Falls, about 15 km upriver (Flemming 1802, unpaginated), but returning to it several times while watering the ship (figure 5.1). They then explored around the mouth of the river, including the future Williamstown area.

James Flemming suggested that the clear land on the "freshwater river" would be the best place for any future settlement because of the areas of good land, good pasture, and thinner timber. He also suggested that the availability of freshwater would be better at other times of the year and that despite the lack of good timber around this place, there was a strong likelihood that better supplies would be available "up-country." He also noted the good sources of brick clay and stone (Flemming 1802).

The first accurate chart of Port Phillip Bay was compiled by Matthew Flinders, who incorporated Grimes' new survey information about the rivers into his own chart (Flinders 1802). This illustrates how rapidly information about Port Phillip was gathered after Murray's discovery.

Following the reports of Murray, Flinders and Grimes, Governor King suggested establishing a settlement on Bass Strait or Port Phillip, to forestall any possible French interests in the region. The British government paid heed, and in early October 1803, Lieutenant Colonel David Collins arrived from England to establish a settlement on Bass Strait.

Sullivan's Bay 1803: A Failed Settlement

In 1803, the British and colonial governments were concerned about possible French ambitions in southern Australia. Napoleon Bonaparte had sent a scientific and survey expedition led by Nicholas Baudin to chart the southern coast of Australia. The French expedition met Matthew Flinders under a flag of truce in April 1803, off the coast of modern South Australia. Flinders later recorded his misgivings about the French motives for their visit and was particularly concerned about Francois Peron's (the expedition naturalist) apparent claim that France had discovered all lands between the Nuyts Archipelago (offshore from modern Ceduna, in South Australia) and Western Port (Flinders, in Flannery 2006, 100–103). Baudin's expedition later visited Sydney to replenish their ships and the relationship was amicable, but the concerns remained.

On October 10, 1803, Collin's small fleet, consisting of HMS *Calcutta* and the chartered transport *Ocean*, which carried convicts and some free settlers, arrived in the bay. The primary account of the short-lived Port Phillip settlement is that of James Tuckey, first Lieutenant of HMS *Calcutta* (Tuckey 1805).

Choice of the Sullivan's Bay Site

Collins spent a week searching for a place to establish his camp. His primary concern for the choice of the site was that it should be easily accessible to shipping, so he focused his search on sites close to the heads. He chose a small bay, which he named Sullivan's Bay (modern Sorrento), despite recognizing that the light sandy soils and lack of water would prevent the cultivation of crops and create other difficulties. The only water available was brackish groundwater that seeped into casks that were sunk into the ground. The remains of at least two of the buried water-filtration barrels were discovered in the 1920s and other artifacts have been found in the area (Parks Victoria 2002, 8–9).

They realized that with a bay that large there were likely to be better sites, so a long search began using two boats led by Lieutenant James Tuckey (Tuckey 1805, unpaginated). Tuckey carried out a reconnaissance of the top of the bay and probably sighted the mouth of Yarra River but was unable to enter Hobsons Bay because of unfavorable sea conditions.

In December 1803, Collins received permission from Governor King to abandon Sullivan's Bay in favor of moving to Tasmania (see chapter 2). The last members formally left on May 20, 1804, but some remained. William Buckley, a convict former soldier, had absconded from the camp with two others on Christmas day and remained behind. He would greet the Port Phillip Association arrivals thirty-two years later, when he would play a part in establishing Melbourne by acting as a go-between for Aboriginals and the 1835 settlers.

Thirty-two years later, John Fawkner, one of the original 1803 British settlers at Port Phillip, returned to the bay after growing up in Hobart Town. With John Batman, he established Melbourne.

Geology and the Failure of the Sullivan's Bay Settlement

The entrance to Port Phillip Bay is guarded by Point Lonsdale and Point Nepean, both formed from calcareous and cemented aeolian sands with some swamp deposits (VandenBerg 1997a). This landscape of sand peninsulas, and shallow treacherous tidal waters is one of the defining influences on the course of the early European settlement of Port Phillip Bay. The difficult entrance was one of the main factors that made David Collins consider the bay to be unsuitable for a major settlement, especially one of a largely political and military nature (to deter French expansion). One of Collins' ships was held up at the heads for three days as it attempted to exit against the strong onshore breezes and the intense swells of Bass Strait. He believed that it would be too difficult to deploy a naval fleet at short notice, in a time of emergency in such conditions. The entrance would also hamper any commercial whaling vessels that may need to exit the harbor rapidly. The soils of the shore were also found to be poor, and unsuitable for agriculture, following the examination by Grimes and Flemming. And there was no significant source of freshwater. David Collins believed that the difficult harbor entrance downgraded the attraction of the bay as a settlement site, and that a population could not be sustained there. He sought permission to move the settlement to the Derwent River after only twenty-seven days at Sullivan's Bay. Governor King agreed with the request after reviewing Grimes' Port Phillip survey (Tuckey 1805, unpaginated).

The British attempt to create a penal colony in Port Phillip in 1803 failed; the Tasmanian attempt thirty-two years later succeeded. However, it would be incorrect to suggest that the failure in 1803 was due to British ineptitude, or lack of Australian experience. David Collins had been integral to the foundation of Sydney and Parramatta in 1788, and after abandoning Port Phillip, he went on to successfully found Hobart. The failure at Port Phillip was probably due to bad luck more than anything. The colony was established in the wrong place. And bad luck, lack of time, and poor sailing conditions prevented Tuckey from entering Hobsons Bay and discovering the entrance to the Yarra River (Tuckey 1805, unpaginated). If the river been discovered in 1803, then Collins may have moved his settlement there.

The Foundation of Melbourne: A Van Diemonian Colony

Melbourne began as an unauthorized private enterprise venture organized and managed mainly by Tasmanians (or Van Diemonians). It was not, strictly speaking, founded by the British. One of the main instigators of the settlement was John Batman, an Australian born in Parramatta, who had later moved to Van Diemen's Land. He visited Port Phillip to explore the ship *Rebecca*, liked what he saw, and returned to set up a base at Indented Head. From here, John Wedge, Batman's surveyor, explored the region in detail. Three months later, John Pascoe Fawkner's expedition arrived on the ship *Enterprize* and established the first permanent settlement where the Melbourne CBD is located.

Unlike the foundation of other Australian settlements, John Batman and his sponsors (the Port Phillip Association) sought to legally purchase the land from the local Aboriginal leaders and presented them with a treaty in May–June 1835 (Presland 2010, 84–85). It is not known precisely where the meeting with the Aboriginal leaders happened, but it was far to the north of the future settlement site.

Batman arrived with his wife and seven daughters and settled at Batman's Hill in April 1836, where he had approximately 20 acres improved by the time his only son was born a year later (figure 5.2). The NSW governor, Bourke, who had nominal jurisdiction over the Port Phillip region declared Batman's treaty illegal and that the settlers were trespassing. However, two years later, 350 people and 55,000 sheep were landed at Melbourne and eventually Bourke was forced to accept the new settlers and to recognize the settlement. Governor Bourke visited the Port Phillip district in 1837 and named it Melbourne, after the British prime minister. The street names were also given by Governor Bourke during this visit.

The Choice of the Site

Unlike previous Australian settlements, the site of Melbourne was not chosen by experienced soldiers or naval officers. It was chosen by amateurs, albeit with long-standing experience of living in Australia. The modern location of Melbourne seems to have been chosen by Fawkner's party. John Batman may have initially chosen a settlement site further west because the surveyor John Wedge marked a place on Batman's survey chart, at the confluence of the Yarra and Maribyrnong rivers, as "Reserved for Township and Public Purposes" (Wedge 1835). This may be the original place that John Batman thought was "the place for a village" but was probably drawn by Wedge before he arrived in Port Phillip Bay (Boyce 2011, 77). The extensive wetlands and tea tree that occupied this area in 1835 suggest that it would not have been proposed as a settlement site by anyone who had ever seen it (Presland (2009, 211).

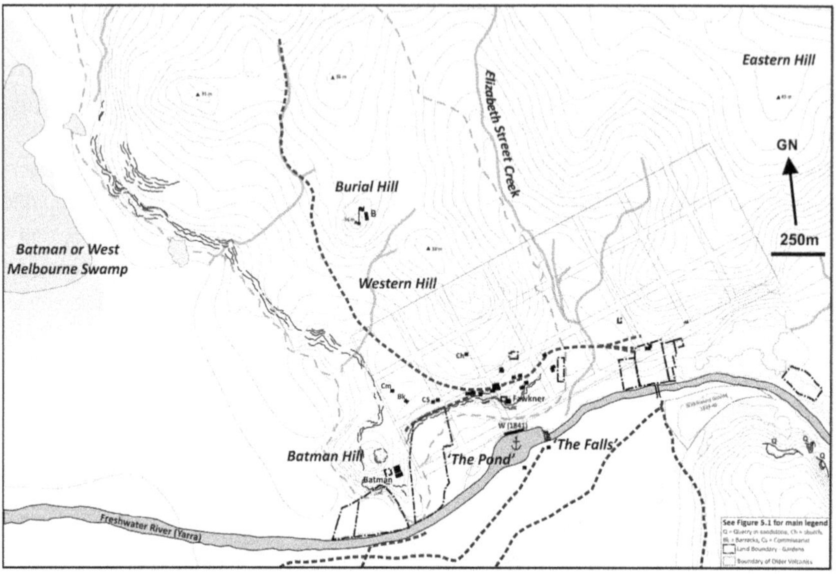

Figure 5.2 The Melbourne Settlement. Topographic reconstruction of the landscape of the Melbourne settlement site at the time of Van Diemonian (Tasmanian) settlement, circa 1836, based mainly on the historic mapping of Hoddle, (1835; 1837), and 3D topographic reconstructions. The first ramshackle village was concentrated near the landing place, on high ground above the north bank of the river. A natural rock bar prevented navigation of vessels further upriver and marked the change from brackish estuarine to fresh water. There was low-lying marshland and swamps to the south and west and the high ground here provided the first site on the Freshwater River that could be reached by small ships. When government authority was finally established, the appointed Superintendent (later Lieutenant Governor) built his residence and offices at the western end of the settlement. The first street grid, surveyed by Robert Hoddle in 1835 is shown. This survey was imposed on the footprint of the original settlement, which was rapidly obliterated by more formalised land allocations. Please refer to the extended caption at the rear of the book for additional information.

Ultimately, the primary reason for choosing the current site of Melbourne was the availability of abundant freshwater in the Yarra River. A basalt ledge formed low falls across the channel and impeded tidal flow. Upstream of this point, the river became a reliable source of freshwater (figure 5.2).

GEOLOGY OF 1835 MELBOURNE

Melbourne lies at the head of Port Phillip Bay, a drowned sunkland basin bounded by major faults and flexures, including the Selwyn Fault (east) and the Rowsley Fault (west). A minor northwest-trending flexure associated with the sunkland margins passes through the western end of the CBD, where it caused tilting and down-warping of the Cenozoic rocks and was important in the formation of the trough in which the Yarra Delta was deposited (Neilson 1992, 224) (see further discussion below).

The Port Phillip Sunkland was a dry riverine plain in the past, and possibly also a salt lake (Holdgate et al. 2001, 2011). It was inundated during the sea-level rise that followed the last glacial maximum to form the modern bay. At highstand, the edge of the bay reached the western slopes of Batman's Hill (now the site of Southern Cross Station) and to just beyond the King Street Bridge (Holdgate and Norvick 2017, 316). Sea levels have varied over time but are now lower than they previously were and began falling around 2,500 years ago (Holdgate and Norvick 2017, 316).

Despite its small geographic area, the site of the 1835 settlement occupies a microcosm of the diverse geology of southeastern Australia. Four major geological terrains come together here. Folded Silurian turbidites form the basement and are covered by two distinct generations of recent volcanics, some of which only formed in the last 15,000 years. It also lies at the head of a Quaternary delta, exposed only recently by a fall in the level of Port Phillip Bay. Most of the important elements of Victorian geology are represented in this small area (except for the granitoids), and each part contributes to aspects of the character of the landscape.

Basement Silurian Sedimentary Rocks

The basement of most of modern metropolitan Melbourne is formed by deformed and weakly metamorphosed Silurian sedimentary rocks. In the 1835 settlement area, the Silurian strata comprise shallow marine siltstones and thin-bedded sandstones of the Melbourne Formation (Holdgate and Norvick 2017, 305; VandenBerg 1997b; Powell et al. 2003). They were originally deposited as mass-flow turbidites on the marginal continental slopes of ancient Gondwana.

The Silurian rocks are closely folded along persistent northeast-southwest (N20–35E) trending axes. The axial planes dip east, and the dip of fold limbs is variable, with the western limbs usually steeper. Minor folds mostly occur on the western limbs. Folds are more closely spaced near the 1835 settlement area. Reverse faults are developed where folding is most intense and are mostly parallel to the strata (Nicholls 1930), including the "crush zone" that Hills (1941) observed further east, between the Johnston Street Bridge and Dight's Falls.

The basement rocks were sculpted into an ancient landscape by erosion, which was then infilled by Cenozoic sediments and volcanics. In the context of the 1835 settlement, the Silurian basement now outcrops as two inliers, which represents the eroded remnants of ancient hills that predate the Cenozoic basalt flows that surround them. The most extensive inlier underlies the linear, northwest-southeast trending Eastern Hill ridge, the highest ground in the CBD proper, and extends from just south of the Yarra at Jolimont, through Royal Park to Brunswick. This is one of the most important geological features of modern Melbourne because it was the location of the first suburban development. The earliest suburbs of Melbourne; Carlton, and part of Fitzroy, spread along this higher ground while the adjacent suburbs of Collingwood, Richmond and Brunswick developed later, on the broad flats formed on the basalt-infilled valleys that surround the Silurian exposures. A smaller Silurian inlier occurs at Clifton Hill and there is another further south, in St Kilda. The largest inlier underlies Preston and reservoir to the northeast, and outside the study area.

A whole ancient topography was formed on the Silurian basement as it was weathered and eroded during the late Cretaceous, and into the Cenozoic. It is now preserved beneath the younger volcanic rocks and sediments as a paleotopography and has been intersected in geotechnical drilling programs, tunnel excavations, and foundation work beneath the central city. Compilation of the subsurface information by Golder Associates (2016) reveals the ancient landscape, and the original valley of the Yarra River, now infilled by basalt and sediments but preserving the ancient river landscape (Holdgate et al. 2007).

A series of northeast-southwest oriented ridges is preserved by the older topography formed on the Silurian basement east of Melbourne. Similar ridges are preserved by the paleotopography, and some project above the volcanic peneplain as topographic highs (inliers), such as Eastern Hill and Clifton Hill. The orientation of these ridges is strongly influenced by the strike of the folded strata and faulting in the basement rocks. East of Melbourne, where the entire landscape is underlain by the Silurian rocks and the ancient drainage system has been largely unaffected by the infilling younger volcanics, the Yarra River and its tributaries are also strongly influenced by the strike of the Silurian rocks and drain toward the southwest.

The strong southwest trend of ridges and the drainage pattern becomes disrupted near Port Phillip Bay and the central Melbourne area, where they abut against a strongly northwest-southeast oriented belt of Older Volcanics, and the basement becomes faulted and warped at the margin of the Port Phillip Sunkland. Newer Volcanic have infilled northwest-trending valleys in the basement topography that also follow this trend. The course of the Yarra River becomes more complex and convoluted as it finds its way around both the Newer Volcanics and across this belt of structural complexity.

Basic Dikes

Basic dikes are common throughout the Silurian strata but tend to be focused along zones of structural weakness, such as fault planes (Nicholls 1930), and zones of small-scale faulting (crush zones) and folding on the limbs of tight folds. Most dikes are deeply weathered to plastic clay. Key examples in the study area occur at the northeast corner of the city grid, where dikes less than 2 m wide were mapped in telecom service tunnels beneath Lonsdale Street (Purcell et al. 1978, Plate 1) and in Spring Street (Brumley 1974a, b; Neilson 1989).

Multiple parallel dikes intersected in tunneling on Lonsdale Street intrude a northeast-trending belt of deformation in the northeast corner of the settlement (between modern Swanston Street and Spring Street). A southwest flowing tributary of the Elizabeth Street Creek has eroded into this belt of deformation and drains the northeast corner of the settlement (figure 5.2). The headwaters of the tributary probably developed in this position because the line of deformed rocks and weathered dikes was more susceptible to erosion. To the east, a 10 m (30 feet)-wide dyke was encountered beneath Wellington Parade during drilling investigations for the construction of the underground rail network (Brumley 1972).

Further dike occurrences outcropped immediately south of the river, in the Royal Botanical Garden (Selwyn 1860), and in Studley Park to the east, where Hills (1941) noted that they varied in width from 30 cm (1 foot) to 3 m (10 feet).

Cenozoic Volcanics and Interflow Sediments

The ancient basement topography was infilled by tuffs and linear basalt flows that emanated from multiple eruption points to the north and west of Melbourne. This volcanic activity was long-lived and persisted throughout the Cenozoic and into the recent past (it may still be active) and ultimately created the grasslands that attracted European settlers.

There were two volumetric peaks at 42–57 Ma (historically defined as the Older Volcanics) and 0–5 Ma, historically known as the Newer Volcanics (Price et al. in Birch et al. 2003). The long-standing terminology of the Cenozoic volcanics has been recently revised, because geochemical and petrological evidence suggests that they belong to the same suite and are a continuous series of related volcanic flows (Price et al. in Birch et al. 2003). The earlier classification is retained in this discussion because it is a useful stratigraphic distinction in the Melbourne settlement area. Here, the fine details of the stratigraphic relationship between the Older and Newer Volcanics, and their intervening sediments and weathered surfaces, clearly mark the time breaks between the otherwise geochemically similar volcanics.

The Older Volcanics were formed as lavas flows, which infilled paleo-valleys formed in the Silurian basement and were then deeply weathered and eroded. The older flows now have a limited outcrop extent because they are often mantled by Cenozoic marine and nonmarine sediments, such as the Brighton Group, and they are also buried beneath the Newer Volcanics. The Older Volcanics mostly outcrop as a northwest-southeast trending linear belt that underlies the prominent landmarks on the western side of the settlement site, including the Western Hill and Batman's Hill, and Emerald Hill, which were all important landscape elements in the early days of the settlement (figure 5.1 inset).

The Newer Volcanics were formed as linear flows. They overlie the Older Volcanics and the sediments of the Brighton Group. Flow bases are often peperitic and can be silcretized and weak geotechnically. So, they are significant in deep foundations and underground excavation, and particularly in major tunneling operations (Schofield 2014).

Pre-Quaternary Sediments

Pliocene and Miocene sediments are intercalated with the Older and Newer Volcanics (Holdgate and Norvick 2017, 301), some of which were deposited when lakes were created by lava flows damming the paleo-drainage. They include the Upper Miocene-lower Pliocene Brighton Group (sandy clays); the Miocene Fyansford Formation (limestones and marls) and the Eocene to Oligocene Werribee Formation (clays, sands, and coaly sediments) lying beneath the Older Volcanics.

Sediments of the Brighton Group mantle the Older Volcanics and are mostly preserved as eroded remnants in isolated highpoints, and beneath the younger volcanics, but were once more extensive (Holdgate and Norvick 2017, 305). Notable high points include modern Royal Park, Parkville, and the Western Hill.

The Yarra Delta

The western limit of the 1835 settlement was defined by a low bluff formed at the boundary between the high ground underlain by the Older Volcanics, and the low-lying Quaternary sediments of the Yarra Delta to the west. The change in the topography at this boundary (an unconformity surface) was one of the most important geological influences on the early development of Melbourne and marked the change in land use from urban and commercial development to (later) industrial and transport infrastructure. The unconformity continues to define the western limit of the modern CBD and still marks the change from commercial and residential land uses to industrial and transport infrastructure.

The Quaternary Yarra Delta Group unconformably overlies the Cenozoic volcanics and sediments. It was formed from sediments deposited by the Yarra and Maribyrnong rivers where they entered the head of Port Phillip Bay (Nielson and Jenkin 1967; Holdgate and Norvick 2017). Deposition in the delta was contiguous with sedimentation in the adjacent Port Phillip Bay and the type of sediment deposited in the delta was coupled with the sea-level fluctuations in Port Phillip Bay that followed the last glacial maximum (when Port Phillip Bay was largely dry), subsequent rises, and the most recent sea-level reduction. Erosional periods occurred between deposition of each Quaternary unit, as river channels cut and fill (Holdgate and Norvick 2017, 301). Wetlands formed on the low-lying Quaternary sediments exposed following the last sea-level retreat and now occupy the southern and western margins of the study area.

The delta covers an onshore area of approximately 20 km^2 and the 45–50 m thick sequence, named the Yarra Delta Group, is comprised of six main stratigraphic units, which from youngest to oldest are (Holdgate and Norvick 2017, 303–4):

1. Port Melbourne Sand—medium- to fine-grained sand, originally formed as sand ridges at surface.
2. Coode Island Silt—dark-gray to brown, variably fossiliferous and pyritic silty clays. It infills river valleys cut into the older Pleistocene and earlier units.
3. Fishermans Bend Silt—yellow to gray stiff clays with sandy lenses and calcareous nodules. The upper part is interbedded with the Burnley Basalt which is dated early to middle Pleistocene.
4. Moray Street Gravels—uncemented gravels and coarse sand beds.
5. Jolimont Clay—red, yellow, or gray mottled stiff slay with minor sandy clay.
6. Burnley Basalt (Newer Volcanics). Series of weathered basalts with numerous soil horizons occupying a former bed of the Yarra River.
7. Upper Miocene-lower Pliocene to Eocene to Oligocene sediments such as the Brighton Group and Older Volcanic lies beneath a basal unconformity.

Active deltaic sedimentation ended between c. 2,500 and 1,000 years BP as base levels fell, and the Yarra and Maribyrnong river sediment loads bypassed the delta because of lower bay levels (Holdgate and Norvick 2017, 316).

The engineering properties of the deltaic sediments result in the area it occupies being underutilized, despite its central position (Neilson 1992, 223). The presence of thick, compressible, and reactive, clays, particularly in the Holocene stratigraphic units, has caused foundation problems west of Melbourne. The poor load-bearing strength and high pyrite content of the Coode Island Silt creates challenging conditions for major structures and concrete foundations (Holdgate and Norvick 2017, 304). In the early settlement, and into the later nineteenth century, the wetlands and the poor drainage of the delta inhibited residential-style developments and the area was often used for noxious industrial purposes, and then a major wharf precinct. Modern planning maps of Melbourne still show how the distribution of inner-city residential development is closely constrained by the outcropping areas of the bedrock geology, while industrial land uses dominate in areas underlain by Yarra Delta Group sediments.

THE MELBOURNE LANDSCAPE IN 1835

In 1835, the landscape of the site chosen for initial settlement was both ancient and young, with elements that had been denuded and weathered since the end of the Jurassic, and other elements that reflected the recent sea-level fluctuations and volcanic activity of the last few thousand years. It was comprised of five distinct topographic elements: the Nillumbik Terrain and its ancient drainage system, the Brighton Coastal Plain, the Lava Plains and their youthful drainage system, the Yarra Delta and its many wetlands, and the modern coastline (Neilson 1967, 12–16).

Melbourne was imposed on many diverse natural environments and crossed the boundaries between them. The settlement was initially focused on the well-drained and dry eastern and northern hills, with their variable cover of open eucalypt woodland (figure 5.1). The impact of the settlement then rapidly spread across the adjacent environments, which included the salt lagoons and mudflats to the west; the tea-tree–lined lower river estuary and delta; the wetland-dominated west and south side of the river, and along the densely forested, fertile soils of the Yarra Valley to the east.

The centrality of modern Melbourne is not fingerprinted by a defining natural geographic feature like the other Australian capital cities. There is no cove or peninsula that defines its place. The significance of the modern Melbourne CBD as a central landscape feature is almost completely a human construct. If there was no CBD, the location would be indistinguishable from its surrounding landscape.

The Yarra and Maribyrnong Rivers

The site of Melbourne is located between two rivers, the Yarra (originally known as the Freshwater River) and the Maribyrnong River (originally known as the Saltwater River). The two rivers originally joined at a confluence approximately 4.5 km north of the estuary mouth at the head of Hobsons Bay.

The Yarra River is a meandering, mainly west to southwest flowing river until it encounters the eastern margin of the tongue of Newer Volcanics that underlies many of the inner eastern suburbs of Melbourne. At this point, the course of the river becomes very convoluted before it swings sharply to the south-southeast to follow the eastern boundary of the volcanics. Near Hawthorn, it then skirts around the southern fringe of the volcanics before again turning sharply westward and passing through the settlement (CBD) area to the delta.

As the underlying geology changes, so does the character of the Yarra River. It is more deeply incised and meandering where it is underlain by the Silurian rocks to the east of Melbourne, but the reaches become somewhat straighter where the channel skirts the southern margin of the Newer Volcanics. West of The Falls (since removed), it became a broad arcuate estuarine channel lined by tea-tree wetlands and saltmarshes as it crossed the delta. Prior to straightening of the channel west of the city in the late nineteenth century, the Yarra River formed a broad inverted "U-shaped" arc as it crossed the delta, before reaching the confluence with the Maribyrnong River (figure 5.1).

The Yarra River is both ancient and recent. Many of its middle and upper reaches were formed by the long denudation of a deeply incised landscape formed on Silurian basement rocks. However, the lower reaches in the settlement area formed as the old river had to develop new courses as lava flows infilled the lower parts of the ancient valleys. This process began in the Cenozoic but continued into the Quaternary and recent. Marine and deltaic sediments were deposited during recent higher sea levels that inundated the lower reaches of the river.

Between the future site of Melbourne and the Maribyrnong River confluence, the riverbanks were low-lying and marshy, rising from below high-tide level to only 1–2 m in places. The north bank of this 5 km stretch of the river was bordered by a dense stand of tea-tree scrub for much of its length. North of the belt of tea tree there was a large saltwater lagoon that would later be named Batman's Swamp (or the West Melbourne Swamp). It would not have been visible from the channel. The south bank was also low, with saltmarshes that merged inland to the south with a crescent-shaped system of linear swamps and marshes interspersed with strips of higher ground with woodland. There was also dense tea-tree scrub on this bank, but it was restricted to places where belts of marshlands and wetlands came close to the river. The largest stand lined the bank at the confluence with the Maribyrnong River (figure 5.1).

South of the wood-choked river intersection, the east and west banks took on very different characters. The west bank was formed by Newer Volcanics basalt and was high and rocky in places, rising from narrow mudflats at the shoreline to form a steep ridge around 5–6 m high in places. There were grassy plains above the bank that extended far to the west, with occasional but widely dispersed stands of trees. There were also some low rounded hills that had more dense stands of woodland at their bases.

Stony Creek entered the main channel from the west, through a broad muddy estuary and from this point to the mouth of the river, the west bank was low, but still rocky, with narrow and muddy intertidal zones (figure 5.1).

The east bank was bordered by narrow mudflats that also merged into an inland wetland system lying parallel to the bank. There were occasionally steep and eroded and cliff-like sections in the bank carved out by flood waters. Dense tea-tree scrub occupied the mudflats and wetlands that lined the first 2 km of this part of the river channel, which would have made it feel closed-in.

Any journey upriver in 1835 was unlikely to have been a silent one because there were many thousands of birds inhabiting the wetlands on either side of the riverbanks, immediately behind the screening vegetation. The noise from the birds that inhabited these wetlands must have been noticeable, if not deafening at times. The first densely populated wetland encountered was probably one that was preferred by geese, on the west bank near modern Footscray, which for a while was known as Gosling Flat.

The Maribyrnong River may be young compared to the Yarra River. It mostly drains a large catchment formed on Newer Volcanics and then follows the southeast-trending belt of Older Volcanics that extends as far as the Western Hill of the CBD. The channel of its southern reaches now closely follows the eastern contact of the volcanic plains that spread far to the west of Melbourne (figure 5.1).

The Yarra River had two important but minor tributaries near the CBD, including The Elizabeth Street Creek and the Moonee Moonee Chain of Ponds. The latter drained southward before dissipating as it approached the West Melbourne Swamp and did not originally enter the Yarra. Both discussed in more detail below.

Hobsons Bay

The Yarra River enters Port Phillip Bay in a small embayment at its northernmost point, known as Hobsons Bay. Hobsons Bay is partly sheltered by the Gellibrand Peninsula and was once the main anchorage for large ships that were unable to enter the shallow river. In 1835, the mouth of the river was guarded by broad mudflats and shoals. The main channel was only 140–150 m wide, and the river mouth was impeded by a shoal, around which the channel bifurcated (figure 5.1).

The Beach

In the 1830s, new seaborne arrivals in Melbourne had a choice of paying a fare and being rowed up the river or being landed on a beach on the northeast shore of Hobsons Bay, at a place near modern Station Pier. They could then walk approximately 3.5 km to the settlement. The fare to be rowed upriver to Melbourne was prohibitive (and was also slower than walking), so many disembarking passengers elected to be put ashore on the beach and walk. They could follow a "well-defined" track through close-spaced tea tree growing between eucalypts and tall grasses near the coast and lagoon, and then "sober-colored" woodlands. The track followed the wooded higher ground as it wove between low-lying marshlands to the west and the higher ground of Emerald Hill to the east, which then gave way to marshy ground on both sides as the track neared the southern bank of the river. Within 1.5 km or so of The Beach, the noise of hammering could be heard before the track emerged in a cleared area on the south bank of the Yarra River, among some scattered houses and an Aboriginal encampment (John Hunter Kerr, in Flannery 2007, 103–104). The track ended on the south side of The Falls, where the new arrivals could cross over to the developing center of Melbourne on the north side. When the first trestle bridge was built further upstream in the 1840s, the track from The Beach terminated there (the bridge may have been an easier and safer crossing). The track that the new arrivals formed in the 1830s from the landing place at The Beach to the river crossing at The Falls evolved into modern Bay Street and City Road and is the oldest arterial road of Melbourne.

Services for new arrivals developed at the landing place near The Beach and the village of Sandridge (now Port Melbourne) developed on an elevated site there (figure 5.1). Piers were eventually built near the beach landing place to offload passengers and cargo from ships too large to enter the river and it became a major port area with a railway connection. Port Melbourne continued to be a significant part of Melbourne's port even after new facilities capable of berthing large ships were constructed on the lower Yarra River in the late nineteenth century. Station Pier at Port Melbourne remains the primary entrance point for seaborne passengers arriving in Melbourne. Many of those passengers are Tasmanians because Station Pier is the terminal for the Bass Strait ferries.

1835 Landscape Features of Melbourne

In 1835, the landforms of the settlement ranged from low-lying marshlands and lagoons to steep hills and deeply incised meandering river and stream valleys.

Several important topographic features dominated the small site and influenced the way that the settlement developed as a village and then as a small town.

The Falls and the "Pond"

One of the results of the interaction between lava and water in the old channel of the Yarra River was the creation of the "falls," a rock bar of Newer Volcanics that was located at the approximate location of the modern Queens Bridge. This was the place where the old Yarra River found a way around (and partly over) the last of the Quaternary lava flows and was slow in the process of wearing it down in 1835.

The basalt ledge of The Falls ultimately gave the Yarra River its modern European name. When Batman's surveyor John Wedge was exploring the Yarra and Maribyrnong river system, he decided to name the channel coming from the east after what he thought was the local Aboriginal name of "Yarrow-Yarrow" (Wedge 1836, 423). Twenty years later, Wedge explained to historian James Bonwick that he had later realized his mistake and that his Aboriginal guides were referring to the flow of the water over The Falls and not its name (Bonwick 1868, 48–49; Sydney Mail 1934).

The natural rock bridge created by The Falls had long been used by local Aboriginals as a crossing point and it became equally important to early European settlers, who also used it to cross the river. John Batman's only son was drowned there in 1845, when he fell into the river while collecting the small fish that were often caught among the rocks as the tide receded. He was buried in the same vault as his father at the Old Melbourne Cemetery (Sydney Mail 1934).

The bar of rock marked the end of estuarine influence by preventing the salt water of the estuary reaching further upstream, and impeded the downstream flow of the river, forming a natural weir. This geological feature is the most important influence on the location chosen for the Melbourne village because it created the reliable source of freshwater needed for the first settlement. The river above The Falls remained Melbourne's primary freshwater source for almost twenty years. Initially, water carters extracted water and distributed it through the settlement. Then a steam pump and filter system were installed to move the water to a central distribution point. This inefficient water system was eventually outstripped by the rapidly developing town and the Yan Yean Swamp, a large wetland on the Plenty River, 32 km to the northwest of Melbourne, was developed into a gravity-fed water supply scheme and operational by 1857 (Laing 1977, 2–3).

The Yarra River became wider immediately below The Falls, where the action of tides and overflowing water had etched out a basin. This was a place in the river where the small ships and boats that had been warped up the river could be turned around for the return journey and so became known as the Turning Basin, or the pond. The first Melbourne jetties developed here, and it became Queens Wharf, the first port to develop (now Enterprize Park).

The needs of the expanding town rapidly outgrew the site and various conceptual plans and studies were undertaken to develop larger docks and port infrastructure, including excavated ship canals from Port Melbourne to the city (e.g., Millar 1860). Such concepts eventually led to the construction of Victoria Dock downstream, and the creation of Coode Island by the excavation of a canal for a new course of the lower reach of the Yarra River west of the city. The Queens Wharf location is still the site of the historic Customs House, now the home of the Immigration Museum.

The Settlement Hills

In 1835, three hills dominated the topography of the new settlement, the Eastern Hill, the Western Hill, and Batman's Hill (e.g., Finn 1841, in Hateley 2010, 31). The Western Hill, Batman's Hill, and Burial Hill (further to the north) formed a well-defined, arcuate northwest-trending ridge developed on a belt of Older Volcanics and their mantle of Quaternary sediments. The western foot of this ridge defined the western edge of the settlement and marked the approximate position of the unconformity between the Older Volcanics and the sediment of the Yarra Delta. This ridge is the main topographic feature of the western end of the city. The western foot of this ridge once formed the shoreline of Port Phillip Bay and marks the approximate position of the uniformity at the base of the Yarra Delta.

At Batman's Hill, the Older Volcanics formed a topographic high that impeded the penetration of the estuary further upriver during the last sea-level rise. It also marked the southernmost point reached by the lava flows that filled the ancient Yarra Valley to form the Newer Volcanics. The modern channel of the Yarra River is deflected around this point before entering the delta.

The Eastern Hill was a markedly different landscape in 1835 and has different geology to the Western Hill. It forms a part of the largest of several ancient hills formed on Silurian basement rocks that protrude through the mantle of Newer Volcanics basalt. They were hills before lava flows inundated the once deep valleys between them and are now inliers within the volcanics.

The 1835 settlement was established on the southern toe of the Western Hill. A spur projected toward the river from the southwestern side of the Western Hill and became known as Batman's Hill after John Batman built his house on a prominent site there. The southwestern slopes of the Western Hill faced the Hobsons Bay anchorage and the early port, so became an important site for early government and administration. The first superintendent's residence was built there, as was an early jail, a commissariat store, and a hospital (figure 5.2). John Batman's former home was incorporated into the new government precinct as temporary offices for the Treasury and superintendent.

The highest point of the western ridge formed by the Older Volcanics (and their mantle of Brighton Formation sediments) was a small hill about 1 km northwest of the settlement. The southern slopes of the hill were chosen for the first burial ground, and it became known as Burial Hill. Several of the earliest settler graves remain there. The crest had panoramic views to the west and southwest, across the estuary toward Hobsons Bay and the mouth of the river, so it also became an important place of communication between the settlement and the anchorage, with signals raised on a flagstaff when ships arrived at the anchorage. The burial ground was soon replaced by a larger one to the east, where the Queen Victoria Markets now stand, and the Flagstaff Hill became the site of an observatory (which was moved to Botanical Gardens in the 1880s). Now known as Flagstaff Hill, the land was eventually reserved as a recreational park at the request of residents, which became the Flagstaff Gardens (now the oldest public park in Melbourne). A beacon was lit on the hill in 1851 to notify the residents of Melbourne that the proclamation creating Victoria as a separate colony had been read at the nearby government offices.

A fourth hill to the south of the river was an important landmark for settlers making their way to Melbourne from the landing place at The Beach. This prominent rise, formed on the Older Volcanics, became known as Emerald Hill because it was originally covered in verdant green vegetation. The raised ground was important to Aboriginals, then became an early gold-rush era tent town before developing into the suburb of South Melbourne. The top of the hill is now occupied by the architecturally significant South Melbourne Town Hall.

Elizabeth Street Creek and Catchment

The Elizabeth Street Creek (originally Townend Creek; Presland 2009, 71; Woodhouse 1888) is a southward draining tributary of the Yarra River that occupies a geologically controlled, moderately steep-sided valley between the Eastern Hill and the Western Hill. It follows the eastern boundary of the Older Volcanics and the unconformable contact with the underlying Silurian basement and valley-fill deposits. The bottom of the channel cuts through the older valley-fill gravels beneath the basalt and into the underlying basement. Faults and weathered dikes in the Silurian rocks form more easily eroded lines of weakness that also influenced the alignment of the channel.

Weathered dikes probably controlled the alignment of a small northeastern tributary gully.

The Elizabeth Street Creek is the primary drainage outlet for a catchment that includes most of the north-central settlement area (and the modern CBD, figure 5.2). In 1841, the course of the creek was described as a "jungly chasm—an irregular broken up ravine" with "shattered tree trunks, displaced rocks and roots" (Edmund Finn 1841, in Hateley 2010, 31). The creek was prone to flash flooding and the channelized course was an impediment to east-west traffic, and footbridges were required to span it (e.g., Woodhouse 1888). Despite such drainage works, it has been a long-standing flooding problem for central Melbourne (most recently in 2010), particularly because the alignment of modern Elizabeth Street mostly follows the creek and lies within the linear valley that defines its course.

The Elizabeth Street Creek originally entered the Yarra River above The Falls. Given its large catchment area within the area of the settlement, considerable volumes of polluted town runoff must have been carried into the water supply. The early importance placed on managing this runoff is shown on Robert Hoddle's 1837 survey plan of the settlement, where the course of a diversion drain is shown as "proposed," and it was in place by 1838 (Woodhouse 1888). This was probably an anti-pollution measure designed to divert the extensive urban runoff away from the freshwater supply above The Falls and into the pond below. Construction of the drain represents some of the earliest public works in Melbourne. Despite the early attempts to maintain the quality of the water, it became steadily more polluted as the Yarra Valley became more populated upriver.

Wetlands

Early Melbourne was bordered by saline and freshwater wetlands. Extensive swamps, salt lagoons and marshes once occupied most of the ground south and west of the Melbourne settlement site, of which the West Melbourne (or Batman's) Swamp was the largest. Marshy regions fringed the riverbanks throughout its lower reaches, and a series of large wetlands and marshes extended along the southern banks and to the southeast, where they were particularly extensive (Albert Park Lake is the only modern survivor) (figure 5.1).

The high ground on which early Melbourne grew was originally adjacent to but above the shoreline. When the sea covered the low-lying areas of the Yarra Delta, the site of Melbourne would have had a similar spatial relationship with the shoreline of Port Phillip as modern Williamstown does today. The arcuate wetland and dune systems occupying the low ground were the product of fluctuating sea levels that were higher than today. The sandy deposits left behind by the retreating bay created a landscape of curvilinear southeast to northwest trending wetlands, lakes, and strips of vegetated higher ground. This landscape preserved the shape of earlier shorelines and sand banks until it was drained and backfilled, or developed into port infrastructure (figure 5.1).

The bordering wetlands were a defining element of the geography of the developing town. They restricted the places where urban residential development could take place, at least until adequate drainage works and reclamation occurred in the later nineteenth century. Land that was subject to inundation was used for industry, transport, and port infrastructure (e.g., Victoria Dock) and as places for sporting facilities and recreational open spaces. These land-use types came to predominate on former wetland areas. In many ways, it is more accurate to consider early Melbourne as developing around the periphery of a single large and complex wetland system, than that there were wetlands on the fringes of Melbourne.

The West Melbourne or Batman's Swamp

The western boundary of the 1835 settlement was defined by a low, and probably rocky, escarpment that dropped sharply from a height of approximately 14–18 m (hill crests) to just above sea level at the foot of the Western and Batman's Hills (figure 5.2). Most of the area to the west was occupied by sea-level flats that extended to the Saltwater River (the Maribyrnong). The saltwater lagoon known as the West Melbourne Swamp occupied most of these flats.

The surface area and shoreline of the West Melbourne Swamp varied seasonally but may have reached a maximum extent of approximately 2.6 km from northwest to southeast, and 1.9 km from northeast to southwest (figure 5.1). There was a permanent central, oval-shaped lagoon that remained as a permanent water body throughout the year. The long axis was oriented parallel to the coastal dunes systems, the Hobsons Bay shoreline, and the linear salt marshes and wetlands of the marshy region stretching from the river to the coast. The lagoon was generally poorly vegetated but was rimmed by pink "pigsface." It was also a haven for many different species of waterfowl and other birdlife (Tovey 1911, 61; Lahey 1994 in Vines 2016, 37–38).

The western escarpment bordering the 1835 settlement can still be recognized in the modern CBD, where it continues to define the western edge of the Melbourne commercial center, despite the extensive railway yards and the other infrastructure of Southern Cross Station. The rail infrastructure was originally built at the base of the western escarpment, but was later expanded to the south, which required the removal of Batman's Hill. The arcuate northern course of the western railway corridor is an imprinted "memory" of the West Melbourne Swamp, which the railway alignment had to avoid when it was first built in the 1850s.

The wetlands and lagoons west of the settlement were incorporated into the integrated late nineteenth century port and rail transport network. Once drained, the broad expanses of flat land became an important industrial region and maritime transport hub.

Moonee Moonee Chain of Ponds

The Moonee Moonee Chain of Ponds was a south-draining water course that dispersed into a marshy area as it entered the north end of the West Melbourne Swamp. It was comprised of a series of water holes and ponds that were only linked by flow during wet seasons (figure 5.2). The alignment was redeveloped as a coal canal when the Coode Canal was cut to redirect the Yarra River and the Victoria Dock was created. The adaptation of this water course as a canal was the beginning of the use of drainage alignments as important transport corridors in Melbourne, with several other creek lines being adapted in later periods as motorway corridors.

1835 Vegetation and Environment

The site of Melbourne straddles the boundary of two major vegetation groups: the western volcanic plains, with their dominant grasslands, and the eastern woodlands and often much older forests that cover the rising ground and hills underlain by Silurian basement and along the old course of the Yarra River Valley. The pre-European vegetation of Melbourne varied according to the underlying geology, with woodlands and forest predominating in the east and grass and shrublands to the west. Small variations in the underlying geology also played an important part in many details of the distribution and density of vegetation (figure 5.1).

The site of Melbourne was lightly to moderately wooded on the rising ground in 1835, becoming more densely wooded to the northeast (Wedge 1836, 423). The Eastern Hill was more heavily timbered, and this belt of woodland extended to the northwest, mostly following the outcrop extent of the Silurian bedrocks. There was a rich forest in Merri Creek and other valleys east of the settlement. The Western Hill was lightly timbered and grassy, and the hill slopes carried stands of mainly she-oak. The low-lying and marshy banks of the river were lined by dense tea-tree scrub in places. The marshes and wetlands seem to have been largely unwooded (Hoddle 1836, 1837; Finn 1841 in Hateley 2010, 31), except at their fringes and on the linear belts of low rising ground between swampy areas.

The 1835 vegetation pattern of central Melbourne strongly reflected the underlying geology, with different plant communities and vegetation densities occupying zones underlain by different rock types or more recent sediments.

The salt marshes and lagoons to the west were largely treeless. She oaks occupied the western hills and slopes of the settlement overlooking the salt lagoons of the West Melbourne Swamp and were underlain by the Older Volcanics and their interflow sediments. Light to heavy eucalyptus and wattle woodlands and forests covered the Eastern Hill formed on the basement sedimentary rocks. Forests of mature trees of variable ages occupied the river valleys to the east and grasslands covered the Newer Volcanic basalt to the north, west, and northwest. Tea tree backed by larger overhanging trees and scrub grew along the low-lying riverbanks west of the settlement, where the river became estuarine.

What seems apparent is that there were much richer and more mature vegetation communities to the east of the settlement, on the older landscapes of the Silurian basement. Very old trees of "every age and condition" inhabited the Yarra River valley and Merri Creek, in the places that were the favorite camping places of Aboriginals (Edmund Finn 1841, in Hateley 2010, 32) and where scarred trees were evidence of Aboriginal woodcraft. It is possible that the large and mature trees of the hill and river valley ecosystems east of the settlement were descendants of the river system ecology that was destroyed elsewhere by Cenozoic lava flows.

The existence of rich grasslands was the primary reason for the interest in Van Diemen's Land in establishing a Port Phillip Bay settlement. The grasslands of the mostly treeless Western Plains stretched away to the north and west and were only broken by the forested hills of the distant You Yangs, and the more heavily timbered water courses. The Western Plains are underlain by the Newer Volcanics and may have remained largely treeless because the geologically youthful and clay-rich terrain formed on the basaltic rocks had not yet developed an effective drainage system. And the forest tree species had not yet adapted to the poorly drained and clay-rich soils.

The Aboriginal Landscape

The archaeological evidence for an Aboriginal presence in the Melbourne region is some of the most ancient in Australia. A cranium was found during sand extraction in Keilor that was dated at 14,500 years (Presland 2010, 121). This means that Aboriginal ancestors probably witnessed the western district volcanic eruptions and the infilling of the old topography of river valleys by lava flows, and the changing environments these events caused. Over hundreds of generations, they must have witnessed the evolution of the landscape first encountered by Europeans in 1803. Aboriginal people were there and adapting to the final stages of the evolution of the 1835 landscape and recent ancestors would have seen the last eruptions only 5,000 years earlier.

Aboriginal Use of the Melbourne Landscape

Aboriginal groups used the water courses as "highways" (Presland 2010) and ranged into and out of the Melbourne area along these routes. Any burning of the landscape was probably centered on the water courses (e.g., the Maribyrnong River and Moonee Moonee Chain of Ponds). Many of the key resources came from the river valleys, including the silcrete, which was used to manufacture implements (see below).

Gary Presland (1994) recorded the locations of camping and ceremonial sites used by Aboriginal people around the Melbourne settlement at or shortly after 1835. It is apparent from this work that the first settlement site was highly utilized by Aboriginal groups before 1835 and continued to be accessed afterward as an important meeting place. Presland's (1994) information was plotted on the landscape reconstructions of Melbourne to evaluate their landscape context (not presented here). This suggested that Aboriginal groups preferred to camp in the meandering sections of the Yarra River to the east of the CBD, a suggestion supported by other historical sources, such as Flemming (1802). They also camped in places to the south of the river, and further to the southeast, where many huts were observed. There seems to have been a preference for camping on high ground and hunting and gathering in the water courses and wetlands.

Aboriginal Use of Geological Materials

Silcrete was an important raw material for stone tools. The most important sources of this valuable raw material were outcrops of the ancient weathering surface underlying the basalt at the unconformable contact between the Newer Volcanics and the Silurian basement. Exposures along the Maribyrnong River, the Moonee Moonee Chain of Ponds (Presland 1993), and on Toolern Creek were the most important.

The Mount William axe quarry to the northwest of Melbourne was one of the most important and valued sources of a greenstone used to make axes that were traded across Victoria. Interestingly, this trading network did not extend eastward to the Gippsland people because of the language and cultural differences (Presland 1993).

Stone was used to enhance natural ledges in the Maribyrnong River to create the fish traps and were seen by Charles Grimes in 1803 (Presland 1993, 18). This bar also served as a river crossing.

LANDSCAPE, GEOLOGY, AND THE EARLY DEVELOPMENT OF MELBOURNE

Volcanism was focused to the north and northeast of Melbourne, and slowly inundated the old Silurian basement landscape. The southernmost limit of the Newer

Volcanics lava flows would eventually determine the location of Melbourne by creating The Falls and diverting the flow of the ancient Yarra River around its eastern and southern fringes. Now, the lowest reaches of the river follow a largely new course that is dictated by the final extent of the volcanic flows. The youthfulness of this riverine landscape is illustrated by the basalt dams that were still present in 1835, where the actions of the new river course had not fully eroded through them (e.g., The Falls, and the north side of Dights Falls).

1835–1837 Batman's and Fawkner's Village

The 1835 settlement developed as an unplanned village and grew organically, spreading northwest along the low but steep and rising ground of the Western Hill, where it formed the riverbank. The village spread upslope to the west, following the line of the bank from the low and gently sloping landing place at the eastern end of the pond. The ground steepened to the northwest as it rose to the Western Hill and Batman's Hill. A single track developed from the landing place, along the crest of this bank to the village and was probably used for unloading goods from the small ships that could be worked upriver, and for the carting of fresh water from above The Falls.

The unplanned first growth of Melbourne shows how original settlers had to adapt to a landscape they could not significantly modify. The first cluster of buildings was focused on the crest of a small southern promontory of Older Volcanics that jutted out southward toward the river and formed the steep riverbank that overlooked the basin anchorage of the pond (figure 5.2). The promontory occupied most of the modern city blocks bordered by King Street and Queen Street, and south of Collins Street. The nose of this promontory is now occupied by the historic Customs House (Immigration Museum). Initially, the Tasmanian arrivals settled into a natural landscape that would have been familiar, but which they were unable to significantly modify, apart from clearing the native vegetation. This situation rapidly changed with increasing wealth and, ultimately, the Gold Rush Boom of the 1850s.

Once the illegal Port Phillip settlement started to grow, a government agent and commandant was appointed by the New South Wales governor to oversee the settlement. A Commandant's residence, commissariat store, and barracks were constructed on a prominent site that is now bordered by Collins Street, Bourke Street, Spencer Street, and King Street (figure 5.2). This was known as the Government Block, and it became the focus of all administrative functions in the early years of Melbourne (Doyle 2011, 26).

First Farming

Melbourne was created as a grazing settlement and large numbers of sheep and cattle were imported or overlanded from New South Wales. Grazing activities rapidly spread through the grasslands to the north and west of the settlement. Arable land was less immediately important and was focused on the needs of Melbourne's residents. The first arable farming was focused along the river foreshore, on the alluvial soils and deposits of the riverbank below the settlement (e.g., Hoddle 1837). John Batman had twenty acres improved as a garden and orchard by the time his only son was born a year after he arrived (near the corner of modern Spencer Street and Flinders Street). Batman's garden occupied the low and gently rising ground on the slope below his house, which stood above the steep shoreline. Several other small garden plots were associated with other dwellings further east along the riverbank and occupied similar positions next to the river (figure 5.2). From this small nucleus, arable farming spread eastward along the Yarra Valley to exploit its fertile alluvial soils.

Fawkner planted the first wheat crop on a lot on the south side of the river, but it was prone to flooding and unsuccessful. The site became the second cricket ground for the Melbourne Cricket Club (MCC) until it had to move to make way for the Sandridge railway line (Doyle 2012).

1837–1850: The Latrobe Era—a Planned Town

A government superintendent was appointed to the Port Phillip district in 1836 and order began to be imposed on the village. The city grid (Hoddle's grid) was drawn up in Sydney without any reference to the landscape (Flannery 2007, 12) and fitted to the site by the Assistant Government Surveyor General Robert Russell and his assistant Robert Hoddle (Hoddle 1837a). The original land survey of Melbourne incorporated the area between Spencer Street and Spring Street (east-west) and initially, from Flinders Street to Lonsdale Street (north-south) (Hoddle 1837a) but it was soon extended to Latrobe Street (Russell 1837) (figures 5.1 and 5.2). Several more town blocks were added on to this alignment to the north, extending the street grid to Batman Street and Franklin Street, and incorporating Burial Hill (now Flagstaff Hill) (Hoddle 1837b). This northern range of blocks was interrupted by the wedge-shaped southern corner of the new burial ground (now the Queen Victoria Markets), which was established on a north-south alignment after the abandonment of the graveyard at Burial Hill.

When referenced against the reconstructed topography (figure 5.2), there appears to be little evidence to show that any significant attempt was made to find a "best-fit" of the street grid to the landscape of the twin hills and the Elizabeth Street Creek, apart from rotating it to the northeast. The east-west streets, such as Collins Street, seem to fit in well with the saddles on the northwest to southeast trending ridge crests created by the Older Volcanics and the valley created by the Elizabeth Street Creek as it eroded the unconformity. Once the street grid was established, the crests of the sharp hills in the settlement had to be cut down to improve road gradients, particularly the linear crest of the Western Hill, where it was crossed by Bourke Street, Lonsdale Street, and Latrobe Street. Reducing the gradients improved the east-west flow of horse transport.

When the city grid was finally laid out, and land allotments surveyed and sold, many of the buildings of the first ramshackle settlement were out of alignment, with some even located in the middle of newly surveyed road alignments (Russell 1837; figure 5.2). The old "organic" village disappeared as the new formalized town grew.

High Ground as Preferred Suburban Locations

The locations of elevated ground were particularly influential on where the first villages were established, and where residential development took place. Some of the earliest satellite villages were located on restricted elevated positions surrounded by low-lying wetlands and marshes. Emerald Hill is a key example, but Sandridge (Port Melbourne) and Footscray also began as small settlements largely confined to elevated sites. Further to the south, St Kilda developed on a hill underlain by the most southern exposure of the Silurian basement on the east side of Port Phillip Bay. As the Melbourne urban area grew, the earliest residential suburbs of Carlton and Fitzroy spread across the elevated sites above the wetland and river flood levels and formed mainly on Silurian basement. When the major urban expansion of the 1850s gold rush happened, the broad flats underlain by the Newer Volcanics to the east and northeast became to suburbs of Collingwood and Richmond.

Not all planned suburbs came into being. Robert Hoddle's 1837 plan for South Melbourne showed a subdivision where part of the Royal Botanic Gardens now lies. There is a note on this plan stating how the river overflowed during flooding. This may be the reason why the planned urban development was reduced in size, and no development occurred at the north end.

Wider Settlement and Growth

On leaving the small area of the Hoddle grid, the natural topography and underlying geology become more influential on urban layout. The best example in the study area is North Melbourne. The streets here were laid out to follow the topography of the western slopes of the arcuate ridge underlain by the Older Volcanics that is the northern continuation of the Western Hill and Burial Hill. The sediments of the Yarra Delta begin at the foot of the ridge, and it forms the shoreline of the West Melbourne Swamp. The railway lines here exhibit a similar geometry to the curved streets of North Melbourne because they occupy the space between the ridgeline and the high-water mark of the West Melbourne Swamps.

The original settlement was hemmed in by rivers, wetlands, and marshy low-lying ground so the amount of good land for residential and commercial development was restricted. The only readily available space for urban expansion was to the east and northeast of the city grid, on the Eastern Hill and the adjoining hills (underlain by Silurian basement), and on the broad flat plateau formed on the Newer Volcanics overlying the basement and bordered on its eastern flank by the Yarra River. The lava-inundated former valleys now formed high, level, and dry land for the earliest Melbourne suburbs. Being coextensive with the original settlement area, the basalt flats formed a logical location for northward urban expansion and were largely unencumbered by wetlands. The flats of the plateau became the locations of the early suburbs of Fitzroy, Richmond, Collingwood, and Brunswick. The only other level ground beyond the wetlands south of the river was Prahran and the high ground of Emerald Hill and Sandridge.

Satellite villages were founded in the surrounding hinterland at various points for a variety of purposes. Sandridge grew at the main landing place on Hobsons Bay. Footscray began where a punt crossed the Maribyrnong River, above the original confluence with the Yarra River, where an inn was built. Later, Footscray became a center of stone quarrying, producing blue stone for the building industry. This was one of the first important industries of Melbourne's western suburbs. Williamstown was founded as a port town at the mouth of the estuary. All the first satellites settlements to grow near Melbourne were located on elevated ground. The urban island of Emerald Hill is the most pronounced example. It was totally confined to the elevated ground formed by the isolated outcrop of the Older Volcanics and surrounded by low-lying salt marshes and wetlands formed on Quaternary sediments.

Europeanizing the Landscape

Large tracts of land on the peripheries of the settlement were reserved for public purposes in the early years of the organized development of Melbourne. This was particularly prevalent during the long tenure of Charles Latrobe (1839–1851), the Port Phillip Superintendent and then first lieutenant governor of Victoria (City of Melbourne 2019). Latrobe showed great foresight in reserving these lands because, although they lay at the periphery of the settlement in the 1840s and 1850s, they now form an inner ring and defining landscape setting for the Melbourne CBD. Many of the city's great public buildings were designed to inhabit these landscaped parkland settings.

The reserved lands were often left as native bushland or were used for utilitarian purposes such as grazing or quarrying (e.g., the Fitzroy Gardens) but were gradually developed into ornamental parklands and sporting facilities. The earliest was the Flagstaff Gardens, but it was followed by Carlton

Gardens, the Fitzroy Gardens, the Treasury Gardens, the Royal Botanic Gardens and Royal Park, and many others. The public land was also used for other purposes, such as the site of Government House and the Domain Parklands, the grounds of the University of Melbourne and its residential colleges, and the Melbourne General Cemetery.

As the reserved lands on the city fringes and in the new subdivisions became more formalized, the Melbourne landscape rapidly changed from an Australian landscape to a European-style landscape, through the introduction of exotic plant species and landscaping. With the wealth of the gold rush that began in 1851, the residents of Melbourne desired to create a great European city to exhibit that wealth. The small unplanned village that was formed in an Australian landscape that many contemporary writers considered remarkably beautiful (e.g., see summary in Hateley 2010, 29–33) was transformed into a city encircled by an increasingly European-style landscape of parklands, recreation grounds, and sporting fields.

The original landscape-based settings of the grand Victorian-era buildings of Melbourne have largely been lost or modified in the modern high-density and high-rise urban environment. Infill developments in the parkland have often ruined the once wide views and the original settings of these buildings. The original aspects have been lost for many buildings, particularly those along Spring Street.

1850–1880—Gold Rush to "Marvelous Melbourne"

Gold was discovered in central Victoria in 1850–1851, and the small town of Melbourne was overwhelmed by wealth, and large numbers of people suddenly coming to Port Phillip in search of riches. As Melbourne developed post-gold rush, the Victorian-era sense of building grandeur and landscape began to dominate the development of the town, particularly the public buildings. The Victorian-era ideals of recreation began to influence the layout of parks and the grand settings in which important buildings such as Parliament House and the Treasury Building were set. Charles Latrobe's foresight at reserving large tracts of public land came to the fore during this phase and they were laid out with gardens and sporting facilities. Grand buildings were set in picturesque landscapes created by the parks that were laid out around them. This era became known as "Marvelous Melbourne," a phrase coined by London journalist, George Sala, in a series of articles he wrote for London's *Daily Telegraph* newspaper following a visit in 1885 (eMelbourne 2019).

Landscape and Infrastructure

Parts of the landscape were used for different purposes throughout the settlement and the underlying geology, usually acting through the topography often controlled those uses. The most enduring impacts are preserved by the locations of transport corridors (road and rail), sport and recreational facilities, and the sites chosen for many of the key landmark buildings.

The underlying geology also came to influence administrative divisions as it acted through landscape features. For example, the eastern and southern administrative boundaries of the modern Melbourne City Council were determined by the locations of major rivers whose channels owe their current geometry to the boundary of the Newer Volcanics.

Roads and Railways

The early road alignments often developed to best exploit the high ground and as transport routes to link elevated locations. The first major transport route formed outside the village developed as a track that skirted the periodically inundated marshes south of the river and followed the edges of the high ground between The Beach on Hobsons Bay and the main settlement. City Road and Bay Street evolved from this track. To the south, St Kilda Road developed as a track from the river crossings (the punts, The Falls, and then the first trestle bridge) to growing southern suburbs. The original alignment followed the western foot of the hill forming the modern Botanic Gardens as it also skirted the edges of the many salt marshes (of which Albert Park Lake was the largest).

The Beach remained an important place for immigrant arrivals in the 1850s and the first railway in Melbourne was built from Sandridge to the town, opening in 1854. The rail alignment followed part of the original walking track to The Falls, and it crossed the Yarra River via a bridge built next to them. This pioneering line was followed in 1857 by the first railway servicing the inland from a terminus on Spencer Street, on the western edge of Melbourne. The original station was placed on the low-lying land at the western end of the street grid, in an embayment in the western slopes of the ridge between Batman's Hill and the Western Hill. As more land was needed for expanded railway infrastructure, Batman's Hill was quarried away in the 1860s to accommodate an enlargement of the station.

The railway line north of the terminus had to skirt around the eastern shore of the West Melbourne Swamp before heading west and northward into the Victorian hinterland (figure 5.1). When the lines were extended to the east, to Flinders Street Station and beyond, the only site available for the rail corridor was the north bank of the river. To create an urban rail network from the central city eastward involved following the banks of the Yarra River and other wetland corridors.

The 1835 drainage pattern and wetlands remain imprinted on the transport network of modern Melbourne, not only through the railways but also through the motorway system. Many freeways follow creek alignments because when they became necessary in the 1950s, these reserves and watercourses were the only available undeveloped corridors along which major roads could be developed into the central city (e.g., the Tullamarine Freeway follows the course of Moonee Ponds Creek, Presland 2009, 19).

Elevated Sites

Both Batman and Fawkner chose prominent, elevated positions overlooking the river on which to build their Melbourne residences in 1835, perhaps as statements of their belief in their importance to the settlement (figure 5.2). Like these entrepreneurs, elevated sites in prominent positions were also chosen for important government buildings. The first Port Phillip Commissioners camp was established on the Western Hill, where it was visible from the anchorage and the Turning Basin. Prime later examples are Parliament House, placed on the rise of the Eastern Hill at the eastern end of Bourke Street, and the Treasury Building, at the eastern end of Collins Street. And the current Government House occupies on one of the highest topographic points south of the river, in an elevated garden setting.

Many other buildings and institutions that were important to the community were also placed in prominent positions. Hospitals were placed on elevated sites because they might benefit from the well-ventilated positions. The Benevolent Asylum in North Melbourne was placed on the ridge that extended north from Western Hill. The University was placed on rising ground to the north of the town in what was then an open rural setting. Princes Park and the Melbourne General Cemetery were laid out on a linear ridge formed on Older Volcanics north of the town. Following World War I, the Shrine of Remembrance would be sited on one of the most prominent hills south of the river, which was formed on the ancient Silurian basement.

Port Facilities

The West Melbourne Swamp and Moonee Moonee Chain of Ponds were adapted to the essential nineteenth-century transport infrastructure of any major city of the period: wharfs, canals, railway yards, and warehouses. This landscape on the western fringe of the city would ultimately be developed into the main harbor facility of Victorian Melbourne. It was in an ideal location for accessing the main railway lines and lay adjacent to the central business district. To construct the new port facilities, the lower reaches of the Yarra River were diverted by a canal, the Coode Canal (named after the designer), creating Coode Island within the former northerly loop of the river. Victoria Dock was created in the lowest point of the former swamp and was surrounded by major dockland infrastructure.

Sporting Fields and Parks

Wetlands adjacent to the river and in the surrounding low-lying marshes were difficult to drain and therefore difficult to develop for residential purposes. Some proposals for early suburbs in the marshlands of South Melbourne seem to have been abandoned after the area was inundated during floods in the 1840s. Low-lying land then often became reserved for sporting grounds and parklands, such as the original Melbourne Cricket Ground and the Flemington Racecourse.

Batman's Hill was a popular recreational area in the 1840s—for picnics and promenades. It was also the location of a racetrack (Doyle 2011), which is marked in this position on early plans. A Botanical Garden was created adjacent to Batman's old house. The first cricket ground in Melbourne was also on the flat ground adjacent to Batman's Hill but had to move to make way for the new railway station. The MCC then appealed to Governor Latrobe for another ground. One suggested site was next to the existing ground, and another was further along Flinders Street, while a third was in the Police Paddock just to the east of the town. The MCC chose the third option because they believed it was flat enough for the ground but sloped enough to avoid floods. This is the current Melbourne Cricket Ground, one of the most important sporting venues in Australia.

Locations of Early Industries and Defense Facilities

Many of the low-lying swampy and marshy areas adjacent to the future CBD were not reserved for parklands and became progressively more industrialized because they were generally unsuitable for residential development. Some of the more noxious industries, such as tanneries and slaughterhouses, were focused on riverside wetlands downstream of the settlement, and their refuse was discharged into the lower reaches of the Yarra River. This was the fate of John Batman's gardens. The land use rapidly became more industrial near Batman's Hill in the 1850s. The former garden became the site of slaughterhouses and other noxious industries that moved into the locations immediately downstream of Melbourne.

The broad but poorly drained western volcanic plains provided level acreages that were ideal for industrial and defense uses, particularly military airfields. Many defense establishments were also located on the plains, such as munitions storage depots and factories (Altona and Deer Park).

Australia's first military airfield was established at Point Cook, west of Melbourne, immediately prior to World War I, and then followed by Laverton in the 1920s. As a major industrial center, Melbourne became a focus of wartime aircraft production at Fisherman's Bend during World War

II, where there was also an airfield. Civil aviation was focused on the broad flats of the western volcanic plains from the 1920s, including Essendon Airport and Melbourne International Airport (Tullamarine).

The volcanic plains became the focus of large-scale industry from very early times. One of the earliest manufacturing industries was of Sunshine agricultural harvesters, and then later, car manufacturing. This type of production-line manufacturing requires large and level factory sites. The plains now provide the spaces required for huge factories and other modern industrial complexes.

Geohazards

The winding course of the Yarra River is prone to flooding during wet seasons. In 1803, James Flemming noticed flood debris at a height of 2–3 m in the trees near the confluence with Gardiner's Creek as he was exploring the river east of the future settlement site (Flemming 1802). And the lower reaches of the river were choked by fallen trees to such an extent that it made upriver navigation difficult for even small vessels (Finn 1841, in Hateley 2010, 31).

The ancient basement landscape infilled with the Older Volcanics and then the Newer Volcanics created a particularly convoluted and winding section of the river immediately east of the settlement. Drainage patterns may still have been evolving in the relatively new volcanic landscapes present in 1835. This would cause many flooding problems in Melbourne in the nineteenth century.

Flooding was a significant problem associated with the Yarra River for several decades, with major flooding events in 1835, 1839, 1842, 1844, and 1849 (Presland 2009, 84–86). Efforts to lessen the impact of flooding have included widening the channel and the removal of bends in attempts to improve flows and lessen the flood risk. The course of the river has been straightened in several places near the settlement site, with "U" bends removed by quarrying in the Botanic Gardens and at the modern Herring Island Reserve.

The landscape was also affected by the recent drop in sea level, leaving barrier dunes that prevented ancient water courses from reaching the sea (e.g., Moonee Moonee Chain of Ponds and the Carrum Swamp).

THE BUILT ENVIRONMENT

The exploitation of building stone began immediately after settlement in 1835. Extractive industries were initially focused on the south bank of the Yarra River, where several small freestone quarries were developed in the Silurian rocks by 1846.

The most abundant building stone near Melbourne was bluestone, the stoneworkers name for quarried basalt, which is the defining element of the built character of Melbourne. This fine-grained, pale gray to dark blue-gray rock formed as lava flows of the Newer Volcanics. Varioles (gas bubbles) are common in the cut stone and formed when it flowed across the landscape and the gas was released from solution by the cooling lava.

The first bluestone quarries were started where the modern Fitzroy Gardens now are, and in Carlton, and then became focused further east, along the western bank of the Yarra River. Other early quarries were developed along the banks of Stoney Creek, a small tributary on the west bank of the Yarra River near its mouth (figure 5.1). The quarried stone from here was used as ships ballast. Large resources of bluestone of very high quality were quarried in the Footscray area and used for many of the historic buildings of Melbourne (and elsewhere, see figure 8.2, panel 2). Bluestone quarrying became a major industry during the early growth phase of the western suburbs. Many of the former bluestone quarries of the western suburbs of Melbourne became local government refuse tips and landfills, with several then being repurposed as playing fields.

Basalt bluestone is the definitive construction material of pre-twentieth-century Melbourne and remains a largely unique aspect of the built environment (figure 8.2). Bluestone is more durable and was cheaper than other freestone (such as sandstone) but was more difficult to work. Its dark color was also considered too gloomy for major buildings (apart from prisons and many churches). An underlying structure of bluestone was often faced with sandstone, such as Parliament House, where it was used to construct the two separate core chambers of the building before they were eventually linked by subsequent additions, and then reclad in sandstone.

Despite the more aesthetically pleasing sandstone facings, the street-level character of most of Melbourne's grandest buildings (and some of its most humble), and much of its grand beauty, is built from bluestone, including most foundations and base courses of major and minor buildings, wall footings, and decorative cast-iron guard railings across the city (figure 8.2 2C–2F). Bluestone also features in many smaller stone-built features, paving blocks in Melbourne's alleys and laneways, as kerbing, and as steps leading into some of the grandest buildings (figure 8.2 2F and 2K). So, in many ways, bluestone forms the foundations of the built environment of the city.

And in spite of the preference for lighter-colored and more easily worked freestone in many public buildings, bluestone was used to construct several historic landmark buildings, including St Patrick's Cathedral; Victoria Barracks; the Victorian Arts Centre, and the Old Melbourne Jail. Bluestone was used to construct many historic schools and parish churches across Melbourne (figure 8.2 2I), and it was used to build the Gellibrand Point

Lighthouse, one of the only surviving convict-built buildings remaining in Melbourne.

Basalt "floaters," unweathered kernels of basalt left in the weathering profile of the Newer Volcanics, were used to build the dry-stone walls and the agricultural fence bases in the western suburbs of Melbourne, and throughout the Western District of Victoria. And most aggregates used in the city are crushed bluestone.

Brickmaking

Brickmaking began with John Fawkner using local clay sources. By the early 1840s, sites marked "Brickmakers" are recorded on survey plans of the north bank immediately south of Batman's Hill, between the slaughterhouses, and a large area labeled as "Brickmakers ground" was recorded on the south bank, to the east of the first timber bridge that crossed the Yarra (near the current site of Princes Bridge). The sources of brickmaking clay were Quaternary River alluvials.

The clay-rich soils formed on the Newer Volcanics are reactive, expanding when wet and contracting when dry (approximately 20 cm per year). The surface heave and settlement as the soil moisture content changed through the seasons could damage solid brick and stone structures built on them. It was this expansive characteristic of the soils that, at least in part, led to the gradual widespread adoption of the brick veneer method of housing construction in Victoria in the late nineteenth and early twentieth centuries. Previously, timber and weatherboard structures have been more successful in withstanding the effects of the reactive soils because they were built on lighter footings and more readily flexed with the seasons.

However, weatherboard did not have the same fashion and social status as brick buildings. Brick veneer construction was developed as a means of imparting the look of a high-status brick building on a structure that had the flexibility and lightness of a timber one. The nonsupporting brick veneer allowed the creation of a more substantial looking house (Cuffley 1989, 141). The construction method was only slowly adopted, with some reticence by banks and insurers to approve loans for the method, but it became more generally accepted in the 1920s (Lewis 2020, 6.04.5). While solid brick homes were more desirable, they were also more expensive than timber structures. This further increased the popularity of brick veneer construction, which provided the external appearance of a brick home without the cost. In addition, brick veneer required less external maintenance, had improved thermal insulation, and services were easier to install (ATSE 2000). Brick veneer construction would be followed later by the floating slab.

Chapter 6

The Woodland Plain

Adelaide, South Australia, 1836

The city of Adelaide, the capital of the modern state of South Australia, was originally conceived as one of the first government-sanctioned "free" settlements in Australia. It would be established without any dependence on convict labor and would be planned from the outset. It was also to be a free enterprise settlement that would be funded by prior land purchases by British investors. The founders of Adelaide achieved their goal, and it became the only Australian capital to be created by a free settler population.

GEOGRAPHIC SETTING AND STUDY AREA BOUNDARIES

Adelaide is located on the south coast of Australia, on the eastern shore of Gulf St Vincent, a 140 km long inlet of the Southern Ocean, at latitude 34° 55' 36.8" S, longitude 138° 35' 57.5" E (Adelaide GPO, Google Earth).

The modern city was established on an inland site toward the southern end of a broad undulating coastal plain in 1836. Bordered by the sea to the west, and a semiarid hinterland dominated by the Mount Lofty Ranges to the east, the plain extends for over 95 km, and is now a prime agricultural region of modern South Australia. The peak of Mount Lofty (727 m) is the most prominent geographic feature of the region. The coastline extends southward, beyond the site of Adelaide, to form the Fleurieu Peninsula (figures I.1 and I.2). The plain is traversed by several meandering, westward draining streams, and small rivers, the most significant of which was named the Torrens River.

This chapter is focused on an area corresponding to the current local government boundary of the city of Adelaide, which incorporates the original site of the 1836 settlement. The defining landscape features are a low, elongate plateau rising to an average height of 50 m above sea level, and the centrally located Torrens River, which bisects the plateau. The original Adelaide town allotments were laid out in two sections. North Adelaide filled the elevated ground of the plateau to the north of the river. South Adelaide (the city "square mile") covered the larger area to the south of the river and is now the modern CBD.

Since 1836, the urban area has grown to cover most of the southern half of the Adelaide Plains. Modern Greater Adelaide now extends from the original center to the Adelaide Hills and the foothills of the Mt Lofty Ranges to the east, and to the former fringing wetlands and the first landing place at Glenelg Beach on the coast to the west. The area discussed remains the economic and cultural center of Adelaide, and the hub of a radiating transport infrastructure, and is now home to over 1.7 million people (Australian Bureau of Statistics 2021).

EUROPEAN DISCOVERY AND EXPLORATION

European knowledge of the southern coastline of the Australian continent began around the turn of the nineteenth century. In this period, Bass Strait was recognized, and the south coast was charted. Most of the significant geographic features of the gulfs were named by British navigator Matthew Flinders during his circumnavigation of Australia in 1802, and by his contemporary, the French explorer Nicholas Baudin.

Flinders was the first to sight (and name) Mt Lofty, the high point of a range that he could see while taking survey angles from Kangaroo Island on March 23, 1802 (Flinders 1814, unpaginated). The eponymous Mt Lofty Ranges take their name from it. Flinders made other observations of the coastline in the modern Adelaide region as he charted the coast, and recorded fires on the ranges that he attributed to Aboriginal burning. Unfortunately, Flinders was detained by the French governor of Mauritius on his return journey to Britain, so many of Baudin's observations and charts were published first. However, most of Flinders' names eventually passed into common usage, while several of Baudin's also remain, including the Fleurieu Peninsula.

The European settlement of the South Australian region began informally. Late eighteenth-century reports of whales made by ships navigating through these southern waters (Clarke 1996, 51) triggered commercial interest. From the beginning of the nineteenth century, unofficial whaling settlements developed in the form of extended ship-based visits and temporary shore encampments. The focus of these temporary occupations was the islands off the southern coast of Australia, and particularly Kangaroo Island. The presence of these settlements led to the first significant (sometime negative) interactions between Europeans and the Aboriginal inhabitants of the region, particularly the Ngarrindjeri of the southeast coast and Murray River mouth (Jenkin 1995, 25–26).

Early Exploration of South Australia

In 1829–1830, Captain Charles Sturt traveled down the Murray River by boat and explored the mouth of the river. He reported the existence of good water and good land in the region (Jenkin 1995, 27–28). He was in London during the early movement to create a free (non-convict) colony in South Australia, which gave impetus to the movement (Davies 1989, 163). Other reports had also become recently available. In March 1831, following closure of the Albany convict station in Western Australia, Captain Collet Barker, the last commandant, explored the region around the Murray River mouth on his return journey to Sydney, at the request of the NSW governor. While examining the region in April 1831, he was the first European to climb Mt Lofty, from where he observed the Port Adelaide River and Sturt River, and he recognized fertile land on the east coast of Gulf St Vincent and along the Murray (Davies 1989, 163; Jenkin 1995, 25–26; Bach 2019).

THE FOUNDING OF ADELAIDE

In the 1820s, British colonization of Australia entered a new phase of organized settlement. Several factors contributed to this change of emphasis, including the early reports of the commercial potential of the well-watered and fertile agricultural land in Australia, social troubles at home, and the lingering concern about possible French territorial ambitions in Australia (Davies 1989, 161). The result was the generation of various discussions and plans to create free settlements in Australia. The first of these was Perth (chapter 4), the second was Adelaide, and the third was the refounding of Brisbane (chapter 3). Melbourne also began during this period, but as an unsanctioned and less organized commercial free-for-all until government control was imposed, after which it was transformed into a planned commercial settlement (chapter 5).

The first proponent of the most significant of the Adelaide schemes was Edward Gibbon Wakefield, but he was ultimately not to be involved in the final approved project. The movement to create a new colony in South Australia went through several iterations before the final successful bid was placed before the British Parliament. It received final approval, as The South Australia Act in 1834 (Davies 1989, 159–163).

Reconnaissance of the Adelaide Region

The first detailed exploration of Spencer's Gulf and the Adelaide Plains was triggered by the colonizing projects and carried out systematically by survey teams with a brief to find the best site for a free and planned British settlement. The first South Australian surveyor general, Colonel William Light, went ahead of the first shiploads of colonists with his survey party. Their aim was to explore the region and evaluate the various potential site options, and to then choose the place for the settlement. They then had to survey the land allocations ready for taking up by the settlers who had prepurchased them in London.

Prior to leaving London, several locations were recommended to Light as potential locations for settlement, including Port Lincoln, Kangaroo Island, and Rapid Bay. Light reconnoitered and rejected them all after careful examination and multiple visits to be sure that he had made the right decision. In the first week of November, the Adelaide Plains were seen for the first time, and the Torrens River was discovered as a series of lagoons by Kingston, Field, and Morphett (a survey party) (Light and Elder 1984, 30; Carter 1998, X). Soon afterward, the southern reach of the Port River was encountered (November 21). After further exploration, Light and his survey teams chose these plains as the place to focus their search efforts, even though it had not received any prior consideration in London (Davies 1989, 168–70).

On Thursday September 29, 1836, Light chose Holdfast Bay as the landing place because of its likely proximity to a suitable site for a capital (figure 6.1). He described the view of the landing place (the future Glenelg) from *Rapid* as a gently shelving sandy beach with low sandhills and a small deep clear stream of freshwater "flanked by an attractive belt of lightly wooded country with the vivid green of grass showing clearly beneath the trees" (Carter 1998). He named Holdfast Bay after the *Rapid* safely rode out severe gales there on October 7–8.

The Choice of the Site

William Light chose the site of Adelaide on the plain for practicality and aesthetics. There was water, in the form of a small river (the largest crossing the Adelaide Plains), but the crucial landscape element was the rising ground. This placed the settlement site above any of the potential flooding that he had found evidence for on the lower ground adjacent to the Port River and along the coast. He also found the setting more aesthetically pleasing as a place for a settlement because "It was on a beautiful and gently rising ground and formed altogether a better connection with the river than any other place" (Light and Elder 1984, 95).

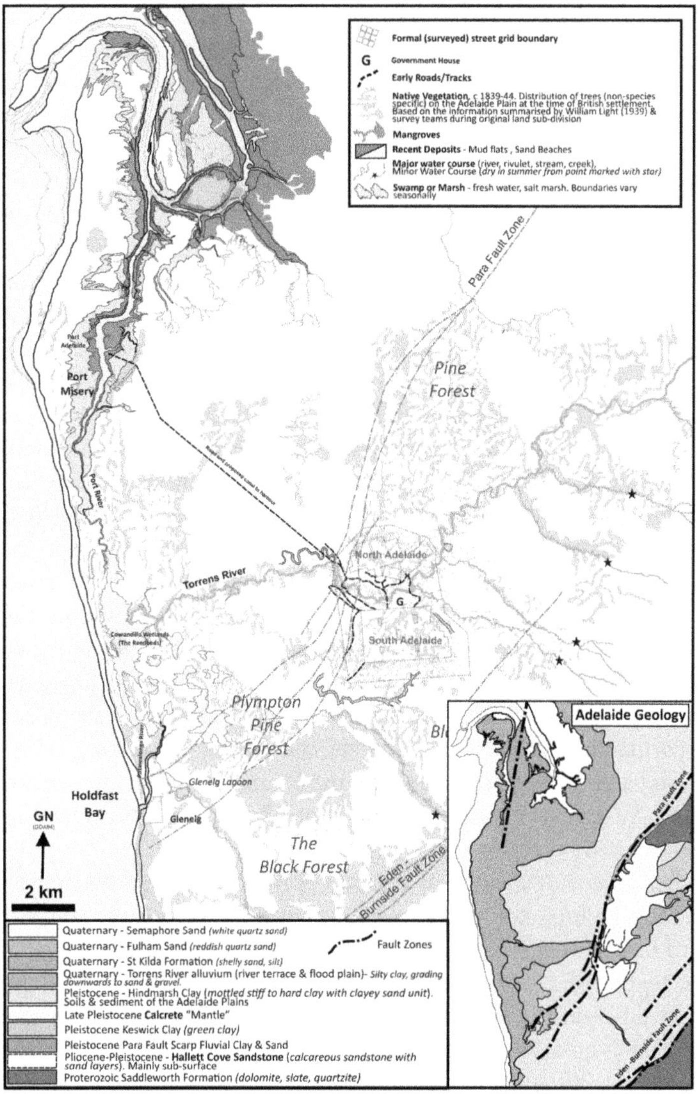

Figure 6.1 The Adelaide Region. The landscape of the Adelaide Plains circa 1836, based on the original survey plans of William Light (published by Arrowsmith 1839), with additions for Port Adelaide by Light and others. The distribution of denser stands of woodland and forest shown is based on the information recorded by William Light in various survey plans of the main settlement and the rural block surveys of the surrounding plains. Additional details are derived from Kraehenbuehl (1996). The Torrens River did not directly enter the sea but entered an extensive hind-dune wetland system of reed beds - the Cowandilla Wetland system - lying parallel to the coastline. The wetlands then discharged to the sea through the Port River and the Patawalonga River. **Inset**. Simplified surface geology of the Adelaide Plain based on Forbes (1980), with additions from Selby (1981). Please refer to the extended caption at the rear of the book for additional information.

Governor Hindmarsh apparently agreed with Light's assessment after he inspected the proposed site shortly after arriving in South Australia. However, he came to prefer a location at the port, or at least closer to it, as the place for the capital and exerted pressure on Light to move it. After initially agreeing to the governor's request, Light re-examined Hindmarsh's preferred site carefully and noted that it was flood prone, and that the river overflowed the mangrove-lined banks, and the low-lying land was subject to flooding. He also realized that there was no freshwater supply there. Ultimately, Light decided that his original site was superior (Davies 1989, 170). Light was ultimately responsible for choosing the site of the capital and not the governor, with the latter only being able to offer advice (Light and Elder 1984, 93). Given that the first anchorage to develop on the upper reaches of the Port River became known as Port Misery (figure 6.1), Light's assessment was correct.

Light had a soldier's and surveyor's experience of reading landscapes, the eye of a skilled artist, and had lived in Mediterranean climates for many years. Constrained by restrictive guidelines, Light read the subtle landscapes of the site he chose and adapted the preplanned design to the topography he encountered. In designing Adelaide, Light went beyond his brief and added a ring of parklands that remain, though modified, one of the defining features of the modern city. Light had to adapt the preordained design brief he received in London to his own practicality and aesthetic, and to the location he chose. Through Light's clever adaptation of the original design to the site, and by adding parklands, the subtle landforms and geological setting of Adelaide have played a defining part in creating the character of this most planned and elegant of Australian capital cities. History has judged William Light this well, as he asked posterity to do (Light 1984, 59).

No other choice of the site of a future Australian capital was preceded by such a thorough regional search. And no other received such careful consideration of how its landscape setting could be used to best advantage to influence the layout of a new settlement. Unfortunately, Colonel William Light died from tuberculosis in Adelaide in 1839 and did not see the settlement develop. But even at the time of his death his contribution to the foundation of Adelaide was recognized. He remains the only person legally buried within the central city, and rests below a monument in Light Square, one of the five open spaces he designed within the modern CBD.

GEOLOGY OF THE ADELAIDE PLAINS

The site of Adelaide is underlain by Neoproterozoic to Middle Cambrian rocks of the Adelaide Rift Complex, a mostly marine sedimentary sequence occupying the eastern margin of the Archaean Gawler Craton (Preiss 2000;

Foden et al. 2006). The sequence was inverted and deformed during the ~500 Ma Delamerian Orogeny (Foden et al. 2006), followed by significant uplift during the Alice Springs Orogeny (Gibson and Stüwe 2000, and Weisheit et al. 2014, in Hall 2018, 3). Tight folding imposed a sinuous north to northwest structural grain on the outcrop geology and landforms of the Adelaide region. This is particularly apparent in the geometries of the Mount Lofty Ranges and Adelaide Hills to the east of Adelaide, as the Flinders Ranges to the north, and as the gulfs and peninsulas of the coastline (figure I.2). This line of ancient rocks forms the central geographic spine of modern South Australia. The sculpting of these ancient basement rocks by erosion has determined much of the landscape character of the wider Adelaide region.

Basement rocks are deeply weathered and eroded and formed a plain on which Cenozoic and Quaternary sediments were deposited. They are now unconformably overlain by a thin mantle of Cenozoic sedimentary rocks of marginally marine-to-marine origin, outwash and alluvium from the Mount Lofty Ranges to the east, and the recent alluvium deposited by the Torrens River and its tributaries (Selby and Lindsay 1982, 18).

Precambrian basement rocks do not outcrop in the study area but have been intersected in drill holes at depths ranging from 35 m, beneath Montefiore Hill in North Adelaide (Selby and Lindsay 1982, 22), to >90 m in Kent Town to the immediate east. Structural contours of the basement unconformity show that it is shallowest beneath modern North Adelaide but progressively deepens to the south below the Torrens River and the modern central business district (Selby and Lindsay 1982, 22–23). To the west, the basement unconformity is downthrown by at least 70 m by normal movement on the Para Fault (figure 6.1 inset).

St Vincent Basin

The Neoproterozoic–Cambrian basement rocks are unconformably overlain by the St Vincent Basin, a restricted Middle Eocene (mid-Cenozoic/Cenozoic) sedimentary basin that formed as the separation of Antarctica and Australia gained momentum during the final breakup of Gondwana (Smith et al. 2015, 41). The 5,200 km^2 St Vincent Basin underlies Gulf St Vincent, the Adelaide Plains, and parts of Yorke Peninsula and Kangaroo Island. It is bounded by the Proterozoic rocks of the Mt Lofty Ranges to the east and the Yorke Peninsula to the west, reaching a maximum width of 50 km (Smith et al. 2015, 24).

The sedimentary sequences of the St Vincent Basin, and the younger superficial terrestrial deposits that mantle them are the most important geological elements of the study area. Of particular importance to the geology of central Adelaide is the late Cenozoic (Miocene to Pliocene) Hallett Cove Sandstone, a fossiliferous marine calcareous sandstone to limestone that was deposited during one of the most recent marine incursions (figure 6.1 inset). The Hallett Cove Sandstone mostly lies at depth in the city area, except where the Torrens River bisects the linear plateau of the Para Fault Block (see below). Here, the Hallett Cove Sandstone was exposed on the south bank of the river (and possibly the slopes of North Adelaide) where it was utilized as an important early source of building stone (see below).

Quaternary Sediments

Several superficial Quaternary units unconformably overlie, and thinly mantle the marine sequences of the St Vincent Basin, including the Poorooka Formation, Hindmarsh Clay, Carisbrooke Sand, and the Burnham Limestone. They formed mainly as outwash fans shed northwestward from the Mount Lofty Ranges (e.g., Twidale 1976, 49–52); as alluvial sands and gravels deposited by the Torrens River and its tributaries; and as coastal wetlands, dunes, and mudflats. The fan-like piedmont deposits dominate the geology of the Adelaide Plain and are intercalated with marine deposits from various Holocene marine incursions (Howchin 1918, 142). Many of the lens-like layers are important aquifers in the metropolitan area.

The surficial rocks and soils were deeply weathered and heavily calcretized during the Cenozoic, a process that was associated with the extensive development of calcrete in the soil profiles. Calcrete deposits are geomorphologically important to the layout of the 1836 settlement and provided a stone source for early building and lime-making (see figure 8.2). It is the subtle landforms of the weathered younger rocks that created the landscapes of the 1836 Adelaide settlement.

The Para Fault Block

Adelaide lies on the eastern margin of the St Vincent Basin, where downfaulting at the boundary created an arcuate series of major faults that traverse the central Adelaide area. Fault movements have disrupted the marine cover sequences and created several fault scarps where the western sides have been downthrown. Of most significance to this discussion are the Para Fault and the Eden-Burnside Fault which form the bounding structures to a fault block on which modern central Adelaide lies (Selby and Lindsay 1982). Most uplift occurred along the Eden-Burnside Fault Zone, while significant uplift also occurred along related structures, such as the Para Fault (Gostin 2014, 3). The rising ground that Light chose for the site of the settlement is formed by a step-like, fault-bounded block of Hallett Cove Sandstone and Hindmarsh Clay to the east of the Para Fault Scarp (figure 6.1 inset). This geological feature formed the first significant rise of ground east of the coast (Light and Elder 1984) and above the potential flooding that concerned Light.

Geological Influences on the Settlement Site

The bounding steep slopes, corresponding to the fault scarps, are the defining elements of the settlement site and the main influences on the layout of the modern Adelaide CBD. They delineate the central city blocks in both North and South Adelaide. The geometry of the high ground areas was a defining factor in the orientation of the street grids, which deviated from the east-west grid of the original conceptual design handed to Light in London. The street grid, and the regular size of the city blocks had to be fitted to the two subtly rising North Adelaide and Adelaide City plateaus of Selby and Lindsay (1982, 21). The high ground geometry also controlled the extent of the parklands, which were laid out on the flat and occasionally marshy land extending outward from the base of the central plateaus.

Hallett Cove Sandstone and the 1836 Landscape

The late Cenozoic fossiliferous and calcareous Hallett Cove Sandstone is the most important geological unit underlying the main north-south plateaus of the settlement and is the foundation of the low hills on which the CBD and North Adelaide are built. It is bounded by the Para Fault Scarp to west and a further scarp to the east (erosional?) and is bisected by the Torrens River. It underlies most of the city area but only outcropped on the southern side of the river, where it formed part of the low and cliff-like southern bank near Government House. This outcrop was developed into a government quarry that extended westward from the grounds of Government House. The stone it produced was used to build the Government House, the Old Adelaide Jail, the armory, the barracks, and the original Adelaide Hospital (Selby and Lindsay 1982, 38; Carter 1998, 199). The former working face now defines the riverside boundary and the rehabilitated sites of the former quarries are now the Torrens Parade Ground and parklands. Quarries to the west have been incorporated into the Entertainment Centre and railway station precincts.

THE ADELAIDE PLAINS LANDSCAPE IN 1836

The Adelaide Plains are a gently undulating erosional landform of low topographic relief. The southern boundary is defined by the point where the plains narrow as the Mount Lofty Ranges escarpment nears the coastline, at the modern suburb of Brighton. The northern boundary is less well-defined, extending north beyond the Wakefield River. The plains are traversed by several small and meandering ephemeral creeks that mostly flow to the northwest before entering the Torrens River, the largest drainage line of the study area.

The site of the 1836 settlement is centrally located on the plain, occupying a northeast to north-trending rise of ground that reaches an average height of 50 m (figures 6.1 and 6.2). This low plateau mostly formed by calcrete and the Hallett Cove Sandstone, on the eastern side of the Para Fault Scarp. The rising ground is bisected by the Torrens River.

Drainage System and Wetlands

The Torrens River is the largest water course of the Adelaide Plains, flowing approximately 85 km southwest across the plain from its headwaters in the Mount Lofty Ranges. In 1836 it was often only a chain of ponds, particularly in the summer months, but could be a raging torrent in winter. The Torrens River did not directly enter the sea, draining instead into an extensive system of wetlands and reed beds that filled the region behind the coastal dunes. This was the largest wetland system in the Adelaide area, and became known as the Cowandilla Wetlands, or simply as the reed beds. The reed beds were in turn drained by the Patawalonga River to the south and the Port Adelaide River to the north (figure 6.1). The extensive wetlands of the reed beds were drained in the mid-twentieth century to build Adelaide Airport and for residential housing. A canal was cut through the coastal dunes to allow the Torrens River to drain directly into the sea.

The Torrens River has several meandering tributaries that flow northwest across the plain from their headwaters in the Mount Lofty Ranges. The most southern is the Sturt River (Warriparri or Warreparinga; Edwards 1964, 184), which now becomes dry in summer, but was originally a permanent spring-fed stream. Several other creeks traverse the plain, including Brown Hill Rivulet (Brownhill Creek), Green Hill Rivulet (First Creek), Hallett Rivulet (Second Creek), Todd Rivulet (Third Creek), Anstey Rivulet (Fourth Creek), and Ormsby Rivulet (Fifth Creek) (figure 6.1). Most of the smaller tributaries have been canalized and covered over to mitigate flooding.

There were abandoned channels of the Torrens River and its tributaries in the settlement area. The longest was immediately to the southwest of North Adelaide and was probably once the main river channel. Another adjoined the northeastern corner of the City Mile and may once have been the location where First Creek (Green Hill Rivulet) entered the Torrens. This small feature was significant because the area became waterlogged in the wetter months and caused Light to truncate the regular street grid along the northeastern corner where the land dropped sharply away to the old channel. He recognized that residential development would not be possible on the

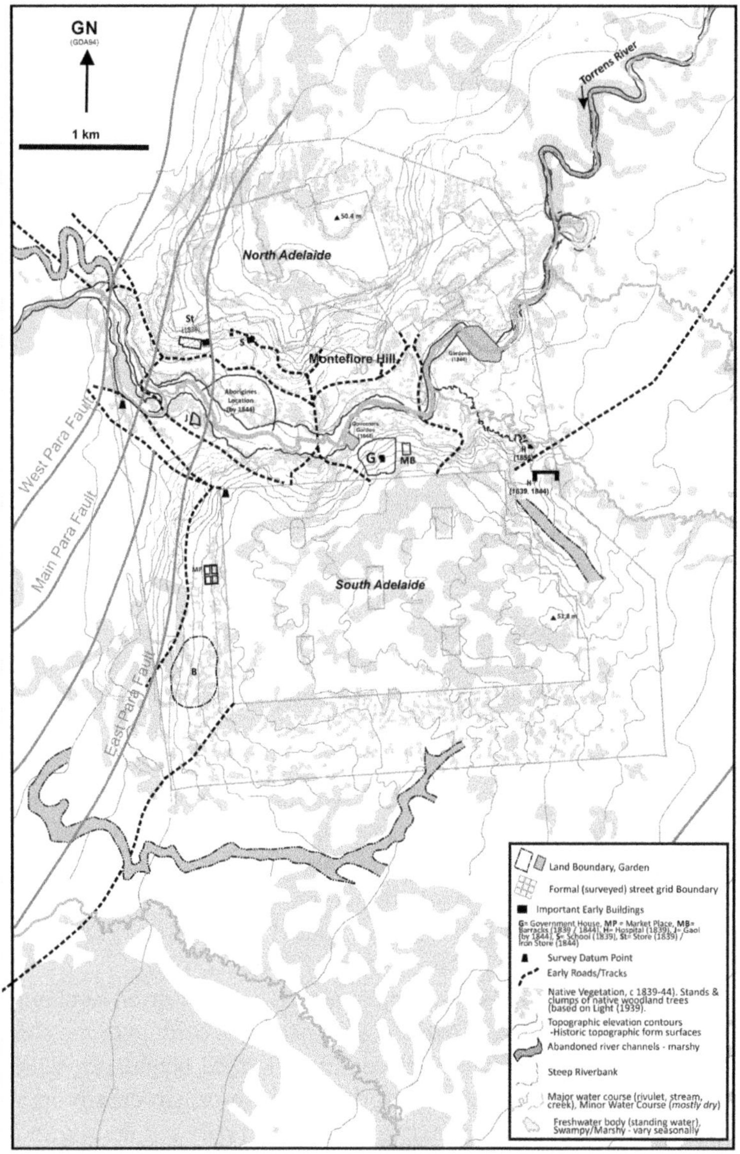

Figure 6.2 The Adelaide Settlement. Reconstruction of the topography of the Adelaide settlement area circa 1836. The reconstruction is based on historic maps and modern topographic data, with reference to written descriptions of native vegetation and contemporary art depicting the area. The scarp east of the Para Fault system is the major geological control on the landforms of the Adelaide settlement. Fresh water came from the billabongs at the western end of the river, where the channel became more complex as it crossed the Para Fault scarp. Sources for this map include Light's map (Arrowsmith 1839, and others). Topographic information is based on 1:50,000 scale topographic mapping, and more detailed topography from Selby and Lindsay 1982). Please refer to the extended caption at the rear of the book for additional information.

periodically flooded ground. A third abandoned channel lay to the southwest of the City Mile and may once have been a large tributary of Brownhill Creek (see Arrowsmith 1839; figure 6.2).

The Torrens River was the most important freshwater source for the young settlement and was the only publicly available early water supply. Water carters drew water from ponds in the Torrens riverbed and distributed it through the town (Carter 1998, 193). The water became contaminated, so many colonists also sunk shallow wells into the river alluvium. However, most wells became brackish in the summer months (Selby and Lindsay 1982, 10).

Vegetation and Environment

The Adelaide Plain was clothed in an open woodland in 1836 (figure 6.1), which became denser in the vicinity of Adelaide, where it inhibited Light's ability to survey the land blocks (Light and Elder 1984, 95). Light's paintings of early Adelaide support the vision of the area as a thickly wooded area of variably dense, mature eucalyptus woodland (Light and Elder 1984, Plate 2). Large fallen trees in the parklands were being rapidly consumed for firewood in 1840 (Carter 1998, 214). No contemporary artworks or written descriptions suggest that the woodlands had been recently burnt and most of the trees illustrated were large and mature.

The vegetation and environment of 1836 Adelaide was dominated by two main vegetation communities; an open woodland community that occupied the plateaus of North and South Adelaide, and the surrounding plains; and an old river red gum forest that occupied the river valley and water courses. The riverbanks had a dense growth of tea tree and wattle (presumably in the wetter areas such as the billabongs and ponds) and various aromatic shrubs (Kraehenbuehl 1996, 76–77; Carter 1998, 192; figure 6.1).

The plateaus were covered in an open woodland comprised of clumps and thickets of trees that varied in density and spacing according to the character of the underlying geology and the geomorphological boundaries within the landscape. The superficial geology of the plain was a particularly important influence on the distribution of native vegetation types and density, possibly because some units functioned as aquifers. Of note are the recent alluvial deposits of the Torrens River, which supported red gum forests. There were higher vegetation densities in the woodlands growing on the calcretized rocks and soils of the North Adelaide and Adelaide City Plateaus (and in the runnels that drained these elevated areas) than on the surrounding plain (figures 6.1 and 6.2).

The Aboriginal Landscape

There is little surviving physical evidence of the Aboriginal occupation of the Adelaide city area and parklands (DASH Architects 2018, 70). While there is relatively little extant archaeological evidence for the pre-European Aboriginal occupation, a campsite on a northeast-facing hillside was discovered at Hallett Cove, approximately 18 km to the southwest, dating to approximately 15,000 years. This campsite probably dates to a time when the sea level was significantly lower. The many stone tools and cores found are of an early form known as the heavy Kartan type (also found on Kangaroo Island) and were manufactured from a locally outcropping siltstone (Flood 1990, 196).

Aboriginal Use of the Adelaide Landscape

In pre-European times there were more than twenty clans occupying the Adelaide Plains, which are now known collectively as the Kaurna (Owen and Pate 2014, 46). Early missionaries and others recorded the name Cowandilla or Kouandilla for the people living near Adelaide (DASH Architects and Bell 2018, 3), probably to the north. Tindale recorded that the clan that occupied the area that is now Adelaide was known as the Tandanja (or Dundagunya), named after the red kangaroo (Tindale, in Aitken et al. 2006, 201).

The people of the Adelaide region lived mostly on the coast in summer, camping near coastal springs and exploiting the mulloway fishery and wetland resources (Tindale 1974, 71). In winter, they moved to wooded inland areas, where they could shelter in the hollows of large red gums along riverbanks (favored as the core of huts), where there was timber for fires, and there was a source of logs and stringybark for building shelters. Stringybark sheets were also transported to the coast for shelters. The Kaurna only accessed the wet sclerophyll of the upper slopes of the hills to the east to hunt small animals and to search for the large insect grubs that bored into the stems of wattles. Gum resin was also an important resource acquired in these higher areas (Tindale 1974, 60, 71). To the east were the neighboring Peramangk, the people of the Adelaide Hills, who were feared and had very different cultural practices, religious beliefs and spoke a different language (Clarke 1991, 60).

The northeast course of the Torrens was probably an important, well-watered and resource-rich route from the coastal wetlands to the higher ground of the Adelaide Hills. Aboriginal groups probably passed through the Adelaide area, perhaps as a part of an annual cycle. The Torrens would have been an unreliable source of freshwater in the hotter months, so it is likely that Aboriginal populations were mostly transitory and did not live there permanently (DASH Architects and Bell 2018, 3).

In some areas, the first settlers found well-worn paths or tracks already in existence when they arrived. One major track ran from the Adelaide Plains southeast toward the Bremer River, continuing to the Murray River (Clarke 1991, 59).

Aboriginal Natural Feature Names

The Kaurna name for the Torrens River is Karrawirraparri (red gum forest river). Several locations within the study area have retained Aboriginal place names, including Tandanya, the name for the city area; kainka wirra (eucalypt forest), a water hole in the Botanic Garden identified by Amelia Taylor in the 1930s and possibly the name of an original waterhole on First Creek that was redeveloped in 1857 (DASH Architects and Bell 2018, 4). The Patawalonga River has retained its name, though it has been extensively modified, and the Cowandilla Wetlands (or reed beds) retained their name until they were drained and redeveloped.

Campsites

The Kaurna occupied semipermanent campsites in the wetland and coastal areas, which they moved between. Debris accumulated at such sites to form distinctive occupation mounds in the landscape, such as the one described on the banks of the Little Para River (Owen and Pate 2014, 46). Mounds were recorded from Glenelg in the south and Middle Beach in the north, with concentrations between Salisbury and Port Gawler. Approximately half of the mounds excavated contain burials and burials have also been discovered nearby (Owen and Pate 2014, 47).

A large campsite was exposed during realignment of the Sturt River in the suburb of Marion, and there was a nearby burial ground uncovered in 1925 when intensive agriculture began in the area (Edwards 1964, 184). Most such occupation sites seem to be focused along the coastal wetland and to the northward beyond Port Adelaide and reflect a cultural occupation tradition stretching back at least 1,500 years (Owen and Pate 2014, 46–47).

Three locations were used as camping places in central Adelaide in the earliest days of the colony, and may have been used earlier. Pinky Flat is a locality on the north bank of Torrens River, which may have been used as a camping place before European settlement. It gained its name (at least in part) from the Aboriginal name for the burrowing bilbies that were common in the area ("pinkies" or "pingkos"). This site extends from below Montefiore Hill to the southwest of modern North Adelaide and reaches as far as the Adelaide Oval (Sumerling 2011, 214). The "Aborigines Location," also on the north bank of the Torrens River and to the west of Pinky Flat, was an area that was later reserved for Aboriginal use (marked on the map by Kingston and

Stephens 1842, see figure 6.2). This location was close to the largest billabongs in the channel of the Torrens (with their central island), which were an important early source of fresh water and were later developed as a weir to create Torrens Lake.

There was another camping area east of the modern settlement, on the south bank of the Torrens in the modern Botanic Park area, which was probably the one sketched by von Guerard in 1855 (Sumerling 2011, 142). Corroborees, ceremonies, and burials were said to have occurred there and a hollow dead red gum was identified as a former shelter (DASH Architects and Bell 2018, 4; Aitken et al. 2006, 201). Other authors suggested that this place was mostly used after European settlement because the native vegetation persisted there longer and provided shelter, hunting, raw materials, and firewood (Aitken et al. 2006, 200). Alexander Murray sketched an Aboriginal encampment adjacent to the first wooden bridge in 1842 (DASH Architects and Bell 2018, 5).

LANDSCAPE, GEOLOGY, AND THE EARLY DEVELOPMENT OF ADELAIDE

The formal European settlement of South Australia began in 1836 and comprised several short phases.

Phase 1: Exploration of Spencer Gulf and Gulf St Vincent to find a suitable settlement site.
Phase 2: Choice of the site of Adelaide, establishment of the first survey camp and creation of the survey datum.
Phase 3: Establishment of the settler camp at Holdfast Bay.
Phase 4: Town survey, establishment of the settler camp near North Terrace, and allocation of land sale grants.
Phase 5: Urban expansion beyond the town survey blocks and parklands, and into the country sections.

Survey parties ranged from the region of Port Lincoln to the south of modern Adelaide as they examined the countryside for locations to form the settlement. Once Light had fixed on the location for Adelaide, a survey base camp was established near the southern bank of the Torrens River, near a supply of freshwater source provided by the billabongs occupying the nearby river bend. The location was between the northwest corner of modern North Terrace and the river, and now on the site of the modern Royal Adelaide Hospital. This campsite became the location of the survey datum for the layout of Adelaide.

While the land grants were being surveyed, the immigrants established a temporary camp at the landing place at Holdfast Bay, now modern Glenelg. It was here that the proclamation of the British Province of South Australia was read by Governor Hindmarsh on December 28, 1836. Soon afterward, the first Adelaide settler encampment was established near Light's survey camp, between North Terrace and the Torrens, and to the west of King William Street. The place became the focus of the first temporary settler dwellings (Carter 1998, 190). In the 1840s, the northwestern corner of the main city grid was still the focus of most urban development within the settlement (see Kingston and Stephens 1842).

Light's 1836 Town Survey

A conceptual street grid was designed in London to provide the required number of presold town blocks and was given to Light before he departed. Light and his team had to adjust the predesigned street grid to fit the local topography, as well as to meet the commissioner's instructions. As subtle as the landscape of the plains was, the sharp slopes at the edges of the central plateaus created by the fault scarps dictated the best-fit scenarios of the design.

In the main city block (South Adelaide), Light included five squares: Victoria, Light, Whitmore, Hindmarsh, and Hurtle squares (figure 6.2). All these squares have survived subsequent development and remain a central focus of the communities that live around them. It was not possible to form a fully rectangular street grid because East Terrace was not as regular as the outer settlement boundaries. Light had to avoid the seasonally swampy area at this corner, so some blocks were left out of the planned grid (Fenner 1927, 210; Arrowsmith 1839, after William Light). The swampy area was later developed into the site of the Adelaide Zoological Gardens and the Adelaide Botanic Gardens, with their ornamental lakes.

The layout of North Adelaide was particularly strongly influenced by the topography of the escarpment and plateau, and the course of the River Torrens. The high ground north of the river was smaller in area than that to the south. Light's original intent was to arrange North Adelaide streets parallel to those of the main city block (see Light's map of February 7, 1837). This was not possible given the topography and the requirement to fit regular and equal-sized blocks into the available space. To fit the town blocks into the topography he had to rotate the grid anticlockwise and create a smaller gridded area on the lower ground to the southeast of the main plateau (Fenner 1927, 210–211). Light created a single central square Wellington Square, which remains a central landscape feature of North Adelaide. Despite all the restrictions of the predetermined elements of the design, the survey of the

1,042-acre town area was completed within two months, by March 10, 1837 (Davies 1989, 170).

The Parklands

When Light surveyed the Adelaide town blocks, he also reserved a ring of parklands around the fringe of the main city block, North Adelaide, and along the Torrens River. This is considered one of his most visionary achievements (Sumerling 2011, ix). The reservation of the parklands provided the space to focus on important cultural and sporting facilities. This distinguishes Adelaide from other Australian cities, where sporting facilities were usually developed on land that could not be used for other purposes because it was too poorly drained (e.g., Brisbane and Melbourne). In Adelaide, there was reserved land available for such facilities, and not just because it had been left vacant because of a detrimental landscape feature.

The parklands became the location for many of the important institutions of the settlement, including Government House and the Domain, the university, the Adelaide Oval, and the racecourse. By 1838 a hospital site had been chosen at the eastern end of the settlement, south of the Torrens. This site remained the location of the Royal Adelaide Hospital until 2017, when it was moved to a new site at the west end of North Terrace (on the former railway goods yards). The parkland also became the site of the main cemetery.

The parklands were also the location of the first farming in Adelaide, in the form of small gardens established on the recent alluvial sediments of the Torrens River. The Governors Garden, on the south bank of the Torrens, and a larger garden on both banks of the river at a site near the modern Zoo are both marked on the map of Kingston and Stephens (1842).

The Country Section Surveys

Once the town survey was complete, the survey parties moved outward to subdivide the Adelaide Plain for rural land grants. The mostly flat landscape of the plains offered little topographic impediment to the regular pattern of rectangular to square land sections. Without major topographic impediments on the plain such as hills, major creeks, or wetlands, which would require diversions, the plains could be divided into a regular grid of mainly north-south oriented blocks, with the divisions recorded on cadastral maps (e.g., Arrowsmith 1839, after William Light).

The main deviation from the regular grid pattern are the blocks surveyed at 90 degrees to the course of the Torrens River (to provide river frontage), and the road corridor to the port. Another significant interruption to the regularity of the rural land surveys was at the place where the Torrens River dissipated in the reed beds, near modern Fulham. But these deviations are rare, and the north-south "chequerboard" grid pattern was not often diverted.

The Evolution of the Road and Rail Network

The original square and rectangular rural land survey boundaries remain the most important influence on the street pattern of Adelaide suburbs. As urban development spread across the plain, particularly along the first arterial road corridors, the pattern of surveyed blocks came to control the shape of urban subdivisions. The regular land tenure grid then became the road network grid of the developing urban areas, and the section surveys strongly influenced the geometry of the system of secondary roads and suburban streets that evolved to serve the spreading suburbs. Urban development has simply filled in the 1837 pattern, and access and arterial roads have mostly honored the land tenure boundaries as new areas have been subdivided. The pattern of those first rural surveys now remains deeply embedded in the modern suburban street pattern of Greater Adelaide.

The first main roads surveyed on the Adelaide Plains in 1837 were straight lines, largely un-influenced by topography, a sign of the low topographic relief. These first arterial roads led to the Glenelg landing place and to Port Adelaide and followed the course of the Torrens River to the northeast (figure 6.2). They were surveyed as alignments from the corners of the central city block grid and North Adelaide, in a hub and spokes pattern focused on the city and parklands. All subsequent arterial roads were laid out in a similar manner. Later road transport corridors have mostly honored the surveyed boundary lines of 1836–1837, perhaps more so than the railways, which have overprinted and transected the grid (e.g., the southwest corridor to Glenelg).

Unlike other Australian capital cities, there were no convenient undeveloped water courses to use as corridors for railway and motorway access to the city center. Only the Torrens River provided a significant corridor of undeveloped land into the central city, so its southern banks became a focus of railway infrastructure, including the main station and the nearby goods yards. Beyond the city center, rail lines utilized the peripheries of the parklands as they exited the western side of the CBD and then turned to the south and north. Private land must have been resumed by the government to create the rail corridors to Port Adelaide (the first railway line in South Australia, opened in 1856), and southwest to Glenelg and the southern suburbs. However, it seems that William Light foresaw the need for at least one major transport corridor between the town and the port because he reserved the wide alignment of Port Road as the location for a future canal during the survey (Arrowmith 1839, after William Light). Port Road remains the main transport outlet to the northwest of the city center.

Beyond the Parklands: Early Suburban Development

The plains surrounding Adelaide were opened for settlement after 1840 and many separate villages developed beyond the parklands. Urban development rapidly spread beyond the parklands and into the country sections, where speculators bought larger rural blocks and then subdivided them into half-acre sections that could be bought by working people. It was cheaper for tradespeople and carriers who could not afford city blocks to buy in such areas, and the locations along major roads were particularly favored (see Denholm 1989, 183). These small settlements became important "nodes" in the evolution of the city because they created links in the developing track and road network on the plain, eventually developing into ribbon developments radiating out from the center (Denholm 1989, 183).

Anchorages

A 29-acre subdivision was also surveyed on the Port River, so that sections could be offered to the settlers who agreed with Governor Hindmarsh and wanted to purchase their town acres on the coast. This compromise resulted from a meeting in which most settlers voted against moving the settlement from the site chosen by Light (Davies 1989, 170).

The first Port Adelaide town blocks were on the upper reaches of the Port Adelaide River (the Port Creek Settlement). The first port facilities were basic, and the place was initially known as Port Misery. It consisted mainly of mangroves and mudflats and the residential allotments were not almost at the same elevation as the mudflats. Passengers had to be landed by sailors who carried them on their backs to shore across the mudflats (Carter 1998, 168).

Governor George Gawler ordered new port facilities to be developed to the northeast of Port Misery, and to be linked to the main road network. The new Port Adelaide facilities grew rapidly from this small beginning to become the main harbor of South Australia and are now a focus of an integrated transport network.

Torrens River Flooding

The superficial geology of the Adelaide Plain is comprised mainly of alluvial outwash deposits, fans, and terraces formed between the Adelaide Hills and the coast. These deposits show that wide-scale and repeated flooding on the plains has a long history (Twidale 1976, 49–52). In 1836, as William Light was exploring the Adelaide Plains, he observed evidence of these processes when he found indications of flooding in the water courses to the north of the future settlement. This observation prompted him to seek a more elevated site. Light's recognition that parts of the Adelaide Plains were prone to flooding was one of the prime reasons why he persisted with the development of the modern site of Adelaide, against the new governor's wishes.

The Torrens River was flood prone, even though it could be reduced to a chain of ponds for much of the year. The large catchment, which extended to the Mount Lofty Ranges, meant that it was subject to dangerous seasonal flash flooding when swollen by winter rains. Crossing the river was necessary to move between the main commercial center and the mainly residential area of North Adelaide and several fords were used. The riverbed was a barrier and river crossings were often difficult or impossible during wet periods. This impeded the economic development of the northern and southern parts of the town. Most of the northwest flowing tributaries of the Torrens were also flood prone and fed further runoff from the mountains into it upstream of Adelaide. They were eventually canalized to minimize the flood risk.

Removal of the vegetation along the banks increased scouring and this undercut the riverbanks, widening the course, and creating steep erosional cliffs. Bridging the river became important but proved to be difficult. The first 1839 wooden bridge was damaged twice and then destroyed by flooding in 1844 (Selby and Lindsay 1982, 10–11). The first successful bridge did not open until 1856, following engineering consideration of the potential effects of scour and the maintenance of bank support (Selby and Lindsay 1982, 10–11).

THE BUILT ENVIRONMENT

Timber suitable for building was in short supply on the Adelaide Plains. What timber there was available to the early settlement was often obtained illegally from the parklands. Decrees had to be passed to prevent further degradation of the parkland vegetation (Sumerling 2011, 142–143). So locally sourced stone was an important building material in the early colonial period and beyond. Local building stone was readily available on the plain and in the nearby hills and four main sources were used; calcrete (limestone); Hallett Cove Sandstone; Tea Tree Gully Freestone, and "bluestone." Stone became a defining element of the built environment of Adelaide and remains the key component of the built heritage of the modern city (see figure 8.2 for examples).

Adelaide is built from the most ancient rock of any Australian city. The most unique Adelaide building stone sources are the low-grade metasedimentary rocks sourced from the Proterozoic basement, which have undergone moderate to strong deformation and regional metamorphism. The "bluestone" and slate that form much of the unique built character of Adelaide were laid down in a Proterozoic ocean that knew only microscopic life. Also important to the early built environment were rocks formed during the several marine transgressions that inundated the Adelaide Plain during the Cenozoic, particularly the Hallett Cove Sandstone.

Calcrete (Limestone)

The first stone buildings in Adelaide were built from calcrete (Jack 1923, 36), a chemically precipitated "limestone" deposited as a weathering product in the soil profile. It underlies most of the original settlement site and formed during the prolonged Holocene deep weathering of the superficial cover sequences of the stable plain landscapes. It was generally found just below the surface but where eroded, it could form low cliffs, such as some of those along the banks of the Torrens River (DENR and Young 2008, 8). It was often found as field-stone rubble, which had to be manually removed to make soil tillage possible. Calcrete was the most important stone of the earliest phase of the growth of Adelaide, and its built environment, because it was so readily available.

The "travertine," as it was known, was extensively developed beneath North Adelaide, where it was probably derived from the deep weathering of Hallett Cove Sandstone and was used in the construction of many of the oldest houses there (Jack 1923, 38, see figure 8.2, 3E). In the city area, it was probably at least partly derived from Hallett Cove Sandstone and quarried concurrently with it in the government quarry, as a mantle, although Selby and Lindsay (1982, 58) state that it transitions downward into Hindmarsh Clay in the city area. It is known as the "calcareous mantle" by modern civil engineers (Selby and Lindsay 1982, 57).

Calcrete was strong and durable, though very difficult to work, and generally produced a rough material in sizes and shapes that were mostly suitable for rubble stone walling (for which it was used extensively, see figure 8.2, 3D and 3E)), though several historically important buildings were built with it, including the Old Parliament House (figure 8.2, 3F) and the boundary wall of the Government House grounds on North Terrace (figure 8.2, 3D).

Rubble construction methods used a large amount of mortar (Selby and Lindsay 1982, 6), but the calcrete could also be burnt for lime, with much of the raw material coming from the wastage of building stone production (Jack 1923, 9, 18).

Bluestone

In South Australia, "bluestone" refers to a sedimentary rock that is usually siltstone, obtained from the Proterozoic basement rocks to the east and south of the settlement, in the Adelaide Hills, and the foothills of the Mount Lofty

Ranges. It is probably the most distinctively "South Australian" building stone, being unlike any stone used in other Australian colonies. The stone became particularly important for walling and flagstones (figure 8.2, 3G–3J), with early sources including shale from Beaumont (1838) and then Glen Osmond (1851) and later, the Tapley Hill area to the south. Bluestone became particularly important during the later colonial phases of Adelaide development, a period after 1855 which has been referred to as the Victorian boom period (DENR and Young 2008, 6).

The stone from a variety of sources was originally used as a main construction material of older houses and many buildings throughout Adelaide (e.g., figure 8.2, 3G), but by the 1920s was being more commonly used for base courses and foundations (Jack 1923, 19–20). Important buildings constructed of Adelaide "bluestone" include the base courses of St Peters Cathedral and as the "backing" behind the dressed sandstone of the GPO (DENR and Young 2008).

Sandstone

Sandstone became more common in the later Victorian period for ornamental public buildings and domestic purposes. An important source was quartzite obtained from interbedded units of the Skillogalee Dolomite (such as Tea Tree Gully Freestone) and it was used to construct key landmark buildings such as the General Post Office (sandstone facing backed by bluestone), the Adelaide Town Hall, the Supreme Court building in Victoria Square, St Peter's Cathedral, St Francis Xavier's, the Stow Memorial, Flinders Street Baptist Church, the War Memorial, the Mitchell Building of the University of Adelaide and the Scots Church, North Terrace. Important buildings were also constructed from Hallett Cove Sandstone, including Holy Trinity Church and Old Parliament House.

Brickmaking Clay

The banks of the Torrens were illegally developed as a series of brickmaking clay and gravel pits. Brickmaking and quarrying were major industries in the parklands in the early years and Governor Hindmarsh had to concede that the workers of these industries could continue living in the parklands, despite the orders for their eviction, because they were too poor to move (Sumerling 2011, 142).

Historic Quarries

Major quarries from which stone was sourced to construct mid-Victorian buildings of the "sandstone and bluestone period" include Glen Osmond, which was the main source of high-quality bluestone; Tea Tree Gully and Glen Ewin, which supplied most of the finest Adelaide freestone (DENR and David Young. 2008, 6). The most important former quarries in the CBD are those on the south bank of the Torrens, at the Torrens Parade Ground and the Railway Station.

Chapter 7

The Northern Outpost

Darwin, Northern Territory, 1869

When the Northern Territory (NT) town of Darwin was bombed in early 1942, many of the already historic stone buildings survived. Even the stone walls of the heavily hit post office remained standing, despite the considerable loss of life. Just over thirty years later, several of those same buildings did not endure the far more destructive onslaught of Cyclone Tracy. Significant structural elements of the Anglican Church, the Town Hall, and the Administrators Office remained only as ruins, but the government residency, Browns Mart, and Victoria Hotel survived, despite the latter being unroofed for the third time by a cyclone. The often simple but strong stone buildings of old Darwin have proven more resilient than many more modern structures through its history and remain as defining elements of the built character of the modern city.

Darwin was founded by the South Australian colonial government in 1869, after administrative responsibility for a large swath of northern and central Australia was transferred from New South Wales. The region was largely unsettled by Europeans and was mostly unexplored. Originally named Palmerston, after the then British prime minister, the main purpose of the settlement was to serve as an administrative outpost of South Australia. The name was changed to Darwin on March 3, 1911, when the NT administration was transferred to the Commonwealth (Federal) Government (Surveyor General 1911, annotation on map originally dated June 1906).

The first two decades of the development of Darwin are the primary focus of this discussion. This period spans the arrival of the first Northern Territory Expedition in 1865, to the completion of the successful land surveys of 1869–1870 and the establishment of Palmerston. These events were followed shortly afterward by the arrival of the Australian Overland Telegraph (AOT) in 1872 and the discovery of gold at Pine Creek. Some reference is made to aspects of the subsequent growth and development of Palmerston/Darwin where relevant.

Modern Darwin remains the administrative capital of the NT of Australia, a region that was directly overseen by the Commonwealth government until it was granted limited self-government in 1978. It has not yet reached full statehood (Australian Government 2019). In 2019, the NT had a population of 245,869 of which over 148,500 were residents of Greater Darwin (Trade, Business, and Innovation 2019).

GEOGRAPHIC SETTING AND STUDY AREA BOUNDARIES

The City of Darwin is located on the north coast of Australia, at latitude −12.461049° longitude 130.841959° (Darwin GPO, Google Earth). It is Australia's only tropical capital and has distinct wet season (November–April) and dry season (May–October). The modern city area occupies approximately 112 km^2, and spreads eastward across the Shoal Bay Peninsula from the shores of Port Darwin, a major southern inlet of Beagle Gulf, which in turn opens westward to the Timor Sea (figures I.1 and I.2).

The focus of this chapter is the area of the town allotments surveyed in 1869–1870, which were laid out on three small terminal promontories at the western tip of the Shoal Bay Peninsula. The mesa-like promontories are bounded by cliffs along the western shoreline and mudflats to the east. In 1869, the land rose from sea level to a deeply incised plateau that reached its maximum elevation of 31 m near the modern Parliament House. The northern boundary of the surveyed town was defined by deeply incised drainage lines occupied by the southwest-draining Horse Gully, Doctors Gully, and the wetlands backing Cullen Bay and Mindil Beach to the west and northwest. These central allotments remain the focus of the modern Darwin CBD.

EUROPEAN EXPLORATION AND THE DISCOVERY OF PORT DARWIN

Macassans, from the Indonesian island of Sulawesi, began visiting the north Australian coast in the seventeenth century in search of sea cucumber, or Bêche-de-mer, a fashionable food item sought by the southern Chinese market. They established seasonal camps at several favored places along the coast and interacted with the local Aboriginals. There were conflicts that resulted in deaths on both sides (e.g., Spillett 1972, 40–41) but relations were also cordial. The trade that took place from such interactions was the first between Australia and the outside world.

The north and west coasts of Australia were the first parts of the continent to be encountered by Portuguese, Dutch, and British explorers, who compiled preliminary charts of the north coast. Modern charting of the north began with James Cook and Mathew Flinders, and particularly by Phillip Parker King, who surveyed the north and northwest coasts in 1818, when he completed the coastal surveys begun by Flinders (Lee 1906, 200–201).

None of the early explorers sought to establish permanent settlements on the north coast until Britain decided to enforce its colonial claim on Northern and Western Australia in the early nineteenth century. British settlements were established to deter perceived risks of Dutch and French settlement, but also with the concept that they might become trading centers with nearby Southeast Asia. There were three early attempts at settlement, which all failed because of poor logistical support and the difficult tropical climate. All were military outposts manned by the British Army and Marines, with Royal Navy input, and some families. Fort Dundas, on Melville Island, founded on September 26, 1824, was an apparently ill-chosen site and the adverse reports of the second commandant led to the decision to close it. It remained open until a new settlement was established on the mainland at Raffles Bay in June 1827 and named Fort Wellington. It remained until August 31, 1829 but was abandoned just as it began to meet the original expectations, such as attracting trade with trepangers.

The third establishment, at Port Essington on the Cobourg Peninsula, and known as Victoria Settlement, existed from 1838 until 1849 (Spillett 1972, 11–18). Despite its relatively short life, Port Essington was an important base for several significant surveying and scientific voyages and overland exploratory expeditions that finalized the charting of the Australian coastline. It was visited by HMS *Beagle* in July 1837, on the third surveying voyage to Australia (after the more famous second voyage accompanied by Charles Darwin), becoming the starting point for the survey of the north coast of Australia during which Port Darwin was discovered and named (Spillett 1972, 48–49). Victoria Settlement was also visited by HMS *Fly*, during the four-year survey of the Great Barrier Reef and Torres Strait beginning in 1842, and by HMS *Rattlesnake* in November 1848, while naturalist and future Darwinism defender Thomas Henry Huxley was on board as assistant surgeon (Goodman 2005, 12–14; 196–198). Port Essington was also the end point of Ludwig Leichardt's overland northern Australian expedition from Moreton Bay in 1845.

Port Essington was finally abandoned when it was realized that the original claims for its potential had been exaggerated. There was no commercial development because it was isolated from shipping routes and had a relatively poor anchorage. Ships rarely called there, and no shipping companies ever intended to use it as a coaling station, or for any other purposes. It was considered unsuitable as a penal settlement, and unlikely to be a haven for shipwreck survivors. The British eventually became secure enough of their sovereignty of northern Australia to no longer consider a physical military presence necessary (Spillett 1972, 164–166).

Discovery of Port Darwin

Port Darwin is a deeply incised and drowned river valley, which was first surveyed and named by John Stokes in September 1839, during the third five-year exploratory voyage of the survey ship HMS *Beagle,* under Commander John Wickham. Leaving Port Essington on September 4, 1839, Stokes discovered Port Darwin as he investigated an unexplored bay sighted by Phillip Parker King during his 1818–1822 voyage. From Shoal Bay, he took a whaleboat and four days' provisions to explore the new bay, landing at East Point on the first night. From the top of East Point plateau (see figure 7.1), in the first light of the September 9, 1839, Stokes and Forsyth spotted "a wide bay appearing between two white cliffy heads and stretching away within to a great distance presented itself to our view. Far to the southward, between the heads rose a small table-topped hill" (Hordern 1989, 209). This was the first European sighting of the future position of modern Darwin.

They proceeded south to the next headland where they climbed to the highest point to get an idea of the full extent of the harbor. This landing was probably at the location of modern Darwin. They rowed to the adjacent headland, subsequently named Talc Head, where Stokes observed geological features that reminded him of his friend and former shipmate Charles Darwin, so he named the new harbor after him (Powell 1986, 1). Observing the extent of the harbor from this place, Stokes decided that he needed the resources of the ship to fully explore it, so returned to HMS *Beagle.* After first grounding at East Point, the ship was sailed into the harbor and anchored just to the south of the headland where they had previously landed. Boat parties were sent out along each arm of the harbor to explore and chart it and Lieutenant Emery went in search of fresh water. Emery chose a depression just behind the beach of a nearby bay and began excavating a well. After five days of digging, sufficient freshwater was encountered at a depth of 10.3 m (34 feet) to water the ship (Hordern, 212). The peninsula was named after Lieutenant Emery as a result (Powell 1986, 1) (figure 7.1).

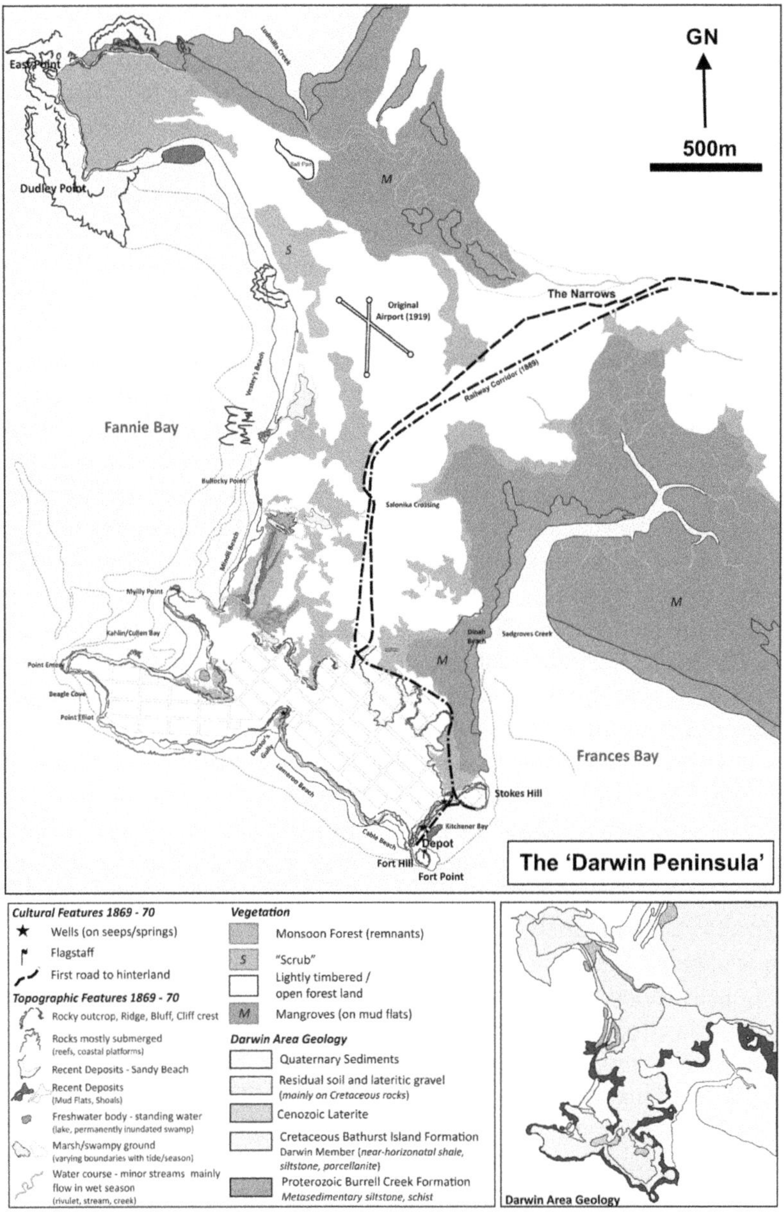

Figure 7.1 The Darwin Region. The topography of the Port Darwin region c 1869-70. The landscape was a deeply incised plateau formed by flat lying Cretaceous sediments that unconformably overlay a deformed and metamorphosed Proterozoic basement. The steep drainage lines of the plateau were infilled by Quaternary and recent sediments and were occupied by wetlands and mangroves along their lower courses. **Inset 1. Geology.** The geology is based on Pietsch (1983). Please refer to the extended caption at the rear of the book for additional information.

Transfer of Administration to the Colony of South Australia

From the 1850s onward, South Australia and Queensland were rivals to become the administrator of the land area that is now the NT. There was a strong belief that a northern port could become a trading outlet to Asia for Australian goods, with the potential to become a second Singapore (Cross 2011, 2). A northern settlement might also become the location of important communication infrastructure, which would directly link Australia to the rest of the world via overland and seafloor telegraph cables. The wealth generated by any such settlement would greatly benefit the governing colony.

Adelaide was an important communications hub by the 1860s, thanks to the construction of a telegraph line to Melbourne. Port Adelaide became the first port of call for ships arriving from Europe, and news was summarized there, and then transmitted to the rest of Australia via the Melbourne telegraph. Any new telegraphic link between Queensland and the north coast would change the status of Adelaide from the first Australian city to receive business news to the last and the South Australian government wanted to maintain its role as an information hub. One way to do this was to gain control of all the land between South Australia and the north coast (Cross 2011, 5–7). The South Australian colonial government had a bold plan to create the Australian communications gateway to the world, with Palmerston hosting the onshore connection via an overland telegraph.

Both colonial governments sent exploratory expeditions to find potential telegraph routes to Port Darwin to strengthen their cases. In 1861, John McKinlay was sent northeast by the South Australian government to search for missing explorers Bourke and Wills and to then explore for a route to the north coast, reaching the Gulf of Carpentaria on May 20 (Cross 2011, 13). A feasible route from Adelaide was discovered after several attempts by John McDouall Stuart in 1862, who was also sent by the South Australian government (reference). Stuart also finally proved that there was no inland sea. Victoria also sent an expedition to the north coast in 1860, led by Robert Burke, which famously failed on the return journey (Cross 2011, 13).

South Australia won the intercolonial competition to govern the vast area in 1863, and the NT was transferred from New South Wales jurisdiction to South Australian administrative control. The transfer of the NT to South Australia was largely the result of John Stuart's successful return journey across the continent in 1862, and the overland route he found for the telegraph across central Australia, from Adelaide to the north coast. This tipped the decision in favor of South Australia (Pugh 2019, 1). The land area became known as the NT when it became as the "northern territory" of South Australia, but only after there was considerable discussion about renaming the entire administrative unit "The Great Central State" (e.g., Cross 2011, ix; 17–18).

The competition continued between the rival colonial governments to become the Australian business telecommunications hub with the outside world. Queensland reached an agreement with the British Australian Telegraph Company (BAT) to connect their line to the Queensland network via Burketown. However, the South Australian government won the race by offering to fund the construction of the overland line from Port Augusta to Port Darwin, across 2,897 km of largely unexplored central Australia, and have it in place by the time the BAT landed their submarine cable from Java on December 31, 1871 (Pugh 2019, 68). This would save the BAT from the expense of laying its own cable from Port Darwin to Burketown, which had previously been agreed to with the Queensland Government (Pugh 2019, 77).

The First Northern Territory Expedition—1865–1866

The South Australian colonial government rapidly set about establishing control of the territory by creating an administrative settlement on the north coast. The model chosen to settle Darwin was like that used three decades earlier, when Adelaide was settled directly from Britain (see chapter 6). Land allotments were presold (there were 1,400 orders from London and Adelaide) and the funds generated were used to pay for the surveying expedition (Pugh 2018, 39).

The initial attempt to create a permanent South Australian outpost on the north coast was known as the first Northern Territory Expedition. It was led by Colonel Boyle Finniss, who had originally come to South Australia with William Light's survey teams in the 1830s. They later formed the business partnership of Light Finnis and Company (Light and Elder 1984, 50, see chapter 6).

Finniss was influenced by Stuart's report on the country around the Adelaide River and chose a site near the river mouth known as Escape Cliffs, on the eastern shore of Adam Bay (Cross 2011, 80; 86–87). Finniss' orders required him to avoid swamps, mudbanks and landlocked harbors, and he felt that the elevated position would provide a better climate than the unhealthy mangroves (Pugh 2018, 27). However, Escape Cliffs turned out to be badly chosen and poorly accessible. The team there was plagued by infighting, which the subsequent government inquiry attributed to Finniss' poor leadership. The site was abandoned, and the team recalled to Adelaide.

In 1865, John McKinlay was again sent north by sea to explore for a better settlement site than Escape Cliffs. He explored the region westward in two stages, and under great difficulties. On his return to Adelaide, he reported favorably on the Port Darwin region (Cross 2011, 13).

The Second Northern Territory Expedition—1869–1870

The aborted attempt to establish a settlement at Escape Cliffs caused delays in the speculative settlement of northern Australia and resulted in litigation by land purchasers. A new expedition to establish the capital of the newly acquired territory was rapidly organized by the South Australian government in 1869. It was known as the Second Northern Territory Expedition and was led by the surveyor general, George Goyder. This successful venture resulted in the creation of modern Darwin, which like Finniss' earlier camp, was named Palmerston (Pugh 2018b). See section below for further details.

Goyder's first camp was at Fort Hill Point, where an orderly tent camp and a series of well-built wood and iron buildings were established on the saddle at the northwestern end of Fort Hill Point (figure 7.2). This camp was to remain the center of the settlement between 1869 and 1872. The camp formed a base for the survey of the first town allotment land surveys that had been presold to investors in Adelaide and elsewhere. Goyder and his team found room on the restricted peninsular site for the required number of 999 rectangular town allotments (Pugh 2018, 39), which set in place the pattern of the subsequent historic development of Darwin and the modern CBD.

Parties then worked systematically outward from this center of operation as they surveyed agricultural allotments in the hinterland, which often required working from temporary satellite camps. Three other satellite town sites were chosen and surveyed by Goyder's party: South Port, Virginia, and Daly. Only Southport grew into a significant settlement, after it began to be used as a landing place to service the developing mines of the hinterland, which greatly reduced the road haulage distance to the goldfield.

The successful land surveys were followed by a period of intense land speculation, and the successful completion of the AOT line. The telegraph line from Port Augusta was connected to the seafloor cable of the BAT at Port Darwin on August 22, 1872, after crossing some of the most rugged and unexplored country of central Australia (Pugh 2019, 77). The link to Java, and then Singapore created the first real-time communications link between Australia and the rest of the world.

Unfortunately, while the telegraph did cement Palmerston's position in world communications, it did not lead to an economic boom. Port Darwin did not ultimately prosper as a trading center to Asia and became a financial burden on the small colony of South Australia. Only 40 of the 999 surveyed blocks were occupied in the first few years and many of the town blocks surveyed by Goyder remained vacant and undeveloped. This was partly because land was tied up with absentee property speculators (one landholder held 100 undeveloped blocks) and this stifled the business and commercial development of the settlement. For many years, the only permanent buildings

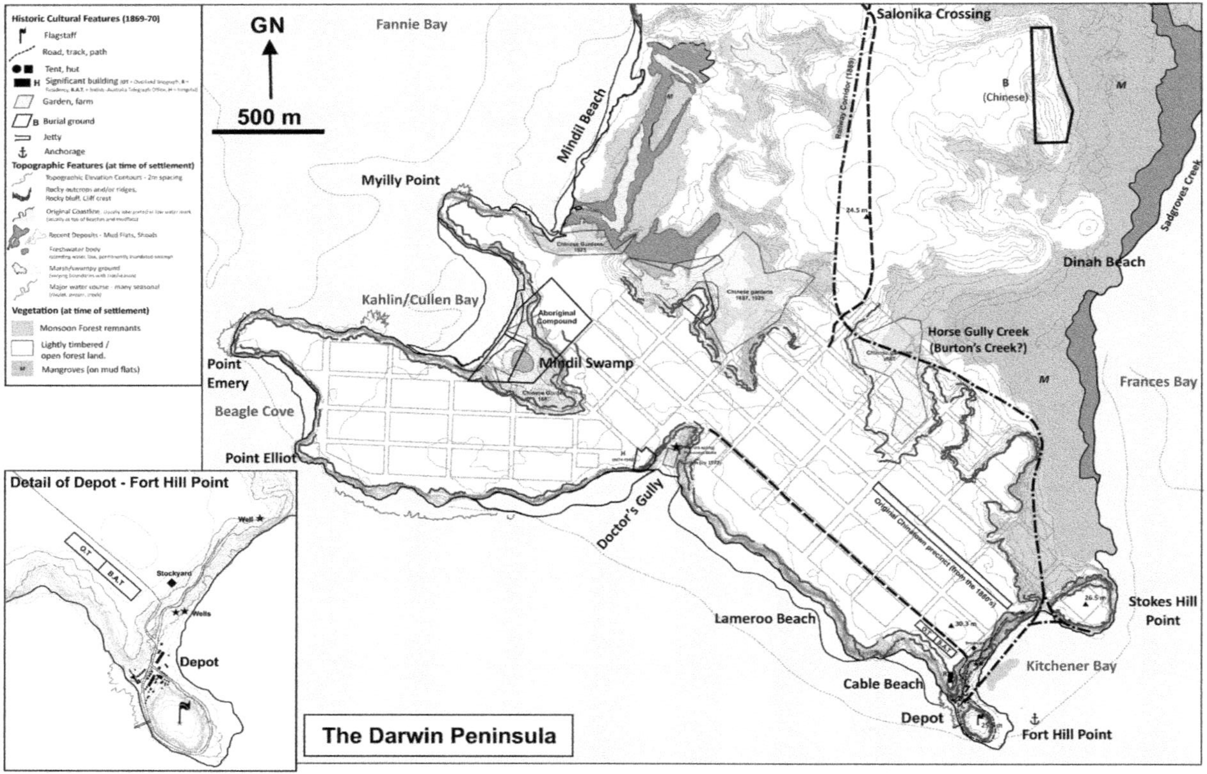

Figure 7.2 The Darwin Settlement. The topography of the Palmerston settlement (Darwin) at the time of South Australian settlement in 1869. The landscape chosen was a drowned mesa comprised of mostly horizontally bedded Cretaceous sediment overlying a Proterozoic basement. This reconstruction is mostly based on the survey plans of George Goyder (1869). The contours are at 2 m intervals and derived from 3D topographic reconstructions. Please refer to the extended caption at the rear of the book for additional information.

in Palmerston were those of the government residency and the BAT headquarters (Pugh 2018b, 110–111). Administration of the NT was officially transferred from South Australia to the newly formed Commonwealth Government on January 1, 1911 when the latter agreed to allocate £6,170,548 to refund accumulated debt, and to purchase infrastructure assets (Zoellner 2017, 45).

THE GEOLOGY OF PORT DARWIN

Port Darwin lies at the northern end of the Pine Creek Orogen, a complex of deformed and metamorphosed Proterozoic sedimentary and volcanic rocks that have remained relatively stable since being folded and uplifted during the Paleo- to Mesoproterozoic (Nott 2003, 48). Following erosion, the Proterozoic rocks were uncomfortably overlain by a mantle of mostly flat-lying Cretaceous sedimentary rocks (Nott 2003, 48) that have since been deeply weathered, leached, and lateritized. The town of Palmerston was laid out on a narrow, table-topped plateau formed mainly by an erosional remnant of the regionally extensive Cretaceous mantle.

Paleoproterozoic Basement

The basement rocks of the settlement area are mainly metasedimentary schists, siltstones, and sandstones of the Burrell Creek Formation that were deposited between c 1,855 Ma and 1,829 Ma (Ahmed and Hollis 2013, 5:5), with those beneath central Darwin mainly comprised of steeply dipping phyllites (McNally, Clarke, and Weber 2000, 36).

Basement rocks are poorly exposed in the Darwin settlement area where they mostly outcrop on shore platforms, and in the bottom of deeply incised drainage lines that have cut through the Cretaceous cover at the edges of the plateau. Basement rocks are exposed in the cuttings on the lower slopes and base of Stokes Hill, where extensive outcrops are visible on the southern cutting faces and along Stokes Hill Road and McMinn Street (where they have been exposed by road and rail cuttings and port works). Proterozoic rocks also formed the lower third of Fort Hill, the western half of which was originally comprised of vertically dipping shale and the eastern half of which was comprised of vertically dipping silty shale and impure sandstone (Gardner and Rix 1963, 9 and Plate 2). South of Darwin, the basement is well exposed and is the dominant influence on the landforms.

The metamorphic grade of the basement rocks varies across the Darwin region. On the Cox Peninsula to the west, the rocks are upper greenschist to amphibolite facies quartzo-feldspathic and mica schist, gneiss, and minor quartzite. To the east, near Gunn Point, they reach lower greenschist facies (Nott 2003, 84).

The Pine Creek Orogen was deformed at approximately 1800 Ma, which generated tight folds with limbs dipping >50 degrees (Pietsch 1983, 1). In Port Darwin, the basement rocks are folded and cleaved along dominant northwest-southeast axes and the geometry of the folded strata strongly influences the landforms of the regions where they outcrop, forming linear ridges on resistant strata and low-lying and swampy areas where they have been deeply eroded. The geomorphology of the upper reaches of Port Darwin, including Middle and East Arms, is largely controlled by the folded geometry of the basement strata. The trends of strata are visible as lineaments in geological and topographic mapping, and in aerial imagery and lidar, and in the geometry of the many tributaries that drain into the harbor. Bathymetric surveys of the head waters show that the same style of tight folding is preserved by the rocks forming the bed of the recently drowned river valley that forms the harbor.

The Angular Unconformity

An angular unconformity separates the near-horizontal Cretaceous cover rocks from the Proterozoic basement. It extends beneath the original settlement area and represents an 1,800- to 2-billion-year time gap (McNally et al. 2000, 36; Nott 2003, 83). The unconformity surface is a gently undulating, mostly low-angle surface that becomes gradually deeper to the north of the CBD toward Nightcliff (Clarke et al. 1999, 2–912). It lies at approximately 2–3 m above sea level at Point Emery but dips gradually below sea level toward East Point, before rising again to 4 m above sea level at Bullocky Point (Nott 1994, 409).

It is exposed near the base of the western coastal cliffs and headlands of the city area and as shore platforms at the intertidal level. It is particularly well exposed in modern road (and former railway) cuttings on the western side of McMinn Street, along Tiger Brennan Drive, and high on the faces of Stokes Hill. It was formerly well exposed by cuttings at the base of Fort Hill, where it dipped shallowly to the west (Gardner and Rix 1963, 9 and plate 2).

Despite being poorly exposed in natural outcrops, the unconformity was an important geological factor in the formation of the 1869 settlement landscape. It strongly influenced the geometry of the twin headlands, the location of the first camp, and the locations of early groundwater supplies and gardens. The sea cliffs that form the edge of the plateau facing Port Darwin and Beagle Gulf are being cut into the easily eroded Cretaceous rocks by tidal action sapping at the level of the unconformity, which outcrops near sea level here. The coastal platforms are formed by the Proterozoic basement below the level of the unconformity surface (e.g., Nott 2003, 83). The 1869 survey camp was established on the eroded unconformity surface forming the saddle at the northern toe of Fort Hill (figure 7.2 inset).

The springs that Lieutenant Emery of HMS *Beagle* exploited to water his ship were probably at the unconformity surface. The first successful well of 1869 was sunk where Goyder's party found water "oozing" from the rocks of the cliffs in a small drainage line that became known as Doctors Gully (Goyder 1869c, 7). This spring was also formed at the unconformity surface. Subsequent wells were sunk at other sites along the cliffs, at or near the unconformity. And many of the seasonal creek that drained the plateau were fed throughout the year by springs at the unconformity. In the Darwin dry season, the groundwater emanating from the unconformity surface was a very valuable resource.

Cretaceous Cover Sequence

The Darwin plateau is mostly underlain by horizontally bedded to gently dipping Cretaceous sedimentary strata and these rocks dominate the coastal landforms. They are comprised of mainly bedded mudstones and siltstones of the Darwin Member of the Cretaceous Bathurst Island Formation, the lowest formation of the Bathurst Island Group. The sediments were deposited in a shallow marine environment with radiolarian claystone, sandy clay stone, glauconite sandstone, and a basal conglomerate (Nott 1994, 408). A basal conglomerate marks the unconformity surface and the beginning of the Cretaceous sequence in most places. It is defined by a narrow quartz pebble conglomerate layer, which is 0.6 m wide where exposed at the former Larrakeyah Quarry (McQueen 1957, 1).

The bedded Cretaceous rocks dip gently north (McNally et al. 2000, 35), where they become thinner. They are well exposed in coastal cliff sections, which were the focus of early geological investigations that identified radiolarians and ammonites (Nott 2003, 84). The Darwin Member also contains common *Ichthyosaur*, belemnite, and ammonite fossils, though the latter are poorly preserved (Skwarko 1966, 54; Murray 1985; Nott 1994, 412).

The physiography of the mouth of the Port Darwin is largely controlled by the weathering and erosion of the Cretaceous cover sequence, and the modern deposition of mudflats and beaches at their base. Palmerston was founded on a partially drowned mesa formed by the flat-lying Cretaceous sedimentary rocks. The Cox Peninsula, southwest of Darwin, and the western harbor shores are all formed in these rocks. Precambrian rocks below the unconformity generally form the shore platforms, but the gentle northward dip of the Cretaceous strata resulted in platforms developing on the cover sequence at Nightcliff.

Deep Weathering and Lateritization

The rocks of the Darwin plateau were deeply weathered and lateritized during the present and past tropical climates. The Cretaceous sedimentary rocks have been particularly affected and the products of weathering processes are widespread and varied. A bioturbated bed in the Darwin Member limits the penetration of weathering in the Darwin area, and where it dips toward sea level, weathering is greater (Nott 1994, 411).

The Darwin plateau and most of the Shoal Bay Peninsula are underlain by a distinctive siliceous duricrust known as porcellanite, which is best developed over Cretaceous strata. Porcellanite is a produced by the leaching and replacement of the pallid zone of the preexisting laterite profile (Piesche 1984). It was formed where the pallid zone of the lateritic profile was exposed and intense silicification took place (Whitehouse 1940). Exposure, weathering, and erosion of this profile gives rise to two products: weathered laterite and porcellanite (Pietsch 1983, 18). Porcellanite is several meters thick and passes downward into bleached and weathered sedimentary rocks and becomes weaker with depth. Porcellanite is patchy elsewhere in the Darwin area and is masked by laterite (McNally, Clarke and Weber 2010, 35).

A detrital laterite formed mainly from reworked material, and cemented by a ferruginous matrix, occurs locally. It outcrops as low benches fringing drainage depressions and marine flats in dissected plains over Lower Proterozoic rocks and is probably early Quaternary in age (Pietsch 1983, 18).

Quaternary and Recent Sediments

The bedrock of Darwin is very ancient, but the modern sequences are very young, with the modern coastal landscapes of Port Darwin only forming in the last 2,500 years, by marine and terrestrial sedimentation (Woodroffe and Grimes 1999, in Bourke 2005, 29).

The occurrences of Quaternary sediments are of limited extent in the Palmerston settlement area, compared to the extensive occurrences in the Adelaide and Alligator River regions to the east. The extensive marine and alluvial deposits record Holocene sea-level rises and stabilization and changing estuarine environments (Nott 2003, 83). On the Darwin plateau, Quaternary sediments occupy structurally controlled, incised gullies on the margins of the Cretaceous mesas where they host drainage lines from the plateau. Despite their limited extent, these drainages were liable to inundation in the wet season and set topographic limitations on the area available for Goyder's original town allotments, so became a significant influence on his street layout design of Palmerston.

THE DARWIN LANDSCAPE OF 1869

The topography of the Port Darwin region is low-relief and mature, with the current physiography formed in the mid- to late Cretaceous, but with a deep lateritic profile formed during subsequent chemical weathering (Pietsch 1983, 1). Port Darwin is a drowned river valley, and the bordering high ground is a deeply incised plateau formed mostly on flat-lying Cretaceous sedimentary rocks overlying Proterozoic basement. Many prominent hills of the settlement area are erosional remnant mesas of the Cretaceous rocks.

The 1869 camp was established at the western tip of the Shoal Bay Peninsula, an intricate landscape of headlands, bays, inlets, and mudflats bordered to the southwest by Port Darwin and to the north by Shoal Bay (figure 7.1). Three prominent terminal points mark the western extremity of the Shoal Bay Peninsula. The most northern is East Point, which marks the entrance to Port Darwin and was named during the 1839 survey of HMS *Beagle* (because it was the most easterly point of the harbor). Two promontories to the south form a hammerhead-shaped peninsula with a western arm that is terminated by Point Emery, Point Elliot, and Myilly Point; and a southeastern arm that ends in Fort Hill Point and Stokes Hill Point. The hammerhead-shaped peninsula is the most historically significant landscape of the 1869 settlement because it is where George Goyder chose to locate the Palmerston settlement. For the purposes of the following discussion, this hammerhead-shaped peninsula will be referred to as the Darwin Peninsula, an informal name used by previous workers such as Wells (2003).

The southeastern arm of the Darwin Peninsula is a plateau formed by a mesa rising to between 18 m and 30 m (60–100 feet) above sea level and bordered on three sides by the ocean. It is deeply incised, and bordered by extensive mudflats to the east, at the head of Frances Bay, and by steep erosional sea cliffs to the west and south, which fall 20–30 m vertically to sea level (figure 7.2). Cliffs also line the shore of the western arm, and continue to the northward, to beyond East Point. They are embayed by the variably sediment-infilled inlets of Doctors Gully, Beagle Cove, Cullen Bay, and Mindil Beach. Coastal erosion continues to undercut the overlying Cretaceous sequence along this coastline and occasional cliff-face failures occur.

The western and southeastern headlands are only joined to their hinterland plateau by a narrow neck of high ground that is constricted at three points by deeply incised drainage lines and low-lying and sediment-filled inlets. The inlets are backed by dunes and wetlands, narrow sandy beaches, cliffs (mainly on the western shore), and tidal mudflats and mangrove forests (mainly on the eastern side). Occasional patches of rainforest occupied the deeper drainage lines entering these wetlands. The most significant constriction is located between the head of the Mindil Beach wetland gully system, draining westward into Fannie Bay, and the head of the inlet occupied by Horse Gully Creek, draining eastward to Frances Bay. This narrow point is now occupied by the road intersection where Daly Street becomes the Stuart Highway (figure 7.2). This neck is used here to define the northern boundary of the Darwin Peninsula.

There is a second narrow neck to the northeast, between the northern arm of the Mindil Beach wetlands and the two easterly draining creeks at Salonika Crossing. Further again to the northeast are two major sediment-filled valleys occupied by Sadgroves Creek and Ludmilla Creek and their rimming mangroves, wetlands, and rainforest. Their deeply incised valleys create a third neck of high ground along which all land transport corridors are focused (figure 7.2).

Fannie Bay and Frances Bay

Fannie Bay and Frances Bay are both major embayments bordering the Darwin Peninsula, with Fannie Bay to the west, and Frances Bay to the east. The shorelines of each bay are distinct. Fannie Bay borders the northwestern shore of the peninsula, and the coastline faces the open ocean of Shoal Bay. There are few mangroves (rare, isolated stands) and the coastline is comprised of sea cliffs, rocky shore platforms, and sandy beaches in sheltered positions. Dune-backed beaches face Fannie Bay, including Mindil Beach and Vesteys Beach, separated by the subdued headland of Bullocky Point. Narrower beaches such as Cable Beach and Lameroo Beach, occur at the foot of the cliffs of the central Darwin and Beagle Cove, and at the mouth of Doctors Gully. Cullen (or Kahlin) Bay occupies the coast between Point Emery and Myilly Point, as a deep-set inlet bordered by their rocky headlands and cliffs. Cullen Beach, Mindil Beach, and Vesteys Beach further to the north are all embayments formed on drowned drainage lines from the plateau. The sandy beaches were originally backed by dunes, behind which were saltwater and freshwater wetland systems that received and filtered the runoff from the creeks draining the plateau along incised drainage gullies.

The sheltered eastern shore of the peninsula is very different to the western shore. It is generally lower and shelves more gently to sea level on Frances Bay and the mouth and estuary of Sadgroves Creek. The shores are very shoaly and occupied by wide mudflats hosting extensive stands of mangroves in their lower levels. There are similar deeply incised gullies draining the plateau as those on the west coast, but they are now infilled by the mudflats.

Historic Landforms of the Settlement Area

The plateau of the Darwin Peninsula originally reached a maximum height of approximately 31 m at the southeastern end, behind the modern Parliament

House precinct. There were several other low rises on the plateau, one near the intersection of Knuckey and Mitchell Streets (28 m) and another low rise on the northeastern flank (22.5 m), which Goyder originally reserved as Cavenagh Square. Many of the high points were reduced and leveled during subsequent development.

Fort Hill Point and Stokes Hill Point

Despite its subtle topography, several landscape elements featured prominently in the first years of Darwin's European history. Of note are two small and isolated hills formed by outlying erosional remnants of Cretaceous sediments that capped the southeastern end of the Darwin Peninsula. They became known as Fort Hill Point and Stokes Hill Point.

Fort Hill Point received its name during the 1864 survey of Captain Hutchison of HMS *Beatrice*, who considered it to be a "fortress-looking headland" (Hutchison and Howard 1864, cited in Wells 2003, 112), a description seemingly later formalized by George Goyder. Fort Hill Point was the site of Goyder's surveying camp, which was built on the lower northern slopes and the neck that joined it to the mainland. Goyder decided that this site on the saddle provided a more amenable climate than the top of the plateau (Goyder 1869). Fort Hill reached a height of 26 m and became the site of the first flagstaff. It was later chosen as the place to bury the first two Europeans to die at Port Darwin.

The narrow, 27 m high peninsula on the mainland side of the neck opposite Fort Hill was later chosen for the site of the government administrator's residence. It was completed in 1871 and is now the location of Government House. The Cretaceous rocks forming the majority of Fort Hill were completely removed for port developments in the 1960s and the level land in the area is now formed on the basement rocks. However, Goyder Park has recently been established to mark the location of the survey camp and will become an important open space in the redeveloped wharf precinct.

Stokes Hill is the table-top hill forming the larger point to the northeast of Fort Hill Point. It was named after the Commander of HMS *Beagle* by Surveyor Patrick Auld, during his short 1865 overland expedition from Escape Cliffs, where he met up with Finniss in HMS *Beatrice* (Pike 1959, 333; Pugh 2018a, 59–60; Pastmasters 2019). Stokes Hill rose to a height of 27.7 m (91 feet) above average high water (1925 plan). The site of the first jetty and camp on Fort Hill Point was exposed to the open ocean of Beagle Gulf. Sheltered anchorages that were close to shore existed in the lee of Stokes Hill Point, which led to it becoming a focus of early port facilities, and eventually to it becoming the main terminus for the North Australian Railway infrastructure. Stokes Hill survives, though its peripheral slopes have been heavily modified by cuttings, to make way for port infrastructure, and for rail access to the wharfs (including a spur line to Fort Hill along the foreshore of Kitchener Bay). The wartime developments in the area had a particular impact, including the fuel bunkers that were excavated into the hill for air-raid protection.

Doctors Gully

Doctors Gully is a shallow, incised drainage line on the southwestern shore of the main peninsula and was the location of the first freshwater well sunk in 1869. Doctor Peele was placed in charge of developing the well, which was excavated via a slit trench in an embankment to a depth of 3.3–4.25 m (11–14 feet) before reaching water (Goyder 1869c, 10). Peel went on to create the first vegetable garden at the wellsite, which was an important source of fresh vegetables and fruit for the survey camp (figure 7.2).

The first Darwin hospital was built nearby in 1874, possibly because of the availability of fresh water from the well and fresh produce from the nearby garden. It was built using funds from public subscription, on a prominent, airy position on the high ground at the base of Point Emery. In the early twentieth century, nearby Myilly Point, which forms the northern headland of Cullen Bay, became an important location for housing senior government officials. These homes now form an important historic precinct.

Wetlands and the Drainage System

The Darwin Peninsula had an extensive and complex wetland and drainage system in 1869. It supported multiple habitats along water courses that stretched from the plateau to the sea. Most water courses were ephemeral, flowing only in the wet season. Permanent surface water only occurred where the creeks were fed by small spring-fed soaks formed at the unconformity surface, and from joints and faults in the Cretaceous cover. These creeks supported important microhabitats hosting small stands of monsoonal rainforest at their heads, but then progressed downward through systems of marshlands, swamps, and mangroves, to the sea. The most important water courses in 1869 were Horse Gully Creek (now Burtons Creek) and Doctors Gully, with the larger Ludmilla Creek and Sadgroves Creek further to the north.

The drainage channels changed in character as they passed from the Cretaceous rocks to the basement. In the Cretaceous cover, the channels were deeply incised and steep-sided, but became broader and shallower on encountering the unconformity surface. Most of the embayments and beaches along the shoreline of the Darwin Peninsula were incised drainage systems that had cut down through the flat-lying Cretaceous rocks of the high ground until encountering the more resistant basement rocks beneath the unconformity surface (e.g., Mindil Beach and the chenier plain behind). All are now filled with Quaternary to recent marine and terrestrial sediments in their lower courses.

Shortly after establishing the camp, a small creek was discovered that drained spring-fed wetlands at the head of a broad, shallow gully on the northeastern side of the plateau. It was the first place where stock could be watered from a surface source and was named Horse Gully Creek. This area became a focus of early horse and cattle grazing (figure 7.2).

To the northwest and north of the main settlement, an extensive system of gullies opened out into a large chenier plain system behind the dunes of Mindil Beach, and onto wetland flats behind the dunes at Cullen Bay. The locations and geometry of these wetland systems defined the northern boundary of the township of Palmerston. Goyder recognized that the gullies and wetlands, with their large timber and thick rainforest undergrowth, were major drainage lines that also became inundated during the wet season. He therefore omitted them from the town allotment survey and reserved the land for other purposes, including cemeteries and gardens (Goyder 1869, 3–4). The reserved land formed the northern limit of the 1869 town survey, and the now drained wetlands continue to define the northern limit of the modern Darwin CBD.

The reserved wetlands of the gullies became the focus of recreational and sporting facilities (cricket grounds), and other open-space recreational facilities, as well as early cemeteries. A particular modern focus of sporting infrastructure is the site of former chenier plain wetlands behind Mindil Beach. The George Brown Darwin Botanic Gardens were established on a site where three gullies draining the plateaus fed the wetlands at the rear of Mindil Beach, one of the largest occurrences of Quaternary sediments in the city area.

1869 Vegetation and Environment

Darwin lies within the wet-dry monsoonal tropics, with 90 percent of the annual rainfall occurring during the wet season between November and April. It experiences spectacular storms and monsoonal deluges alternating with dry-season droughts (Van Oosterzee 2017, 1). The coastal environments are very recent. Formed mostly in the last 7,000 years, some coastal features, such as the dune-backed beaches and wetlands, are as young as 700 years (Van Oosterzee 2017, 3).

The Darwin settlement area has an environment, habitats, flora, and fauna that is typical of much of the northern NT. It is located within a belt of Eucalypt-dominated woodland and open forest that extends from the

Kimberley in the west, to eastern Queensland (Van Oosterzee 2017, 19). It is not homogeneous however, and there are stands of monsoonal forests composed of species that have arrived from nearby Asia and adapted to the wetter areas and seasonal climate. These stands also contain occasional species remnants of the formerly widespread temperate Australian rainforests that were present before the continent began to dry in the Holocene (Van Oosterzee 2017, 3).

The vegetation of the Darwin Peninsula was mostly open eucalypt woodland, described by George Goyder (1869, 3–4) as being comprised of paper bark (some large), palms, fan and fern, screw pines, iron bark, gum, stringy bark, fig, cedar, cotton, and various other trees and shrubs. The character of the woodland remained similar as far east as Freds Pass. Locally, the density of vegetation increased in a 5–18 m wide belt fringing the clifftops facing Shoal Bay, where it formed a dense thicket of "various sized timber, matted together by bamboo, convolvoli, volvoli, and a variety of other vines and shrubs." There was a thick grass cover in the understory. The woodland was more open in some areas and particularly on Point Emery and Point Elliott, where the clifftop thickets were absent (Goyder 1869, 3–4). Microhabitats existed in the wetter drainage gullies and in the wetlands forming their lower reaches near the dunes and mangroves at the shoreline. Stands of rainforest occupied the more deeply incised gullies and one remnant still exists as the Duke Street rainforest, at the head of Burtons Creek.

The vegetation and environment of Darwin at the time of first European contact is well recorded because, unlike other Australian settlements, Mr. Schultze, a professional naturalist, accompanied Goyder's expedition and sent large collections of plants, seeds, and animal skins to Adelaide. Goyder was also a skilled observer and provided accurate descriptions of both the geology and vegetation of the country in the vicinity of Fort Hill Point in his reports to the SA Parliament. He also verified Litchfield's earlier discovery of alluvial gold (Goyder 1869, 4–5).

The Aboriginal Landscape

When Goyder's survey party arrived, he estimated that there were about sixty people living near the camp (Goyder 1869b, 3). The Shoal Bay Peninsula was the home of the Larrakia, or Saltwater People, originally known by their language group name of Gulumirrgin. Their lands extend from the Finniss River and Fog Bay northeast along the coat to Gunn Point west of the Adelaide River and inland to a point 10 miles north of Rum Jungle (Tindale 1974, 230). The Larrakia knew the Port Darwin region as Garramilla, a reference to the distinctive white rock of the cliffs and shoreline (Larrakia Development Corporation 2019; NT Place Names Register 2019a). To the east were the

Wulna, based around the Adelaide River, and closely linked by intermarriage to the Larrakia. To the west were the Wogait, and to the south were the Awarai (Parkhouse 1895, 1; Tindale 1974, 221 and 238).

Group territories were small compared to those further inland because the people prospered on the easily gathered and rich food resources of the Darwin region (Pugh 2018a, 2–3). The boundary between the Larrakia and their neighbors was a "neutral" zone approximately 8–10 miles wide in which no camps were created or food gathering took place. Groups could travel through it with permission and use its resources as they traveled (Parkhouse 1895, 2).

Within this landscape, individual families owned their own territories and a camp on Lameroo Beach was occupied by one family for generations (Parkhouse 1895, 2). After 1869, and probably before, family groups congregated at Port Darwin during the wet season but dispersed to other camps throughout their territory during the dry season.

In the 1890s, several camps on the Darwin Peninsula were focused near the wetlands at the northern end of the town area, where they were organized into tribal and family groups. The main Larrakia camp was at the head of Cavenagh and Smiths Streets and comprised several circles of three to seven wurleys (egg-shaped domed huts). Another camp, 400 m to the west was the home of the Daly family. And approximately 150 m to the north of the main Larrakia camp, there was another group of three circles of wurleys where Wulna people related or allied to the Larrakia lived, with each circle the home of one family. Another Wulna camp was in the scrub on the northeast side of Cavenagh Street and was home to people unrelated to the Larrakia (Parkhouse 1895, 2). The focus of encampments around these wetland areas predated European settlement. Captain Howard encountered a camp at the wetlands here (probably those behind the Cullen Bay beach) in 1865, a short distance inland from Emery's well and Gundalagwa. A living and ceremonial place on Point Emery, which lay on a Dreaming track Ngayin.gilmak to the Cox Peninsula, was described by Wells (Howard 1865, cited in Wells 2003, 57). This place would become the location of the Kahlin Aboriginal Compound and Chinese market gardens (figure 7.2).

There were few interactions between the Europeans and the Larrakia people in the first days, partly because Goyder directed his teams to have no contact with the local Aboriginal people. However, John Bennett, who had been a part of the first settlement expedition of Finniss, was fascinated by the people and compiled a place name list and dictionary of the local Larrakia and neighboring Wulna people's language (Pugh 2018a, 67–73). As the survey teams fanned out into the hinterland to measure the rural blocks, they established remote base camps. Sadly, at one of them, Bennett and a companion were attacked, and Bennett died soon afterward. Despite the attack, Goyder understood Aboriginal culture and realized that while the attack was unprovoked by his party, it may have been revenge triggered by previous Aboriginal interactions with Europeans. He understood that to retaliate would only result in further revenge killings (Goyder, Diary Transcript, Report to Parliament). Fortunately, John Bennett's vocabulary of the Wulna language was published shortly after his death (Bennett 1869).

Aboriginal place names were recorded in the settlement area in John Bennett's (1869) Wulna vocabulary. Police Inspector Paul Foelsche had a copy of Bennett's vocabulary, which he hand-annotated with several other local Aboriginal place names (on page 28 of the State Library of New South Wales copy, see Bennett 1869). It is not certain if the place names recorded by Bennet (1869) and Foelsche on the Darwin Peninsula are all Larrakia names or if some are Wulna names. They include:

- Mayrre = Fort Hill
- Stokes Hill = Laylillar
- Ningee = Peels Well
- Fannie Bay = Beedījik-benning
- Native camp = Lay-marro
- Kings Table [hill to the SW of Darwin, on western shore of harbor?] = Latungo
- Fannie Bay [G J?] Camp = Beri-jool
- Point left of Cullen [Bay] = Goondul
- Bay right of Hospital Point = madlarmanning
- Small beach near hospital = murrarmurroo

Other place names recorded by Bennett (1869) include:

- [The] Narrows, [the] district around = Lo-eil-wil
- Aulds Lagoon (Freds Pass) = Guā-kee
- Escape Cliffs = Pātter Pūrrer

In 1914, the superintendent of the Kahlin Aboriginal Compound (on the shores of Cullen Bay) was requested to document the locations of tribes on a map of the Darwin area for use by the Department of Lands; he also recorded the Larrakia names for local landscape features. Mayrere was the name of Fort Hill; Delila was Stokes Hill and Ninyji was the name for Doctors Gully (in Wells 2003, 123–124). A spring issued from the beach at the cliff base at Lameroo Beach, northwest of the modern war memorial. It was developed into a well that became known as the Overland Telegraph Well, but the name of the beach was adapted from the Larrakia people's knowledge of the location of damoe-ra, a freshwater spring. In the Larrakia language the primary meaning of damoe-ra is eye, however, it also has the secondary meaning of spring (Wells 2003, 125).

Aboriginal Use of the Darwin Landscape

The people were highly mobile and moved rapidly and efficiently through their country. In 1839, John Stokes recorded that a group of Aboriginals previously originally encountered in Shoal Bay visited the well site on Point Emery. He interpreted this to show that the Aboriginal groups migrated along the shore in this region (Stokes 1846, 18; Hordern 1989, 213). Aboriginal pathways crisscrossed the landscape and Goyder and his teams used these as they carried out their surveys (Goyder 1869, in Bourke 2014, 45).

In the first days of the 1869 camp, Goyder's party did not recognize any particular Aboriginal group as the main inhabitants of Darwin. They were visited by people who arrived from the south, west, and east. Some local people had been friends with members of Goyder's party who had been at Escape Cliffs, and they visited Goyder's new camp. By 1870 however, it was recognized that the Wulna were linguistically distinct from the local Larrakia and mainly lived to the east of the Darwin Peninsula and that there were two quite distinct groups (Wells 2003, 88–89).

The evidence of Aboriginal occupation of the Port Darwin region is of great antiquity, despite the relatively recent emergence of the current coastal landscapes. Pre-contact and contact Aboriginal archaeological sites are widespread throughout the Darwin region but none are recorded in the relatively confined and heavily developed limits of the 1869 settlement area. There are several important sites immediately east of the Darwin Peninsula, on the shores of the head of Frances Bay. They are located mainly on high ground, at the inland margin of the mangroves, and near water sources. These sites contain flaked stone artifacts and worked bottle glass, showing that their use persisted after European settlement (Bourke 2005, 60).

Shell mounds and middens are widespread on the coast in the Darwin region. While many are formed by natural processes and supplemented by scrub turkey nest building (e.g., Stone 1995), others are probably, at least in part, true Aboriginal middens formed by many years of occupation, and the use of favored camping sites. Shell mounds were present in or near the settlement area because they were quarried and burnt for lime in a kiln built shortly after Goyder's party arrived (Pugh 2018b, 37). Shell mounds also provided useful reference points as Goyder's party surveyed the region (Bourke 2014, 54).

Two registered sacred sites are in the study area. Stokes Hill is a place inhabited by a spiritual ancestor known as "Chinute Chinute." He is said to sometimes manifest as a Tawny Frogmouth (a native bird resembling an owl) and to act as a guardian of the wharf area of the peninsula (Darwin Waterfront 2018). The second is damoe-ra, the freshwater spring above Lameroo Beach. It is now the site of a shaded foreshore park (Bauman 2006, 38).

LANDSCAPE, GEOLOGY, AND EARLY PALMERSTON

Like the Larrakia name for Port Darwin, Garramilla, its European name recognizes the local geology, if in a roundabout way. It was named for Charles Darwin in 1839, then known as a geologist, when the outcropping rocks at Talc Head reminded John Stokes of HMS *Beagle* of his old friend (Powell 1986, 1). The local geology probably also led to the choice of the Darwin Peninsula as the site for a settlement.

Choosing the Site of Darwin

During the short stay of HMS *Beagle* in the harbor 1839, the need to water their ship led to Lieutenant James Emery's search for freshwater in the area, which he discovered seeping from springs at the base of the cliffs. Thirty years later, Emery's spring may have been one reason why George Goyder chose Port Darwin as the most likely place to establish a new settlement, before he had even left Adelaide (Powell 1986, 1). Finniss had rejected Port Darwin, despite its excellent anchorages, because he felt that it was too landlocked and poorly ventilated, and that the bordering mangroves made it too malarial (Finniss 1865, cited in Wells 2003, 112). A subsequent examination of the harbor by James Manton, Finniss' second in command recommended Point Emery as the best location, describing the excellent nearby anchorages, abundant building stone, access to coral for lime, and the groundwater resources. Prior to his departure from Adelaide, Goyder reviewed all the information available about the previous settlement sites, Port Darwin and Stokes' 1839 account. Opinions such as Manton's probably also helped to influence his final decision (Manton 1866, cited in Wells 2003, 112–13; Reece 1989, 308).

Goyder chose the site of modern Darwin carefully (Pugh 2018b, 34–35), preferring the low saddle between the mainland and the low hill of Fort Point for the survey camp because he believed it was cooler than the top of the plateau above. He found the best landing place was the beach near Fort Hill (where a jetty was constructed), and then immediately set about exploring the area. Members of the party discovered freshwater springs "oozing from the rocks" in a gully halfway between Fort Hill Point and Point Emery that was soon excavated as a government well, and water was carted to the camp (Goyder 1869). The site proved to have sources of reliable freshwater (mainly groundwater), good anchorages, and elevated ground above the mangrove flats.

Early Development

The topography of the confined Darwin Peninsula site defined how the settlement evolved and has heavily influenced land use, and the location and placement of both historic and modern infrastructure. It was a complex problem to impose a regular street grid on a long and narrow site bordered by cliffs and mangroves. Goyder solved the problem by orientating the main streets parallel to the long axes of the arms of the hammerhead-shaped peninsula and establishing shorter streets normal to them.

Goyder chose the southeastern arm for the main settlement, possibly because of its proximity to the sheltered anchorages of Frances Bay and any future port infrastructure. However, the topography and elongate geometry of this restricted site required considerable adaptation of the standard street grid to provide the requisite number of presold allotments for land speculators (999, mostly half-acre allotments in an area of 1,015 ha, Pugh 2018, 39). Here, he arranged the streets northwest-southeast but there was still insufficient area available, so he heavily modified the street layout to create a separate system of east-west allotments on the western arm. Despite fitting the surveyed allotments to this complex site, the demand for land in Darwin was never great enough to fill them. Most of those on the western arm remained undeveloped and they are now largely contained within the Larrakeyah military area.

Following the foundational survey, Palmerston became an important node for international telegraphy. The BAT and AOT headquarters and infrastructure were located at the point where the undersea cable had been brought on shore, on a small shelving beach originally known as Cable Beach (now reclaimed land). The offices were at the top of the cliff above the beach and would later become the site of the more substantial stone-built Darwin Post Office. Subsequently, this site became the location of the NT Parliament House.

The rest of the settlement grew slowly and was mostly focused on the plateau of the southeastern arm of the peninsula. It was here that the most significant developments of the first two decades took place. The small administrative settlement focused on the southern tip gradually spread to incorporate other elements of the surrounding landscape. Of note are the locations of market gardens, cemeteries, early extractive industries, and the extensive railway and port infrastructure that grew to service the burgeoning mining industry from the 1880s. These landscape elements became important to the pattern of Darwin's early growth and settler land use and remain important in the modern CBD (figure 7.2).

Freshwater Supply

As soon as Goyder's team landed, parties were sent out to search the nearby area for potential freshwater sources and useful land. Burton discovered a small rivulet that provided a reliable source of water and good grazing a short distance to the north of the camp. The creek became the first place where stock was watered, and it became the primary location for horse and cattle grazing. The creek drained the higher ground of the plateau and the deep gully that it occupied would form one of the topographic boundaries of Goyder's first town survey.

Surface freshwater is rare during the northern Australian dry season and the Darwin Peninsula and its hinterland are dry for much of the year. Springs were the most important freshwater sources in the first decades of European settlement (even Horse Gully Creek was probably partly spring-fed) and were mostly developed into wells. The most important was the well in Doctors Gully, but several others were also excavated on springs and soaks near the camp. Not all were successful, and one excavated on the saddle found salt water, while another to the east of Fort Hill ended dry in rock at a depth of 1.2 m (Pugh 2018b, 32; Goyder 1869c, 10).

Wells remained the most important sources of drinking water for Darwin residents into the late 1930s, supplemented by rainwater storage tanks. Wells were unreliable however, because of the seasonal rainfall recharge, and this led to the need for a more reliable and permanent water supply, particularly with the influx of wartime defense personnel. Manton Dam was constructed between 1939 and 1941, and reticulated water became available to residents for the first time (Ling 2011, 59).

Early Gardens and Graveyards

Although restricted in area, low-lying patches of fertile Quaternary and recent sediments were particularly important for early agriculture and grazing. The drainage lines they occupied were usually spring-fed, so water was available year around. Peels Well was the site of the first gardens but by 1887, Chinese market gardens occupied many low-lying, marshy areas on the northern fringes of the surveyed town center (as squatters). Chinese market gardeners became the major suppliers of fresh produce in Palmerston (NT Chinese Museum 2020) when they began to exploit these patches of well-watered and fertile soils at the heads of the gullies (particularly the former wetlands feeding into Horse Gully Creek and the Mindil Beach wetlands). These sites became the most important places for early agriculture in Darwin and the Chinese market gardens remained there well into the twentieth century. The places that the market gardens once occupied are now the locations of the modern Kahlin Oval, a golf course, the Botanic Gardens, and the developing greenspace around Burtons (Horse Gully) Creek.

The first burials were placed by Goyder on Fort Hill, adjacent to the flagstaff but later pioneer cemeteries were established on the northern fringes of the surveyed town area, on the upper margins of the wetlands. A Chinese cemetery was located on the western shore of Frances Bay, to the north of the main settlement.

After the Development of the Pine Creek Goldfield

An unexpected result of the construction of the overland telegraph was the discovery of alluvial gold by line workers in 1871. The place, about 100 km south of Palmerston, became known as Pine Creek. The discovery triggered an alluvial gold rush, for which Port Darwin became the main supply base and port. The alluvial rush eventually led to more capital-intensive hard rock underground mining. In Palmerston, which had languished economically, the Pine Creek Goldfield led to industrial and commercial expansion, including better port facilities, railway infrastructure, horticultural developments, and more permanent government and commercial buildings.

The gold mines were hampered by their isolation and the high cost of labor and freightage, with many proving uneconomic despite their high-grade ore. It would take fifteen years, but in 1886 the South Australian government approved the construction of the 3’ 6” (1.06 m) Northern Australian Railway to service the goldfields from Port Darwin, in the hope that it may assist the economic viability of the mining industry. To provide the rail line corridor access to the Darwin port facilities required extensive land reclamation along the western shore of Frances Bay and extensive quarrying of the promontories formed from Cretaceous sedimentary rocks (figure 7.2). While the railway helped the mines somewhat, it experienced its busiest period during World War II, and operated with various extensions until being closed in 1978 (North Australian Railway History).

The Chinese began to arrive in large numbers to exploit the gold in 1874, often on two-year indentured contracts. Once the contracts had expired, many remained to build roads, work on the overland telegraph, work as tradespeople, and later, 3,000 would do the tracklaying for the railway line to Pine Creek. Many became successful businesspeople and by 1892 owned most of the successful Pine Creek mines and crushing batteries (Blainey 1989, 91–96). By 1892, the Chinese population of the goldfields outnumbered Europeans by more than twenty to one (Blainey 1989, 96). Cavenagh Street, in central Darwin, remained a center of the Chinese community until World War II.

The gold boom also led to the much wider settlement of the NT than may otherwise have occurred (Blainey 1989, 96). While the gold-mining industry provided some short-lived development impetus in the early 1870s, Palmerston’s true long-term viability came from the establishment of the pastoral industry, which would provide the economic backbone of the territory and lead to the wider settlement and development of the interior (Pugh 2019, 212).

The construction of the AOT made one other significant contribution to the spread of European settlement into the NT interior. The telegraph repeater stations established along the line between Port August and Port Darwin (usually at water holes) became important European settlement hubs across central Australia. One of those stations has since become Alice Springs, the third-largest city of the NT.

Harbor Facilities

All of Darwin's original port infrastructure was located at the southwestern tip of the peninsula where the first significant harbor facilities developed around Stokes Hill Point, near the sheltered anchorage at the mouth of Frances Bay.

When the decision was made to build a railway to service the mines of the growing Pine Creek Goldfield, there were only limited places where a rail corridor could be brought in to access the Stokes Hill and Fort Hill wharfs south of the plateau necks. A corridor was developed that skirted the northwestern shoreline of Frances Bay and followed the inland margin of mangroves at the shoreline of near the mouth of Horse Gully Creek. Two steep-sided bluffs formed by Cretaceous sedimentary rocks, with small drainage gullies between them, were partially removed and the quarried material was probably used to reclaim low-lying land. Subsequent port developments included excavation for defense purposes, particularly the construction of blast-proof recesses for naval fuel oil tanks excavated into the Cretaceous rock. Post-war, Fort Hill was completely removed to provide a space for an iron-ore export terminal in the 1960s.

Development of the Road Network

The modern road system follows an initial track network that radiated outward from the camp on the neck at Fort Hill Point. Some of the first tracks probably followed long-established Aboriginal paths. In the main settlement, a steep road was cut up the cliff from the landing place at Fort Point to the plateau and is now modern Hughes Avenue (Pugh 2018, 39). Another short track was cleared from a landing site to the soak in Doctors Gully for the well diggers to use. When water was found, it was initially transported to the main camp by boat using tanks or casks but was taken by land once the Esplanade was cleared (Pugh 2018, 32).

The locations of the tracks leading to the hinterland were also dictated by the topography of the plateau, particularly the narrow necks bounded by deep gullies. The narrow neck of high ground between the backing wetlands of Mindil Beach and Horse Gully (Burtons) Creek provided the only restricted dry land access to the hinterland from the settlement. All roads leaving the settlement had to use this corridor, as the Stuart Highway still does.

Further north, a second topographic constriction of the plateau occurs between Ludmilla Creek and Sadgroves Creek, north of East Point and this area constrains the transport corridors leaving the Darwin Peninsula. It was known as the Narrows from the early survey days and this name passed to the post-war suburb that developed there (figure 7.1).

The outer rural locality of Freds Pass gained its original European name from the gap in the Daly Ranges through which Patrick Auld's overland expedition from Escape Cliffs to Port Darwin passed in 1865. Frederick Finniss accompanied the expedition, and it was named for him (Pugh 2018a, 59–60; NT Place Names Register 2019b). Freds Pass Road later became the main arterial road outlet of Greater Darwin, but the original pass has since been bypassed.

The railway corridor of the 1880s created a second land transport access into the Darwin CBD. While it was originally created for the railway, this transport route has since been redeveloped into a modern road network on the eastern side of the city, and redeveloping wharf precinct. It now forms part of the city end of Tiger Brennen Drive, a major traffic outlet. The integrated port and rail infrastructure has been removed and is now focused at East Arm.

The Government Residency and Administrative Precinct

The timber was cleared from the top of Fort Hill by cutting and burning (Wells 2003), and this prominent location became the site of the settlement flagstaff. Later, the official residence of the South Australian government representative was built in a prominent position nearby, on a narrow bluff of the tableland overlooking Fort Hill. This site was highly visible from the harbor and had a sweeping 270-degree view of the harbor and anchorages, and the developing wharfs. The official flagstaff was subsequently moved to a position in front of the government residence.

The site of the 1869 camp and several of the original buildings became government offices, such as the lands survey office, which was developed from Goyder's original stables. Nearby, more permanent government buildings were built on visually prominent sites above the cliffs of the Esplanade and overlooking the port area. These important buildings were constructed from local stone and included the Police Station, Court House, and Government Office complex at the corner of Mitchell Street. The Post and Telegraph Offices, the Cable Office, and the Postmasters Residence grew from the original AOT and BAT structures.

This part of central Palmerston Darwin became a government and administrative quarter, which remains in modern Darwin. The Police Station and Post Office were badly damaged during the Darwin bombings and demolished, but the Court House became the Darwin Naval Headquarters, until largely wrecked by Cyclone Tracy in 1974. Following its rebuilding, it became the Northern Territory Government Administrator's Offices in 1981. The site of the Post and Telegraph office complex is now the site of Parliament House.

Defense

Darwin became a major focus of defense activity from the 1930s onward and particularly during World War II, when it became a base of operations for Australian and U.S. forces operating in the Pacific Theater. The landscape was utilized for many defense purposes. East Point was developed into a series of important fixed defensive positions, with many concrete gun emplacements and other structures built using porcellanite aggregate. Gun emplacements were also installed at Point Emery. Various other fixed defenses were built along the shoreline, and in the immediate Darwin hinterland, to service aerial defense capabilities. There were also air-raid shelters, anti-aircraft batteries, trenches, and munitions storage.

After the first Japanese air raids targeted fuel storage facilities, new tanks were built into excavated recesses in the cliff faces of the Cretaceous rocks bordering the harbor, and on Stokes Hill. They were linked into subterranean storage facilities by buried fuel transfer pipelines.

When Goyder laid out Palmerston, he reserved Point Emery for defense purposes (presumably as a possible site for a gun battery) and it later became the site of a navigation beacon. Urban development did not proceed as originally hoped and the Point Emery and Point Elliott allotments were still mostly vacant by the 1930s. In the lead up to World War II, the entire western arm was turned over to military use, a function that it still fulfills. It is now the most important Australian defense base in northern Australia, incorporating Larrakeyah Barracks and HMAS *Coonawarra*.

Darwin Airfield and Early International Aviation

The original Darwin airport was located near the shore of Fannie Bay, on the flat terrain of the plateau. It was constructed quickly in 1919 as the first Australian landing site for the England to Australia air race. The first successful flight from Britain to Australia was completed by Ross and Keith Smith in a Vickers Vimy bomber when it landed in Darwin in 1919. The memorial commemorating the first successful flight still stands at the original site, near the head of the old runway.

The small Darwin airstrip would go on to play an important part in opening Australia to international air transport and travel. It continued as the first landing place for pioneer aviators developing the route from Britain, such as Amy Johnson, the first woman to fly solo from Britain to Australia in 1930, and for routes across Australia (e.g., De La Rue 2014, 140–141; 177–180). Darwin also became the handover point for the first international airmail arriving and departing Australia, in a service partly pioneered by Sir Charles Kingsford Smith and the Darwin postmaster (Bowden 2016, 22). The old civil airfield was closed in 1946 and the site was eventually subdivided for housing. The alignment of the original runway was preserved as Ross Smith Drive.

Landscape features were of particular importance during World War II when suitable locations had to be found across the NT for large air bases for both defensive and offensive capabilities. The Royal Australian Air Force (RAAF) initially used the civil airfield until the new air-force base was constructed on broad-acre and partially marshy tablelands to the east of the narrow neck of ground between Ludmilla Creek and Sadgroves Creek. The RAAF base is now also the Darwin International Airport.

Maritime patrol aircraft were also a key part of the war in northern Australia, performing reconnaissance, anti-submarine, and search and rescue functions. Darwin Harbor was a hub of seaplane activity and suitable coastal sites for bases were required. One of the most important seaplane bases was operated from Doctors Gully and was the home to Catalina flying boats. The site of the base is now a popular fish-feeding tourist attraction.

Geohazards

The properties of the weathered and leached Cretaceous substrate is one of the most important geotechnical factors considered for footing conditions of any construction project in the modern city. The material is also easily eroded by the coastal wave action along the shores facing Shoal Bay. The constant erosion of the cliff faces causes collapses and cliff-face recession and there have been recent cliff-face failures at the Dripstone Cave cliffs. Ongoing coastal erosion has significantly modified the shoreline at Escape Cliffs since the 1866 abandonment of the settlement. The cliffs have receded over 15 m and several of the original iron water tanks that once remained on the site have been lost (Pugh 2018, 156).

Floods and the inundation of low-lying land during the wet season have been overriding influences on the European exploration and settlement of the Darwin region. Wet season flooding and the saturation of soil and alluvial plains greatly inhibited the passage of early European explorers such as Leichardt and impacted the general movement around the country during Finniss's and Goyder's land surveys. Flooding also seriously impeded the progress of the construction of the AOT. Wet season inundation and associated climatic conditions continue to be one of the most important climatic influences on NT human culture. In addition to floods, cyclones have molded the development of the built character of Darwin by sweeping away many of the buildings that defined earlier phases of the development.

While not a geohazard, any discussion on environmental conditions in the Darwin region affecting the ongoing stability of the built environment would not be complete without discussing termites (also known as white ants). As simple as these communal insects may seem, they are key elements of the landscape and destroyed all the earliest built heritage of the place. They continue to influence the design and management of the built structure of modern Darwin. Native cypress pine (*Callitris intratropica*), a light and useful structural timber, was found to be resistant to termites. It grows mainly on the Cretaceous cover sequence forming the tablelands and became the preferred construction material, leading to increased exploitation (Walpole et al. 1968, 9–11).

THE BUILT ENVIRONMENT

The first camp at Fort Hill Point was a group of tents and prefabricated corrugated iron and timber buildings constructed from imported and local timber, and with bark. Local Aboriginals assisted with construction by supplying large sheets of bark for roofing. White ants attacked many of these early makeshift buildings and they did not last. Immediately upon establishing the 1869 camp, a lime kiln was built, and lime was made by burning coral, and from shells quarried from nearby coastal shell mounds (Pugh 2018b, 37).

As the settlement became more formalized and developed beyond the first camp, more substantial buildings were required for permanent accommodation. Several simple structures were built from local stone to house administrative and communications functions, including a government residency and the buildings of the BAT station. The most readily available stone was a surficial carapace of silicified weathered Cretaceous sediments that capped the higher ground of the plateau. This material became known Darwin porcellanite. Porcellanite is a local term for the typically white, cream, or light gray colored building stone (though usually stained by iron oxides to varying degrees creating shades of yellow orange and brown) sourced from the duricrust developed on the surface of the Cretaceous sedimentary rocks that underly central Darwin. It is less altered than true silcrete, containing around 90 percent Si compared to the more typical 97 percent of true silcrete. It is inhomogeneous and becomes weaker with depth. The duricrust formed at or near the water table (Clarke et al. 1999, 2-911–2-212). Modern Darwin is the only Australian urban center that experiences foundation problems caused by a silcrete profile (McNally 1995, 539).

The first porcellanite freestone was sourced from weathered Cretaceous rocks in coastal exposures, presumably near the central settlement at the southeastern tip of the Darwin Peninsula. Considerable collapsed rock was observed on the shoreline near the camp when Goyder's party arrived in 1869 and material had to be cleared for the development of the first jetty (which Goyder marked on survey plans). The cliff rubble is a likely source for the stone used for this jetty because it is marked on later Admiralty Charts as a "stone-built jetty" (figure 7.2). Remnants of this structure can be seen in early aerial photographs, but the location is now buried beneath land reclamation.

Porcellanite was then sourced from several local quarries that were developed in local area. The Stuart Park Quarry was developed during the World War II as a source of crushed porcellanite for exposed aggregate in gun emplacements, observations posts, and other fortifications at East Point and Casuarina Beach. The site is now the grounds of Stuart Park Primary School. In 1957, St Marys Catholic Cathedral was built from porcellanite quarried on the southern shore of Emery Point (west of Doctors Gully). Finlay's Koolpinyah Porcellanite Quarry was the source of facing stone for the modern KPMG building and for many modern feature walls.

Brick was not a significant building material in early Palmerston. The impetus for a brickmaking industry developed when bricks were needed to build mine infrastructure at the Pine Creek Goldfield. Brick clay was extracted from pits to the northwest of the town center, and clay was transported by a narrow-gauge tramway to works established where the modern Darwin yacht club hard stand is located. The former clay pits are now an ornamental lake known as Lake Alexander (Fannie Bay History and Heritage Society 2016).

Landmark Historic Buildings

Despite the ravages of war, cyclones, and termites, central Darwin retains several historically important stone buildings that remain defining features of the modern streetscape (Doyle 2001, 59–60). All are built, or are faced, with local porcellanite sourced from the coastal cliffs, occurrences in the settlement area, or from quarries in Larrakeyah and Fannie Bay.

Government House (The Residency) remains a central feature of the southern tip of the CBD. The current building was completed in 1879, with much of the work done by Chinese masons, and jail inmates, but incorporates parts of the original building. Nearby is the Administrators Office (the former Court, Police Station and Naval Headquarters), which was reconstructed on the same ground plan, and in the same style, as the original building following the destruction of Cyclone Tracy. The historic significance of porcellanite to Darwin was recognized when stone salvaged from the ruins of the original structures was used to rebuild them.

Other important historic stone buildings include parts of the Fannie Bay Jail complex, now an important museum site, the Victoria Hotel—opened in July 1890—and originally named the North Australia Hotel (renamed in

September 1896), the Anglican Church (the portico and front façade predate the cyclone), Browns Mart, the Town Hall ruin (now an important cultural precinct and performance area) and the Sue Wah Chin Building (Stone Houses). The latter is the last remnant of the extensive Chinese district of the settlement that was focused on lower Cavenagh Street for much of the late nineteenth and early twentieth centuries.

Porcellanite stonework continued to define the built character of Darwin into the twentieth century and remains an influential character feature of the modern city. The BAT Company Residence (known as Lyons Cottage after a later tenant), was built in the 1920s. Crushed porcellanite quarried on the site of the modern Stuart Park Primary School was used for exposed aggregate in the wartime fortifications at East Point and Casuarina Beach (Doyle 2001, 60). In 1957, St Marys Catholic Cathedral was built from porcellanite quarried on the southern shore of Emery Point (west of Doctors Gully). And when KPMG built their new offices in the central city in the 1990s, porcellanite was used as the facing stone of a building designed in a colonial style. Porcellanite is also a part of many feature walls in central Darwin, and blocks are used as kerbing in many of the oldest streets.

Chapter 8

Geology, Landscape, and Australian Cities

The seven Australian state capital cities are not particularly large by world standards, ranging in population from around 148,000 (Greater Darwin) to 5.3 million (Greater Sydney). Only Sydney and Melbourne are listed in the top 100 largest cities (e.g., World Population Review 2021). Yet the fact that they were comprehensively documented from their earliest days through maps, charts, artworks (including photographs), and written records means that they provide valuable examples of how landforms (and their underlying geology) can influence the ways that cities develop through successive stages. This chapter draws together the examples presented in the preceding chapters to identify which aspects of a landscape can most impact the development of settlements, from the initial choice of the site to the patterns of their growth as they spread into their hinterlands. A short overview of the contribution made by locally available building materials to the distinctive built environment of each city is included. The last section will briefly speculate on the possible future evolution of these Australian cities within their landscapes.

This book presents a largely historical review, but the results have applications beyond historical studies. In the Australian context, there are applications of the historic landscape reconstructions to the modern fields of historical archaeology, landscape archaeology, and Aboriginal archaeology because they provide original landscape contexts with which to interpret the distribution patterns of finds and archaeological sites. They also have applications to the reconstruction of lost pre-European environments and ecosystems. An understanding of former landscape elements can also inform the wider fields of urban geology, geohazard prediction and management (including flood prediction and possible sea-level rise modeling), and engineering geology because they allow a deeper understanding of the subsurface geological environments of urbanized areas.

Beyond Australia, other cities may benefit from a better understanding of the effects on growth of geological and geomorphological features. The factors that most influenced the growth of Australian settlements are likely to have operated elsewhere in the world and might provide guidance for the study of other cities, particularly those founded by other settler societies. And there may be applications to the study of cities founded in earlier eras, when the interaction of landscape, geology, and the foundational years were not recorded. Interpreting modern urban patterns through a historic landscape lens may help to explain why parts of other cities exhibit the land-use patterns that they do.

CHOOSING THE SITE FOR A SETTLEMENT

Four geological factors most influenced the choice of the sites of the pioneering British settlements in Australia: (1) they had a reliable freshwater source; (2) there was fertile agricultural land; (3) there were elevated, well-drained habitation sites; and (4) there were nearby deepwater anchorages close to shore. Each of these factors is elaborated on below.

Fresh Water

The foremost requirement of each new settlement was an adequate and reliable source of fresh water. The availability of freshwater was a key aspect of why a site was chosen. Sydney Cove had the Tank Stream, Sullivan's Cove the rivulet, the Brisbane Peninsula had Wheat Creek, Point Fraser had the lake system, the fresh water above the Yarra River falls decided the location of Melbourne, the Torrens River billabongs supplied Adelaide, and the Darwin Peninsula had many spring-fed soaks and some small creeks. The basalt falls on the Yarra River dictated the point where it became freshwater and some of the earliest drainage works in central Melbourne were to prevent polluted runoff from the town area from entering the river upstream of the falls, where drinking water was drawn. The Torrens River billabong was the location of William Light's first survey camp, and the subsequent settler's camp. The nearby western end of Hindley Street was the first part of the town to see significant urban development. The small but important landscape features from which fresh water came were the primary reasons why each settlement was located where it was.

Groundwater was also important, particularly in the early years of the Swan River Colony. When James Stirling chose the site for Perth, the abundant fresh water available in its marshes and lakes, and ultimately sourced from the groundwater aquifers, was probably the key factor. Several springs on the banks of the Swan River acted as outlets for the groundwater. One spring was chosen as the site of the first brewery while another was developed into a public water source. Wells and springs remained the primary source of potable water for the population of Perth until the first reticulated water supply scheme was built in the 1880s. Despite modern access to surface water sources, the groundwater aquifers of the Swan Coastal Plain remain one of the most important supplies of freshwater for the modern city.

In Sydney, population growth increased the demand for water beyond the capacity of the Tank Stream and led to the exploitation of nearby groundwater resources. Marshlands that had previously deterred the British from settling at Botany Bay were also extensive freshwater aquifers. This resource was developed into a town water supply scheme that drew water from the aquifer-fed Botany Wetlands via the Busby's Bore tunnel.

While the other Australian settlements were not as rich in groundwater resources as Perth and Sydney, they still benefited from them. In Hobart, the debris fields and scree deposits on the eastern slopes of Mt Wellington and its foothills acted as subterranean water reservoirs, which fed the rivulet from springs, creating a reliable water source for the early settlement. In Darwin, groundwater accessed through wells was the only source of water for the 1869 camp and remained the most important source of fresh water until the 1930s.

Despite the importance of the early and often limited freshwater sources, it is remarkable how quickly they became unfit for human consumption through pollution and abuse. Their yields also became inadequate for growing populations. Most were rapidly superseded and are now mostly buried beneath city development. They all survive however, as part of each city's subterranean drainage system. Only the Hobart Rivulet remains a visible part of the modern CBD.

The supply of potable water to ever-increasing populations has remained one of the key development issues of all Australian cities since their foundation. It is as important now as it was at first settlement and may one day become a limiting factor on the population growth of this largely arid continent.

Early Agriculture

Fertile land for cropping and grazing, or a nearby hinterland that could provide it, was fundamentally important to the first penal settlements. Settlers were on the far side of the world from their primary supply base, and so had to quickly become self-sustaining. Yet agriculture began in a very small way in each settlement.

The first places to be farmed were the limited patches of fertile soil near the camps. At Sydney, the first gardens were on small areas of sediments at the heads of the coves. Gardens were established around the administrative district at Sydney Cove, but the first significant farming was to the east, at Farm Cove, where a creek entered the harbor through a wetland system behind the sandflats. Similar wetlands to the east and west of Sydney Cove were sources of the workable timber and rushes used for the first buildings. Farm Cove was later reserved as part of the Government Domain, which led to its preservation as the modern Botanic Gardens. At Sullivans Cove, there were no areas of recent sediments that could be farmed. Gardens were created above the main camp on the central ridge next to the rivulet until a larger government farm was established on the soils formed on Cenozoic basalt and alluvial sediment near New Town Bay (which became the main Hobart cemetery in the 1920s). At Moreton Bay, convict gardens were established in an area of low-lying and often flood-prone (but fertile) sediments mostly located on the fringes of the inside bends of the Brisbane River before spreading inland to the marshlands on more expansive Quaternary deposits around Eagle Farm. In Perth, the first government garden was planted on the flat ground near the water's edge, at the foot of the steep bank below the first government camp. The recent lacustrine deposits immediately north of the settlement became a focus of market gardening in the 1840s and 1850s (Bolton 1989, 152). The first gardens of Melbourne were on the alluvial soils of the north bank of the Yarra River, below the settlement, and further eastward along the riverbank. Arable farming spread up the Yarra Valley to exploit the fertile alluvial soils while grazing rapidly spread across the grasslands to the north and west. In Adelaide, small gardens were established on the river flats near the Government Domain and to the east. The first fruit and vegetable gardens in Darwin were laid out on sediments deposited at the base of a gully fed by a spring. Later, Chinese market gardens exploited large patches of alluvial sediments in gullies on the northern fringes of the settlement.

The production capacity of the small patches of fertile sediments in each settlement were quickly outstripped by increasing populations and broader expanses of arable and grazing land were sought as an exploration priority. Once located, the new agricultural areas would have a profound effect on the subsequent growth of the settlements. Richer soils to the west of Sydney triggered the spread of settlement toward Parramatta and the Hawkesbury River Valley while the poor sandstone soils to the east, between Sydney and the coast, remained largely undeveloped. Arable cropping near Perth spread along the Swan River Valley and the lake systems of the coastal plain.

The availability of large expanses of good agricultural land was even more important to commercial settlements like Melbourne. The recognition of the extensive grazing lands of the volcanic plains of southeastern Australia was the trigger for an illegal mass settlement that the NSW colonial government was powerless to stop. The subsequent discovery of gold in the nascent pastoral regions triggered a mass migration that largely overwhelmed the natural landscape of Melbourne within two decades.

Wetlands and Settlements

The topographic features that most influenced the spread of the early settlements were wetlands. As the camps evolved into villages, and then towns they spread beyond their original bridgeheads into their local environs. As they did, it was the wetlands that most influenced the patterns of this growth, including the places used for residential or industrial purposes, the placement of transport and other infrastructure, and the location of sport and recreational facilities.

Elevated and well-drained places for dwellings were important to reduce the risks of flooding and disease. The coastal locations of the first settlements meant that there was sometimes a lack of such places. The wetland-, saltmarsh-, and mangrove-bordered shores of Botany Bay led Arthur Phillip to select an alternative site in 1788, mainly because of the potential for disease if a permanent camp was established there. The lakes on the northern borders of early Perth initially inhibited its northward spread, but the fertile lake soils soon became favored places for low-density agriculture. The lake geometry led to a modification of the standard street grid. The limited availability of such sites in early Melbourne was a dominant influence on early land use as urban growth spread outward into the surrounding marshy country.

The location of low-lying marshes and swamps in areas of Quaternary to recent sediments were important influences on early urban expansion. Wetlands and flood-prone areas remained undeveloped or were reserved because they could not be used for residential developments. Many were adapted as industrial precincts (particularly noxious industries), such as the Botany Swamps in Sydney, the West Melbourne (Batman's) Swamp, and York's Hollow and the extensive Quaternary flats to the east of Brisbane. Residential development in Sydney was unable to spread southward beyond the drainage divide because of the sediments of the Botany Basin and the wetlands, and the area became mostly agricultural and then industrial, with a focus on noxious industries. Low and flood-prone land at the mouth of the Hobart Rivulet hindered residential development there and it became the focus of noxious industries (e.g., tanneries and slaughterhouses), and later evolved into the Wapping industrial precinct.

Wetland areas were critical influences on suburban development in other ways. In the case of Melbourne, Perth and Brisbane, residential development was initially restricted to high and well-drained ground. Where low-lying areas were eventually developed for residential purposes, such as Wapping in Hobart, Frog Hollow in Brisbane, and Northbridge in Perth, they became the places where less well-off people lived. Cheap housing was often located in the more flood-prone areas near industry and the wharfs. In an era of cesspits and water wells, this also made them more disease-prone places. Low-lying areas could create socioeconomic divides within urban areas.

The Melbourne settlement site was located adjacent to the northeastern margin of the low-lying and marshy landscapes formed on the sediments of the Yarra Delta, and the the locations of well-drained higher ground were particularly important for the first suburban growth. In 1835, extensive salt lagoons and marshes, such as the West Melbourne Swamp, occupied most of the land south and west of Melbourne, and the southern side and banks of the river were also occupied by marshes and lagoons. These wetlands were a defining element of early Melbourne because they restricted the places where residential development could take place west and south of the river, at least until adequate drainage works and reclamation occurred in the later nineteenth century. Only elevated places such as Emerald Hill, Sandridge, and later, St Kilda, could be used for housing. And when rapid suburban expansion occurred in the 1850s, after the discovery of gold, it mostly spread across the basalt plateau to the north of the original settlement.

However, land that was subject to inundation could be used for transport and port infrastructure, such as the integration of the West Melbourne Swamp into the late nineteenth-century development of Victoria Dock, the major harbor facility of Victorian era Melbourne. It was in an ideal location for accessing the main railway infrastructure of the developing town and for the creation of an integrated transport system. The northern railway corridor linking Melbourne to the rural hinterland had to circuit the West Melbourne Swamp when it was first built in the 1860s. Though it has now long been incorporated into port facilities, and is largely backfilled and drained, the imprint of the West Melbourne Swamp remains preserved in the route of the modern railway, and in the mainly industrial and transport uses that predominate west of the modern CBD.

Later, some of the poorly drained areas of Melbourne would become important recreational precincts.

Wetlands adjacent to the river and in the surrounding low-lying marshes were often reserved for sporting grounds (cricket grounds and horse-racing tracks) and parklands, while others became the locations of industry (particularly noxious industries like tanneries). Many of the most important open spaces and sporting facilities of inner Melbourne exist today because of the wetlands that were there in 1835.

The northern boundary of early Brisbane was marked by several curvilinear ridges with deep valleys in between, such as Spring Hollow and Yorks Hollow. The valley floors were swampy, with occasional chains of ponds, and rivulets. They could not be developed because of their poor drainage and so were reserved for water supplies and recreational purposes. Wheat Creek, a minor but well-incised drainage line on the northwestern fringes of the convict settlement was also one of this series of arcuate valleys and became the first reservoir. Early reservation of these low marshy valleys meant that the land remained largely vacant when the need developed for railway corridors (and later motorways) to enter the inner city.

The recognition of potential geohazards, such as flood-prone land, could also deter new arrivals from choosing sites for settlements. Arthur Phillip abandoned Botany Bay partly because the marshy coastline would be too unhealthy for a permanent town. The low-lying land near the Port River and seacoast deterred William Light from creating an administrative capital there, despite it being preferred by Governor Hindmarsh. And the seasonal wetland occupying an abandoned tributary channel of the Torrens River at the eastern end of the main city block forced him to modify his regular street grid to fit the town blocks to the available high ground. The result was the irregular geometry of East Terrace (Fenner 1927, 210; Arrowsmith 1839).

More than any other Australian CBD, the landscape and urban development of inner Perth has been directly influenced by wetlands and lakes. Though the influences were subtle, the lakes and marshes of central Perth were a defining element of the first decades of European settlement. Their locations influenced the street grid geometry, land use, the placement of transport infrastructure (roads and rail), and the locations of early agriculture. In Hobart, Perth and Brisbane, the locations of flood-prone residential areas led to socioeconomic divides in the later towns.

Anchorages, Wharfs, and Other Transport Infrastructure

Natural outcrops forming rocky shorelines became convenient natural landing places in Sydney and Brisbane. In other places, beaches and low points in high riverbanks were chosen as landings (e.g., Hobart and Perth). The first landing sites were usually developed into jetties and wharfs, with anchorages nearby. First landing places such as Sydney Cove (now Circular Quay), Hunter Island and the rivulet in Hobart, Queens Wharf in Brisbane, and the Turning Basin in Melbourne became important transport hubs in the new settlements and their locations became a focus of road systems and influenced their street layouts. Most remained important for decades but were mostly superseded by new infrastructure and became backwaters (Sydney's Circular Quay is an important exception). In the modern cities, these historically important locations are now being redeveloped as important inner-city precincts, such as the Docklands in Melbourne, the Queens Wharf development in Brisbane, and the Wapping urban renewal in Hobart. Others have retained their original function, such as the Barrack Street Jetty in Perth (modified by land reclamation) and Circular Quay in Sydney.

Harbors and anchorages were essential transport infrastructure for newly arrived settlers dependent on the maintenance of seaborne supply, trade, and cultural links with the outside world. The sites of the first landing places became fixed points in the geography of most settlements, such as Queens Wharf in Brisbane, the government jetty at the Domain in Perth, and Queens Wharf in the Turning Basin of Melbourne.

It was not always possible to find places that combined the key features of an ideal settlement with a good harbor. Even Sydney, with excellent harbor and anchorages required the establishment of Parramatta before large-scale farming became truly successful. Hobart exists on the western shore of the River Derwent Estuary now because there was little potential for agriculture at the earlier-established Risdon Cove settlement on the eastern shore. The lack of good agricultural soils and a reliable freshwater source at the mouth of the Swan River Estuary led to the separate foundation of Fremantle as a port settlement for the Swan River Colony while Perth became the administrative capital.

To Stay or to Move

Once a settlement was founded, a decision often had to be made about whether to remain there, especially if it was found to have shortcomings, or a potentially better place was found. Yet no primary administrative settlement was ever moved to a new site in colonial Australia once it had been formally established as the seat of government. The reasons why settlements remained on their original sites were usually political, economic, and logistical. They had gained "momentum" which deterred any location change (e.g., Statham 1989, 14–15, 26–9), and often despite any shortcomings of the site that resulted from the topography.

When David Collins arrived in the Derwent River in 1804, he had already abandoned one settlement on Port Phillip Bay for largely geological reasons, and then chose a new site, rather than accepting the disadvantages of the existing camp at Risdon Cove. Similar discussions about moving took place in Brisbane shortly after it was established by Henry Miller (however successive commandants agreed with Miller's choice), but particularly later, when there was a debate about whether South Brisbane should be the main commercial and administrative center, given its proximity to the main anchorage. However, the susceptibility of South Brisbane to flooding ended the discussion. And the argument was settled at Adelaide because the surveyor general had been given the authority to choose the site above the governor and had such faith in his choice that he refused to move it. There were later discussions about moving the administrative center to the port, but it never occurred because the governor could not persuade the settlers to bear the cost of moving and could not convince the home government that it was necessary (Statham 1989, 15).

So, while short-term geological reasoning was employed when initially choosing a site, other factors tended to take over once it was firmly established, and geological reasoning became much less important. Sydney is an important example. Almost from its beginning, the location was recognized as being generally unfavorable for agriculture. Once Parramatta was established on richer soils, Sydney could have become the port hub for the colony and not its administrative center. However, this never occurred because the infrastructure and commercial activities had already become entrenched in Sydney and the effort and expense of moving was a deterrent.

GEOLOGY, LANDFORMS, AND EARLY GROWTH

Significant changes in the geology, or major variations in landforms (such as areas of wetlands) often dictated the broader land-use patterns of the modern cities, particularly in the locations of large-scale industrial and manufacturing uses versus sporting and residential zones. However, it was often the small-scale geological and landscape features that were the most significant on a local scale. Elevated positions, such as hills or ridges, were often chosen as the locations for significant administrative buildings, or infrastructure such as military barracks, gun batteries, flagstaffs, and signal stations, and for later hospitals, churches, and elite housing. When the administrative heart of Melbourne was transferred from the original site on the Western Hill, Parliament House was constructed on the prominent lower slopes of the Eastern Hill, facing down Bourke Street while the almost equally important Treasury Building faced down Collins Street. On the other hand, low-lying places prone to flooding were generally considered to be unhealthy, and less desirable for housing, so became industrial areas (particularly noxious industries), low-cost worker housing (e.g., Hobart, Perth, and Brisbane), or sporting and recreational facilities.

The terrain also dictated the locations of tracks and roads, and outcrops could control the alignments of suburban streets (e.g., The Rocks in Sydney). Of particular importance was the siting of bridges. In Sydney and Hobart, the placement of the first bridges across narrow points in water courses, such as the Tank Stream or Hobart Rivulet, shaped the largely organic growth of the land transport infrastructure as it radiated outward from these key water crossings. A natural basalt bridge across the Yarra River at Melbourne resulted in the first significant track being formed beyond the original village to Sandridge. Fords across the Torrens River controlled the position where tracks were formed to link Adelaide and North Adelaide (later becoming the sites of bridges). The natural crossing point of the Swan River at the Heirisson's Islands became the site of the Perth causeway, which focused the eastward spread of the road system.

Many early roads were created as tracks to access raw materials such as timber, brick clay, stone, and the raw materials for lime mortar. The earliest road extending beyond the Sydney Cove camp was to the brickfields, and the first tracks north of Hobart were to timber resources and the brickfields. In Brisbane, the first tracks from the main settlement led along the river to Limestone Station (Ipswich) and to Eagle Farm. The road networks of Perth and Adelaide, both inland settlements, developed to link them with their anchorages and ports, and to inland centers (e.g., Guildford). Later, corridors had to be found for railways, then motorways, and then finally tunnel portals to bring commuters into the city centers.

Drainage Lines and Transport Corridors

The imprints of many drainage lines and water bodies remain embedded in modern urban landscapes because of their controlling influence on earlier land uses. Creek lines, marshy valleys, and lake systems could not be developed for housing so would later provide undeveloped corridors for railways and motorways to reach into the CBDs. Transport corridors continue to be developed along water courses (e.g., Bibra Lake in Perth) but most available surface corridors into central cities are now fully utilized, so tunnels are being developed. Yet even the locations of modern road and rail tunnels are, at least partly, controlled by the locations of existing (or former) drainage lines because they link into earlier surface road and rail infrastructure that originally conformed to the surface landforms.

Simple Choices Leading to Functional Divisions

The often short-term and apparently trivial decisions made by the first settlers could be of great significance to the subsequent evolution of the settlements into towns and cities. Decisions as simple as, on which side of the creek the settlement administrators might set up their tents could establish patterns that remain over two centuries later. In 1788, the spring-fed Tank Stream and its valley decided the broader administrative and commercial divisions of the modern Sydney CBD. The administrators chose to be focused to the east of the stream while the convicts and Marine guards were camped to the west, where the road network came to be controlled by the outcrop geometry of the thickly bedded Hawkesbury Sandstone. Government functions are still concentrated on the eastern side of the Sydney CBD, including the NSW Parliament Building, which occupies a wing of the former colonial "Rum Hospital." The hospital was probably built on this high and originally prominent position east of the Tank Stream to take advantage of the health effects of the better ventilation it offered.

Most government, cultural and health facilities in central Hobart are still located on sites that reflect the layout of the 1804 camp. The Royal Hobart Hospital is near the place where the surgeon first established the camp hospital on the far side of the rivulet. Many local and state government functions are still focused on the central city blocks where the settlement administrators placed their tents on the high bluff above the shoreline.

In Brisbane, the low rocky shoreline of the western side of the Brisbane Peninsula provided an easier landing place than the eastern side and became the wharf, and a focus of the penal settlement. Parliament House and Old Government House were placed on the crest of the spine of the peninsula, and overlooking the first main anchorage that evolved there, on the South Brisbane Reach of the river. Government buildings and functions are still focused around the Queens Wharf area. In Perth, the modern Barrack Street Jetty is still located near the place where James Stirling first landed near a gully that gave easier access to the high ground above the bank, and the Government Domain and Supreme Court (and gardens) occupy the location of his first camp.

Quaternary Deposits and Airfields

The broad expanses of low-lying and level land underlain by Quaternary to recent sediments often provided the necessary space for the twentieth-century development of airfields. The airports of Sydney, Brisbane, and Adelaide are all located on former wetlands. One of Melbourne's earliest airfields was located on Yarra Delta sediments at Fishermans Bend, while modern Moorabbin Airport, southeast of Melbourne, occupies Holocene swamp deposits (Vandenberg 1981). Hobart airport is built on Quaternary marine sediments deposited as a sand bar at the mouth of the Coal River (Pittwater). The first permanent aerodrome in Perth was constructed on the Pleistocene sand flats of the Maylands Peninsula in the 1920s.

Geological Boundaries and Modern Land-Use Zones

The underlying geology still strongly influences how people utilize the landscapes of Australian cities. The boundaries between different land-use

Figure 8.1 Photo Panel 1-3 Sydney, Hobart, Brisbane.
Examples of the uses of characteristic local stone in the built environment of Sydney (Panel 1), Hobart (Panel 2), and Brisbane (Panel 3). Please refer to the extended caption at the rear of the book for additional information.

Figure 8.1 (Continued)

Panel 1 Sydney

1A Partly modified outcrop of Hawkesbury Sandstone on the eastern shore of Farm Cove. Many original outcrops survive in this area.

1B Fig tree on quarried sandstone face, with ornate carved stone fence pillar.

1C View of Admiralty House, Kirribilli Point, north shore, Port Jackson. An example of a grand harbourside home built from local sandstone within the context of natural and modified sandstone outcrops and landforms.

1D Stone steps built into the quarried sandstone face of Bennelong Point, opposite the Sydney Opera House. The vertical face is known as the Tarpeian Wall.

1E Quarried face in sandstone with a sandstone block retaining wall fitted neatly onto the upper surface. 'Hungry Mile' – Hickson Road.

1F Former warehouses built from sandstone, with upper storeys added in brick. The Rocks, western shore of Sydney Cove.

1G Sandstone Guardhouse of Hyde Park Barracks (1811-1819).

1H Finely carved sandstone royal crest on the façade of the Land Titles Office (1912-13), Albert Road, central Sydney.

1I Hawkesbury Sandstone portico, St James Church (1824), Sydney.

Panel 2 Hobart

2A View of Davey Street looking eastwards from Franklin Square, central Hobart.

2B View of the city end of the Brooker Highway, on the eastern side of the Hobart CBD, looking south from the Old University Building. The valley occupied by the highway in the middle distance was formerly the course of the Park Street (Domain) Rivulet.

2C The channelised Hobart Rivulet where it is crossed by Barrack Street. East of this point, the rivulet becomes obscured by city developments, until reappearing at the Collins Street Canal.

2D The steep incline in Harrington Street looking north from the intersection of Macquarie Street.

2E The former Governor's Private Secretary's Cottage seen from Davey Street, and now within the modern Tasmanian Museum and Art Gallery complex. The cottage once stood high above the north shore of Sullivan's Cove, within the grounds of Old Government House.

2F The 1810 Commissariat Store building, which originally stood at the shoreline of Sullivans Cove and at the head of the sand spit that joined Hunter Island to the mainland.

2G Sequence of buildings forming part of the State Treasury Department facing east across modern Franklin Square. This complex of buildings occupies the western end of the former Georges Square.

2H Sandstone mile marker on the Harrington Street bridge (1844) across the Hobart Rivulet.

2I A rare building constructed mostly from dolerite blocks in Salamanca Square.

2J An example of a modern red gravel (weathered dolerite) driveway to a garage in suburban Hobart.

Panel 3 Brisbane

3A A remnant outcrop of the Neranleigh-Fernvale Beds on Queens Wharf Road, central Brisbane.

3B Detail view of the water-level outcrops of Neranleigh-Fernvale Beds at the base of the Riverside Expressway, near the site of the former Queens Wharf. The surviving exposure may be a remnant of the outcrops that spurred Henry Miller to choose this site as his main landing place.

3C View to the east, from the Spine of Brisbane (now the site of modern QUT) and across the site of the former convict farms and gardens of the penal settlement.

3D Detail view of a corner of the riverside boundary wall of the Commissariat Store (1829) of the Moreton Bay Penal Settlement.

3E Panoramic view of part of the Kangaroo Point Cliffs, a former stone quarry near the southern base of Kangaroo Point.

3F The soaring structure of St John's Anglican Cathedral, Brisbane; with external walls built largely from Brisbane Tuff, and internal walls of Helidon sandstone

3G Retaining walls and steps built from Brisbane Tuff, leading eastwards from QUT to the low-lying ground of the City Botanic Gardens.

3H The Brisbane General Post Office building (1872), built from local Triassic sandstone (quarried near Breakfast Creek) and stone from Murphy's Creek (near Toowoomba) and tuff.

3I The ANZAC Square Shrine of Remembrance (1930). It is built mainly from Helidon Sandstone and shows the remarkable structural qualities of the stone.

Figure 8.2 Photo Panel 1-3 Perth, Melbourne, Adelaide.
Examples of the uses of characteristic local stone in the built environment of Perth, Melbourne, and Adelaide. Please refer to the extended caption at the rear of the book for additional information.

Figure 8.2 (Continued)

Panel 1 Perth

1A Face of former limestone quarry on Mounts Bay Road, on the eastern slope of Mt Eliza (Kings Park). The building complex in the foreground is a former pump house for a freshwater bore.

1B The limestone rubble wall on the southern boundary of the Government House grounds, near the Supreme Court buildings in central Perth. This wall was once on the original shoreline.

1C Steps in the grounds of Government House, which traverse the steeply rising ground from the original Swan River shoreline to the level of St Georges Terrace.

1D Limestone block wall of an Edwardian era house in West Perth. Such limestone walls were (and are) a common feature of the suburban environment.

1E Limestone block wall with variably weathered and etched surfaces that reveal the eolian cross bedding.

1F Detail of Tamala Limestone base courses of a legal chambers in central Perth.

1G The Old Perth No 1 Fire Station (1901).

1H Ornate limestone-built house with brick detailing, Mount Street, West Perth.

1I Intricate and detailed limestone masonry of the first section of St Mary's Catholic Cathedral, East Perth.

1J The northern façade of the 1904 Parliament House, Perth, comprised of intricately carved Donnybrook Sandstone.

Panel 2 Melbourne

2A Modern view of "The Pond" or "Turning Basin" of the Yarra River, looking east from near the King Street Bridge.

2B View southwards down Elizabeth Street, central Melbourne, looking from a position just north of LaTrobe Street. This was once the course of the main drainage line of the settlement area.

2C Bluestone (basalt) paving stones of a laneway in central Melbourne.

2D Intricately ornamented bluestone (basalt) being used as a base for iron railings in the central city area.

2E Bluestone foundation stone and base courses of the Town Hall, Swanston Street, Melbourne.

2F Ornamental bluestone base courses, steps and a basement public lavatory entrance at the Melbourne Town Hall (facing Collins Street).

2G Bluestone former wool store on the corner of William and Bourke Street. Bluestone was less commonly used to construct complete buildings.

2H Detail of the blue stone construction of the former Methodist Church in Lonsdale Street.

2I The Wesley Uniting Church in Lonsdale Street, central Melbourne.

2J Bluestone forming the southern bridge pylon and the river embankment at the south end of Queens Bridge.

2K Weathered and worn simple kerbing and gutters formed from both roughly shaped and cut bluestone blocks, in a side street near the Queen Victoria Markets.

Panel 3 Adelaide

3A The former Hallett Cove Sandstone (known as limestone) quarry on the south bank of the Torrens River, at the boundary of the Government House grounds.

3B View eastward along the Torrens River from the 1879 Albert Bridge (Frome Road), which is east of the modern CBD.

3C View of the Torrens River to the west of the main settlement area, now occupied by the weir walls that form the modern ornamental lakes.

3D Detail of the stone wall bordering the grounds of Government House on North Terrace, Adelaide. The upper courses are constructed from irregular blocks of pale-yellow calcrete ("limestone").

3E Calcrete rubble and brick wall bordering St Laurence's Catholic Church, corner Buxton and Hill Streets, North Adelaide.

3F Old Parliament House on North Terrace Adelaide (1843, 1857). This original legislative building of South Australia was constructed from Hallett Cove Sandstone and calcrete quarried from along the banks of the Torrens River.

3G Typical double-storeyed and verandaed stone building constructed from shaped blocks of Adelaide bluestone.

3H Detail view of the Allan Campbell Buildings of the Adelaide Children's Hospital, King William Road. Bluestone ("slate") forms the base courses and foundations of this building, while the upper part of the structure is built from faced sandstone (freestone).

3I Detail of stone building façade of a corner building on Pirie Street, Adelaide. This example shows well-shaped rectangular Adelaide bluestone blocks and fine mortar pointing integrated with brick detailing of window and door arches.

3J Adelaide bluestone used as the foundations, base courses, and arched entrance steps of this 1920s building on Kintore Avenue, which was formerly a Teacher Training College.

types shown on modern zoning maps still frequently coincide with those of the subsurface geology. For example, the borders between industrial and residential use zones in both inner Sydney and Melbourne closely follow the geological boundary between Quaternary sediments and basement rocks. In Melbourne, this is the contact between the Yarra Delta sediments overlying the Silurian basement and Cenozoic volcanics. In Sydney, it corresponds to the unconformity between the Triassic Sydney Basin rocks and the overlying sediments of the Botany Basin. Similar relationships are seen to varying degrees in the other cities.

GEOLOGY, LANDSCAPE, AND THE BUILT ENVIRONMENT

Building stone was one of the first resources that European settlers sought when they landed in Australia. As local stone resources were identified and exploited, the local geology began to immediately influence the character of the built environment. The oldest and most substantial historic buildings were usually built from the most readily available stone. Sydney had the Hawkesbury Sandstone, Hobart had the Triassic sandstone (and the often-neglected dolerite, and its gravel weathering product); Brisbane had the tuff (porphyry) and Ipswich Formation sandstone; Perth is most characterized by the Quaternary Tamala Limestone (and Donnybrook Sandstone); Melbourne is built (most characteristically) from Cenozoic basalt (bluestone); Adelaide is built from calcrete, slate, and bluestone (metamorphic rock); and historic Darwin (Palmerston) is most defined by porcellanite (silcrete/laterite) (figures 8.1 and 8.2).

Locally sourced stone imparts character to both humble and great historic buildings alike, and to the smaller elements of the built environment, such as retaining walls and kerbing. It was the local sources of suitable building stone that had the greatest impact on the built environment of each settlement. Unique and characteristic local stone was sourced for early buildings and other structures that now define the distinctive built character of each city.

Stone-built structures form unique elements of each capital city when concrete structures look the same everywhere. In many ways, Australian capital cities only maintain their individuality because of the surviving buildings and other architectural elements that were created from local stone.

Brick Clay and Lime Mortar

The use of stone triggered a need for lime mortar, which led to the search for limestone resources. The short-term need was met by using the relatively poor-quality materials that were initially available, such as coastal shell deposits. In 1788 Sydney, lime could only be produced by burning intertidal shell deposits (Philip 1879), some of which may have been Aboriginal middens. Raised coastal shell beds were important lime sources in early Hobart (Banks and Leaman 1999, 47). Other early sources of limestone were the products of Quaternary to recent weathering, such as the chemically precipitated calcrete formed in soil profiles near Adelaide, and on basalt west of Brisbane. The Quaternary Tamala Limestone supplied most of the early stone and lime resources needed in Perth and Fremantle.

The need for building materials provided much of the initial impetus for new settlements to expand into their hinterlands. The discovery of clay resources suitable for brickmaking led to the establishment of the Brickfields village south of Sydney Cove. Timber getting and then brickmaking (and the location of limestone resources) drove the first northward expansion of Hobart. The discovery of limestone deposits west of Brisbane led to the foundation of Ipswich. Brick clay was first found in East Perth, but the pits were rapidly exhausted. More extensive deposits were later found in the alluvial clays and silts of the Swan River Valley, where the abundant supply made up for the lack of good building stone, so many buildings in Guildford were made of brick. Brickmaking in Melbourne began using local clay sourced from Quaternary sediments and weathered Silurian basement rocks on the southern banks of the Yarra River, opposite the settlement.

WIDER APPLICATIONS OF LANDSCAPE RECONSTRUCTIONS

Most ancient European and Middle-Eastern cities originated because of "local geological conditions," and the sites were not randomly chosen. The choice of location stemmed from the application of "very elementary geological reasoning, even though it was not recognized as such." Ancient settlements could begin as the locations of "mines, good building stone, passes through mountains, rocky crags, good water supply, natural harbors, river mouths, [or] fords" (Legget 1973, 17). The elementary geological reasoning Robert Legget recognized in the ancient world was also applied by the British across Australia when they chose a particular place for the foundational settlements of the late eighteenth and early nineteenth centuries. Geological reasoning was also fundamental to every decision made about the location, layout, and land-use divisions within the villages, towns, and cities as populations increased and they expanded beyond their foundational footprints.

The historic impacts of now lost or modified landforms can be recognized because Australia was colonized in the late eighteenth and early nineteenth centuries, at a time when accurate surveying and scientific methods were being developed and applied (e.g., the ability to accurately calculate longitude, Sobel 1996). The survey maps and charts that documented Australian landforms in this era are usually very accurate and provided a reliable source of information for reconstructing lost or modified topographic features and the natural environments that once occupied them. Other recent workers have also found the accuracy of the maps and surveys of this era to be a valuable resource for the developing field of historic landscape and environmental reconstruction. The 1782–1783 British Headquarters Map of Manhattan was used as the foundational data for the Mannahatta Project, the reconstruction of the 1611 topography and natural environment of Manhattan (Sanderson 2009, 53–65).

Australian cities exhibit the surprising and subtle ways in which geology and landscape can shape the growth of a settlement. The small beginnings of Australian capitals are well documented as maps, factual and narrative descriptions, official historical documents, and as artworks. Each stage of the transition from natural places to intensely urbanized landscapes is usually well-recorded. The ability to see these places as they were at first settlement, and then to view their subsequent early growth shows the effects that a variety of geological environments and landscape features had on multiple urban centers founded by similar people. Even though the examples discussed are all Australian, there are applications in the wider field of international urban geology.

The documentation of Australian settlements differs from those in Europe and the Americas, Africa, and Asia because of their accuracy. Portuguese, Spanish, Dutch, British, and French colonial settlements on other continents were largely founded in the seventeenth or early eighteenth centuries, and prior to the development of modern survey methods. The Australian examples provide a framework for interpreting how less well-recorded cities evolved in their landscapes, under geological influences, during their foundational years.

As Australia was being colonized, there was a concurrent groundswell of professional and amateur interest in the natural world. This resulted in detailed scientific documentation of the flora, fauna, and geology of the newly settled regions. The scientific databases (physical collections and artworks) created during this period are also a valuable resource for reconstructing lost environments in city centers, when evaluated within the context of the restored landforms.

Cenozoic (Holocene) Environmental Changes

The settlement landscapes of all seven Australian capitals show the importance of Cenozoic and recent geology and processes, and the effects of the sea-level and climatic changes of the past 15,000 years. These recent geological events have left indelible imprints on the urban environments of each city.

The landscapes of Australia's city centers reached their current form during the Cenozoic, through the interactions of varying marine environments with an evolving coastline, marine interactions with basement rocks, active volcanism, erosion, and sometimes deep tropical weathering. The effects of these processes during the Quaternary, encompassing the last 2.6 million years of the Cenozoic, were the most important. It was a period when the Earth's climate was strongly influenced by bipolar glaciation, which in Australia ranged from periglacial processes to deep tropical weathering and leaching. It is comprised of two epochs, the Pleistocene and Holocene (Pillans and Gibbard 2012, 979). As should be expected from seven cities that are located on the fringes of a large continent that extends from the tropics to cool temperate latitudes, the effects of the Cenozoic and recent geological processes were very different in each city.

Deeply incised coastal river valleys were drowned and the estuaries on which Sydney, Hobart, Melbourne, and Darwin lie were created. The deeply incised erosional valleys formed along the fault and fracture systems in the Hawkesbury Sandstone of the Sydney region were partly inundated during the last postglacial marine transgression between 20,000 and 6,000–7,000 years BP (Harvey and Caton 2010, 83). The lower eastern reaches of the Parramatta River (Port Jackson), Georges River, and Cooks River (Botany Bay) formed most of the important landscape features of Port Jackson (such as Sydney Cove) but left the headwaters occupied by relatively minor rivers and mudflats. An extensive icecap covered the Tasmanian central highlands in the late Cenozoic (Kiernan 1991), with the first and most extensive in the early Pleistocene. The ice-cap runoff created the Derwent Valley (Vicary and Mazengarb 2019, 17) along the structural weakness of the Derwent Graben (Stacey and Berry 2004), creating an often deeply incised river system. The lower reaches of this system were invaded by the ocean from 20,000 to 6,500 years BP, forming the River Derwent Estuary (Harvey and Caton 2010, 83).

Subsequent sea-level retreat of several meters in the last 5000 years created the landscapes that Aboriginal people colonized, and which Europeans first encountered. Sea levels have fallen since the mid-Holocene, exposing sand bars and marginal sedimentary sequences at the current shorelines that are now within the limits of the central city areas. Many of these recently exposed sediments were important locations of food production, freshwater aquifers, and had an influence on the locations of the first suburbs, the location of industry, or transport infrastructure.

Early Cenozoic uplift in the Brisbane region led to extended erosion and local continental sedimentation, including the Ipswich Basin coal measures, and associated sedimentary rocks. Late Cenozoic faulting near the present coastline, coupled with sea-level fluctuations strongly influenced coastal physiography. An extensive superficial cover of Quaternary and recent sediments was formed during the erosion of the current landscape and was deposited in low points in the landscape. The extensive, low-lying areas became the site of the extensive coastal marshlands, wetlands, and drainage lines along the shores of Moreton Bay, and also formed alluvial flats on the inside bends of the Brisbane River.

The sea reached much further inland at the site of Perth . On the Swan Coastal Plain, the Darling and Gingin Scarps marked the most eastern extent of Cenozoic and Quaternary marine erosion and deposition (Davidson 1995, 5–6), with the result that the landforms of central Perth have been more strongly influenced by Quaternary geology and geological processes than any other Australian capital. The dune systems formed as the ocean retreated across the plain created the landforms of the urban area. Depressions between dunes hosted freshwater wetlands and the dissolution and re-deposition of calcium carbonate within the sands formed most of the materials used in early buildings.

Cenozoic volcanics partially covered the site of central Melbourne at 42–57 Ma, (Older Volcanics) and 0–5 Ma (Newer Volcanics, Price et al. 2003). The fault-bounded Port Phillip Sunkland basin was inundated by the sea at the end of the last glacial maximum. Sea levels fluctuated but are now lower than their peak, when the shoreline lay along the low cliffs formed mainly by the Older Volcanics at the western edge of the settlement site. Lava flows of the Newer Volcanics mostly inundated the ancient basement topography, creating the western volcanic plains. Deposition of Quaternary marine deltaic deposits at the Yarra River mouth occurred as the sea level fluctuated. The deltaic sediments were exposed during the last retreat of the ocean, which left the site of early Melbourne bordered by wetlands formed on recent estuarine and marine sediments, particularly the West Melbourne Swamp, but also by the swamps of the river and those extending to the south-east. Melbourne bordered these wetlands in 1836, and they were controlling factors in the early growth of the town.

The geology of the Adelaide Plain is dominated by a sequence of Cenozoic piedmont deposits that unconformably overlie ancient basement. They are intercalated with marine deposits from various Holocene marine incursions (Howchin 1918, 142), of which the most important is the late Cenozoic Hallett Cove Sandstone, which underlies much of the original settlement. Superficial Quaternary units unconformably overlie the older Cenozoic sediments and were deeply weathered and heavily calcretized. Calcrete is important geomorphologically to the layout of the 1836 settlement and was a stone source for early buildings and lime-making. A thin mantle of younger sediments overlies the marine sequence, formed mainly as alluvial outwash fans shed north-westward from the Mt. Lofty Ranges (e.g., Twidale 1976,

49–52), and as alluvial sands and gravels deposited by the Torrens River and its tributaries. There are also coastal wetlands, dunes, and mudflats (Twidale 1976, 49–52), which would later become the location of the airport.

Cenozoic geological processes have most strongly influenced the coastal sites of Australian settlements (particularly during the Holocene to Recent). The British arrived by sea and chose places in active erosional and depositional environments that were subject to the effects of recent sea-level fluctuations and climatic variations. Australia remains a coastal nation, with more than 90% of the population living within 100 km of the coastline (Australian Bureau of Statistics 2019). These processes still have the greatest impact on the places where most Australians live.

Landscape Reconstruction and Urban Planning

Reconstructions of pre-European landscapes provide useful information for urban planning: in the areas of geohazard management; heritage assessments and the interpretation of historic land-use patterns; urban and built character; and historical, archaeological, and environmental investigations in urbanized settings.

The examples of landscapes described in the preceding chapters are all retrospective. They describe how landform and geological factors influenced past decision-making. Despite this fact, the observations of the past can provide guidance as to how these same factors may affect future developments, particularly in the fields of geohazard prediction, urban planning, heritage assessment, and the aesthetics of the built environment.

The geological evolution of the settlement landscapes has not ended, and they are constantly changing. Old landscape features are being modified, and new ones are formed by ongoing natural processes, often in the form of so-called geohazards. Landscape reconstructions can be used to predict potential geohazard locations, including potential locations of flood risk, unstable slopes and mass flows, coastal erosion, and of acid-generating soils (which can occupy the locations of former wetlands).

Pre-settlement wetlands and drainage systems are reclaimed by nature during exceptional storm events. Recent intense rainfall overwhelmed the storm-water systems and triggered flooding on the Brisbane River (2011), Elizabeth Street in Melbourne (2018), and in Hobart (2018). The flood events restored many of the pre-European water courses to their ancient purposes. The Brisbane River flooding reclaimed many of the marginal wetlands created over thousands of years of landscape evolution. The Hobart and Domain Rivulets reclaimed their lower reaches and original estuary, and the Elizabeth Street Creek became a torrent once more. In each case, a comparison of the modern flood maps with the historic landscape reconstructions shows that the parts of pre-European landscapes that were previously wetlands and creeks were reclaimed during these floods. Most municipalities now provide flood-prediction maps for planning purposes that could be improved with better knowledge of preexisting landforms.

Coastal and beach erosion are active along many urban coastlines, and it is useful to identify erosional shores prone to coastal and beach erosion to inform better planning and development approvals. Depositional shores are also important to define, particularly regions of constantly shifting sand banks and mudflats, which may require constant dredging to keep rivers and ship navigation channels open, or for deepening to reduce potential flood bottlenecks.

A better understanding of the city center landscapes at settlement provides a contextual framework for interpreting the archaeological evidence of pre-European Aboriginal occupancy, and the historical archaeology of each colony. Knowing more fully where an archaeological site fits into its wider original context may allow patterns in site distribution to be recognized that may not have been previously apparent.

Being able to place the information preserved in scientific collections within the reconstructed landscapes could provide a firm context for understanding the pre-European natural environments of the modern city centers. It may also assist in the reconstruction of original landscapes and environments. And it could help to assess how climate and sea levels have changed since the turn of the nineteenth century.

Public Domain Urban Geology Information

The research for this book involved an extensive search for publicly available geological information about the city CBDs. It was surprisingly hard to find in Australia, despite the fact that subsurface infrastructure projects began very early in some Australian cities, with Busby's Bore started in Sydney in the 1820s. Since then, the exploitation of the subsurface of cities has greatly increased and there have been many large excavations for deep foundations and basements, car parks, underground railway station, tunnels, service conduits, and sewers. All such projects are dependent on detailed geological and geotechnical knowledge. The information that is publicly available is mostly two-dimensional government geological mapping at a range of scales, including 1:25,000 (Hobart), 1:50,000 (Perth and Darwin), 1:31,680 (Brisbane), 1:100,000 (Sydney, Brisbane), and 1:63,360 (Melbourne). Most maps predate the 1990s (except Hobart) and some predate the metric era.

More detailed geological maps focused on the central cities are rare. Published 1:5,000 scale (Brumley 1974b) and unpublished 1:5,000 scale geological maps (Neilson 1989) of the Melbourne CBD exist. There are cross sections in technical reports by Golder Associates (2016); and with the Adelaide maps in Selby and Lindsay (1982). All the publicly available geological maps show the distribution of geological units rather than the material properties of substrates, with the important exception of the 1:50,000 scale environmental geology maps of Perth and Fremantle (Gozzard 1983, 1985, and 1986). Structural information is almost non-existent.

Very little other publicly available information has been created from the many geotechnical drilling investigations, excavations, and tunnels that have been developed in most city centers since the 1970s, with Grubb's (1984) unpublished Brisbane thesis, as a further exception. Most of these sources are now over two decades old.

Modern engineering geology maps that integrate geological observations with geotechnical drilling and materials testing programs to generate engineering geology units would be a useful planning tool for all urban areas. The sources of information would include geotechnical, geological survey, and civil engineering reports about harbor works, railways, roads, sewerage, service ducts, and tunneling. Generally, only summaries are presented in the public domain (e.g., Pells 1993; Och et al. 2004). For example, in Sydney, detailed information about the variability of bedding orientations across the CBD is particularly lacking, as is knowledge about the geotechnically significant dikes that traverse the city center. Public information about the latter is still largely dependent on the work of Pittman (1903) and Willan (1925), supplemented by Chesnut (1983, 1979) and Och et al. (2004). Some useful information has been published about the location (as surface traces), width, character, and orientation of faults and joint swarms that cross central Sydney (Chesnut 1979, 1983; Pells et al. 2004; Och et al. 2004, 2009), but this publicly available work is also fairly generalised.

The lack of availability of modern information for Australian cities is surprising, given the number of large-scale excavations that have occurred both above and below the surface in CBDs since Robert Legget wrote in 1974. In Australia, Legget's message about the need for geoscientific information for city growth was heeded for individual construction projects (by preliminary geotechnical investigations), but the importance of making the information more generally available through local and state governments repositories was not (Legget 1974, particularly chapters 3 and 9). There must be vast amounts of information available, yet very little has entered the public domain where it could be available to local government authorities, civil construction consultants, and academic researchers. In the places where there is the most need for detailed geological data, it is not available.

And there are only a few specific references to the geology of Australia's capitals, including, *Notes on the Engineering Geology of the Sydney Area* (Chesnut 1976); *Engineering geology of the Sydney Region* (Pells 1985);

The Engineering Geology of Melbourne (Peck et al. 1992); the *Engineering Geology of the Adelaide City Area* (Selby and Lindsay 1982); and the unpublished *Engineering Geology of the Central Business District of Brisbane* (Grubb 1989). Gozzard (1985, 1986), and Commander (2003) in Perth. There are also Pell's additional writings on Sydney engineering geology (e.g., Pells 1993), and isolated papers elsewhere in the literature (e.g., Och et al. 2004). The published geological information about inner cities is often subsumed into a regional geological context on geological maps and in map explanatory notes (e.g., Geological Survey of Victoria 1967). With the information that is available, it is particularly difficult to separate specific information about the small but infrastructure-rich CBDs from descriptions of the surrounding regions. While the regional context is important, this tends to diminish the focus on the important details required to understand the small areas of the inner cities.

There are no publicly available 3D models of the complex geology beneath Australian cities. Report figures presented by Golder Associates (2016) and Holdgate and Norvick (2017) show that such models exist for central Melbourne and the Yarra Delta areas, but they are not open source. Two open-file drill-hole databases were located during research for this book: for Perth (Stewart 2003, made available by the Australian Geomechanics Society) and Melbourne (Holdgate and Norvick 2017, Yarra Delta drill-hole information provided as a supplementary table). Both these data sets were used to successfully create 3D geological drill-hole models. There is enough drill-hole information for Hobart in Hofto, Sloane and Weldon (1991) to also create a database but it was not completed due to time constraints. Three-dimensional geological models would be very useful planning tools for the major cities in Australia, as the foundation for major infrastructure projects and planning decisions. This may involve regulations that require all geological and geotechnical information collected during civil projects to be made available to government organizations, in the same way that all exploration data must be provided for exploration leases. The state geological surveys would be the logical custodians of such an important resource.

The analysis of historic information undertaken for this study has emphasized how important small geological and landscape details were in the past. It seems likely that understanding such details into the future will be required to manage urban development. The geology beneath our largest cities is poorly understood at the fine scale required to tunnel and directional-bore, and to create service tunnels and underground structures, rail tunnels, sewers, and the foundations of ever-larger skyscrapers within diverse and variably weathered geology. It is increasingly excavated and tunneled and the interactions between all these works can only be understood if the details of the geology are known, at very fine levels of scale and detail. There is a need to understand the variability of rock types, weathering, structure, and fresh rock mass properties, particularly how they will respond to the long-term effects of the ever-increasing weight of structures that must remain upright above subterranean infrastructure excavations. More cost-efficient building design may come with greater geological and geotechnical knowledge of the substrate. There is a need to document and understand the subterranean geology of these once small settlements (more than ever before).

What is needed are consolidated geotechnical and geological drill-hole databases and 3D geological and rock mass block models of the subsurface geology and subsurface infrastructure. Those models should include features such as weathering profile, soil type, acid-generating soil, and detailed structure (joints, faults, cleavage), the variability of rock properties (e.g., a rock mass block model) and the excavation properties. Detailed geological knowledge at very fine scale should be gathered for all urban areas, at densities and quality comparable to that collected in large underground mining operations for resource estimation and geotechnical management.

The need for geological knowledge about the substrates of cities is probably greater than that needed to define the value and to safely and efficiently mine a great orebody. The data available should all be consolidated into one central repository, including all geotechnical drilling and geotechnical testing, excavation records (wall mapping and geological feature measurements), groundwater information, and rock mass properties. An example of the detail required is the recent investigation of the weak basal zone of basalt flows beneath Melbourne, which highlighted such horizons as a geotechnical risk during tunneling (Schofield 2014).

THE FUTURE OF AUSTRALIAN CITY LANDSCAPES

The consideration of how the seven foundational settlement sites will continue to be developed is essential. Each site remains the dominant urban center of an enormous hinterland; home to most of the population, and the focus of transport networks that radiate outward from their CBDs. They continue to serve as the centers of government, business, and retail activities, as they have done since their beginnings. And they remain the focus of cultural activities, as the locations of the most prestigious universities, libraries, museums, art galleries, and sporting facilities. The focus of so many activities in these remarkably small areas, must lead to continuing development pressure and change.

Landscape and Character under Threat

The individuality of the Australian cities is now under serious threat from the homogenization wrought by redevelopment, and the loss of the relationship with their landscape settings. The topography of urban landscapes is being obscured by suburban consolidation, particularly by high-rise apartment blocks and office towers. Where once the natural undulating form of the landscape remained recognizable under low-density buildings, with large yards and space for trees, there are increasing numbers of angular and jagged skylines where the landforms are obscured.

There is an ongoing loss of the relationship between the built environment of the inner cities and their natural landforms. This is most typified by the "walling-in" of rivers, parklands, and coastlines by high-rise office towers, apartment blocks, and hotels hungry for any view, so placed around their margins. The result is that the vistas that once existed from the first settlement sites are now lost, and the rivers turned into canals bordered by buildings.

There is an ongoing loss of the relationship between historic landmark buildings and the built environment and landscape contexts they were originally designed to occupy. Historic buildings are now often preserved as isolated artifacts when the surrounding buildings, or non-heritage-listed extensions, are removed, or when infill developments are constructed in former open spaces that once surrounded them (such as gardens, yards, and lanes). Unsympathetic extensions to many buildings are also common (e.g., the Melbourne GPO). This process can leave a building that was once an integral part of its urban landscape as an isolated artifact set in profoundly changed surroundings, with no historical context. This process is particularly evident on St Georges Terrace in Perth, where the Barracks Arch, the Cloisters, and the former Perth Technical School seem out of place in their modern surroundings. At its worst, this process results in the historic building becoming an anachronistic remnant in its redeveloped setting.

Buildings designed to be set in a landscape have been partly or completely divorced from their intended setting, and the relationship obscured. The open spaces that once surrounded key buildings have been partly infilled with unsympathetic developments (e.g., the Exhibition Building and former Treasury Building in Melbourne). In Brisbane, historic government buildings that once had grand views of the river and were designed to impress from the main anchorages are now obscured and overshadowed by surrounding buildings.

Massive, out-of-scale developments are often crammed into heritage precincts and historic suburbs (e.g., Queens Wharf in Brisbane). The worst examples are those where monolithic structures are constructed over the top of, and enveloping historic buildings, which even cut them off from the sky. Examples include the Shot Tower in Melbourne, and the Hibernian Hall (a gutted shell stripped back to its basic structure, with the historic patina erased) and the new museum complex in Perth. In such cases, the historic buildings have been preserved, but at the cost of almost every aspect of their historical landscape context. This process can go even further, with only the façade of an original structure being left attached to a modern tower. Other world cities build new complexes to house museums and galleries while leaving their historic centerpiece buildings preserved (e.g., the Smithsonian Institution in Washington and the British Museum in London). The centrality of Australian cities and the focus of so many land-use demands into such small areas seem to generate the need to cram buildings onto the same sites.

The once unique built environment of each city is also being homogenized and the unique architectural character is being lost. The same types of materials are used in all modern cities, including imported facing stone that is cheaper than local materials. The original influences of local geology on construction materials and architecture are being lost from the city environment. The cities will no longer be a physical part of the places they occupy.

Land-use patterns that were established in the earliest days of the settlements are also changing. Land previously considered unsuitable for dwellings and reserved for industry has been converted to residential precincts (e.g., the Melbourne docklands). While zones formerly reserved for transport infrastructure, such as railway yards, wharfs, and airfields, have been subdivided for residential developments. Health precincts have been repurposed. The site of the former Royal Adelaide Hospital is being partly adapted to university and inner-city residential uses, while the new hospital occupies former railway yards that were originally excised from the parklands in the nineteenth century. Former defense sites have also been redeveloped, particularly if they have river or coastal locations.

Rarely, historic open spaces and occasional landscapes are preserved, such as the Parramatta Domain, the Sydney Domain, the Queen's Domain in Hobart, and the Adelaide Parklands. Brisbane has the City Botanic Gardens, where there were once convict farms. Melbourne still has the Botanical Gardens and Royal Park (among others), and Perth has Kings Park (Mt Eliza). Significant areas of native vegetation remain in some inner cities, such as the Queens Domain (Hobart) and Kings Park (Perth). Re-establishment of native ecosystems has occurred in some places, as part of urban pollution management, such as at Frasers Point in Perth, where wetlands have been reintroduced as part of the parklands.

Nevertheless, such open spaces are under constant development pressure, particularly at their peripheries, through incremental encroachments. They provide convenient locations for transport infrastructure, such as tunnel portals and ventilation shafts, car parks, railway stations, and motorway flyovers. Ever-larger sporting and entertainment facilities, and the infrastructure to service them, and public art installations, further degrade the landscapes that survive.

FINAL STATEMENT

The modern CBD of each Australian state capital was an insignificant place before European settlement. They merge into their backgrounds and become parts of much wider natural landscapes and environments. They were not stand-out locations until Europeans arrived and made them so. They were also probably not particularly special places for local Aboriginal groups either, because they were only small localities within the far wider, rich, and diverse natural environments of extended territories. The small areas of the future CBDs were traversed, favored places were used at times for camps, and some parts provided resources that were exploited seasonally, but they were probably not permanently inhabited.

There are many historical similarities in the origins of the seven Australian state capital cities. This might be expected, given that they were founded by people from the same European nation, and within a very short period of time. Even their original purposes fell within two narrow categories: they were either established as military penal establishments or as planned civil administrative and commercial settlements. Regardless of their original functions, they all rapidly evolved to serve identical purposes when they became the commercial, transport, administrative, and cultural centers of their respective colony (and then state). The relatively homogeneous British cultural origins could have led to greater uniformity in layout and architecture than occurred. Yet they are not homogeneous. Why? The answer seems to be that each settlement developed its own unique character because of the geologically diverse coastal or riverine landforms of their locations, and because of the local natural materials that were available to build them. Each city was rapidly imposed on a virgin landscape, but the landscape was immediately embedded in each city, as a fundamental aspect of its unique character.

The basic geological reasons that underlay the foundation of the Australian capital cities remain the overriding considerations for their continued urban growth, inner suburban consolidation, and outward urban expansion. The supply of fresh water; the provision of transport and communications, and the delineation of places where safe housing can be constructed are still the fundamental requirements of the populations of each city. And the fundamental need for a system of food production and distribution remains. Modern urban planners still consider the supply of fresh water, the development of transport corridors and infrastructure (roads, railways, and airports) and undertake geohazard risk assessments before approving any residential housing development. And the provision of retail centers is a part of any suburban subdivision. The fundamental requirements for successful permanent settlements recognized by the British founders of Australia's modern capital cities remain largely unchanged after more than two centuries of growth and social change.

Appendix

Figure I.1. The evolution of the British colonisation of Australia. Each panel shows the foundation year of each colony and the location of the principal foundation settlement discussed in this book (shown as crosses). The boundary of the area claimed by Britain at that time area shown as a dotted line. South Australia, Victoria, and Queensland were separated from New South Wales in the years shown in brackets in the lower two panels. The lower right figure also shows the final configuration of the Australian colonies and territories, which became the modern Australian states. This figure is based on the front and endpaper maps of Statham (1989) and a topographic base from Geoscience Australia.

Figure I.2. Many places beyond the main settlements were important during the first years of each Australian colony. The locations of the most important places mentioned in the text are shown on the inset maps.

Figure 1.1. Topographic reconstruction of the landscape context of Sydney Cove in c1770–1792, with an emphasis on the region between the southern shore of Port Jackson (Sydney Cove) and Botany Bay. It was in this relatively small area that most important historical events took place during the first year of European settlement.

The landscape was an extensive and interconnected wetland system developed on the Quaternary sediments of the Botany Basin. Linked wetlands drained all the land south of the Oxford Street drainage divide, passing through a succession of freshwater swamps and marshlands, coastal saltmarshes, mangroves, and then mudflats before entering Botany Bay. Often, drainage lines passed through systems of coastal sand dunes before entering the bay. The extent of coastal mangroves, mudflats, saltmarshes, and wetlands surrounding Botany Bay was a factor that quickly deterred Arthur Phillip from creating a permanent settlement there in 1788. This reconstruction is based on multiple, cross referenced historical charts and survey plans dating from 1770 to the early twentieth Century.

Figure 1.2. Inset 1. The geology of the Sydney region, simplified from Brunker and Rose (1967), with additions from Chesnut (1978) and Och et. al. (2004). Note the significant expanse of Quaternary sediments immediately south of Sydney.

The boundary of the Botany Basin Quaternary sediments is also shown in the main figure to illustrate their spatial relationship between distribution of wetlands, swamps, salt marshes, mudflats and mangroves, and the Quaternary deposits.

Inset 2. Reconstruction of the pre-European vegetation of the Sydney region. Modified after Benson and Howell (1990).

More detailed information about mangrove and saltmarsh distribution on Botany Bay is shown in the main figure. This is also based on Benson and Howell's (1990) work but supplemented by additional information from charts and maps ranging in age from the 1830s to the early 20th century. The succession of historic maps and charts (e.g. Dixon 1827; Willis 1859) show that mangrove distribution at the headwaters of Botany Bay remained stable throughout most of the nineteenth century, particularly at the mouth of Shea's Creek and in the lee of Towra Point. Mangroves were well-established along the southern shores of Botany Bay by 1827–1830 and occurrences remained largely unchanged throughout the nineteenth to twentieth centuries. The map sources have been supplemented by the contemporary written accounts of Cook (1893), Tench (1789), and Phillip (1789).

Figure 2.1. Reconstruction of the topographic and cultural features located between the 1804 settlement at Sullivans Cove, and the free settler land grants at New Town Bay (Stainforth's Cove). The need to travel from the main settlement to the free settler's farms resulted in the development of an early track and road network that eventually evolved into Elizabeth Street and Main Road. The bridges (and probably earlier fords) required to cross the deeply incised, eastward draining rivulets formed pinch points in the road network. Several cultural features from later periods are also shown, including the brickfields, and the limekilns and the lime deposit on Limekiln Hill. The occurrence of these important primary resources spurred much of the early northward development of Hobart, which followed the first timber felling immediately north of the rivulet.

The approximate boundaries of several of the 1804 New Town free settler land grands are shown. The linear geometry of these rectangular blocks, with their narrow frontages along the New Town Rivulet, still largely control the modern street network of New Town and the post war subdivisions of Moonah. The location of Meehan's 1811 Hobart Street grid is also shown.

Inset. The geology between Sullivan's Cove and New Town Bay. Hobart spread northwards along a lower-lying valley formed on Triassic sandstone. The valley was bounded on either side by tilted blocks of Jurassic Dolerite. Cenozoic basalt surrounding New Town Bay and extending to Moonah produced fertile soils that became important agricultural land, including the first government farm, and was later the site of Chinese market gardens. The location of the Cascades Fault Zone, marking the western margin of the Lower Derwent Graben, is in the lower left corner. The geology is based on that of Forsyth and Clarke (1999).

Figure 2.2. Topographic reconstruction of the 1804 landscape of Sullivan's Cove. The layout of the initial camp is shown, based on Harris (1804) and Bolt (2004) with additions from Mault (in Walker 1889). The first camp was focussed along a lightly timbered and rocky central east-west ridge mainly formed by sandstone. The woodland became drier, stonier, and more sparsely timbered as the land rose gently to the west. The convicts were housed in tents and huts along the southern side of the deeply incised rivulet, where a belt of thick "brush" (scrub and remnant temperate rainforest) occupied its course between the steep sandstone banks. Important officers and civilian staff were located along the crest of the steeply bluffed shoreline, overlooking the anchorage.

James Meehan's 1811 surveyed street grid is shown as an overlay. This remains the streetscape of the oldest part of the modern Hobart CBD. The 2 m contours shown are derived from the 3D topographic terrain reconstruction.

Dolerite hills near the camp, such as Battery and Macquarie Points, and Barracks Hill, would later become sites of important military infrastructure, such as gun batteries, barracks, and semaphore stations, and the new Government House.

Figure 3.1. Topographic reconstruction of the landscape surrounding the Moreton Bay Penal Settlement on the Brisbane Peninsula circa 1825 to 1844. The hinterland of the settlement was a landscape of curvilinear valleys and bounding ridges with variably marshy, low-lying valley floors, marshes, and wetlands. The drainage pattern and ridges were influenced by the foliation trends in the Neranleigh Fernvale Beds and Bunya Phyllite, and by the contacts and joint patterns of the Brisbane Tuff, a highly resistant linear unit of ignimbrite that infilled the older basement topography. The Normanby Fault Zone was a significant control on drainage direction on the western margin of the peninsula.

A general reconstruction of the original native vegetation circa 1825–1844 is shown. This is largely based on the survey mapping by Henry Wade (1842, 1843, 1844). With additional details from various other maps, and written descriptions of John Oxley and Edmund Lockyer.

The geometry of the land blocks first surveyed by Henry Wade, to the northeast of the main Brisbane settlement in the early 1840s, is preserved by the urban landscape of the modern suburb of Eagle Farm.

Inset. The geology of the Brisbane region. The geological units are simplified from Houston and Tucker (1965), with additions from Rands (1887), Grimes, Cranfield et al. (1986), Grubb (1989), and Bennett and Norbert (2014).

Figure 3.2. Topographic reconstruction of the Brisbane Peninsula and the Moreton Bay Penal Settlement, circa 1826–1830, based on historic mapping and a 3D terrain reconstruction model. This plan view shows the layout of historic buildings and land use boundaries, waterways and wetlands, rocky bluffs, and cliffs. Also shown is the modern street grid of the CBD surveyed by Henry Wade in 1842–1844, as Brisbane was opened to free settlers. Wade was tasked with imposing a grid on the 'organic' settlement that had grown as a military style penal settlement.

The first farms of the penal settlement were established on areas of well-watered but low-lying Quaternary sediments. These sediments mostly occur on the inside bends of the river and were mostly marshes and swamps (and some rain forest) in 1825. The boundary of the Quaternary sediments is shown as an overlay (simplified from the Geological Survey of Queensland, 1965). The locations of the original penal settlement farms remain as the City Botanic Gardens.

The feature labeled 'Jetty/Wharf' indicates the likely first landing place chosen by Henry Miller. This is the location where a rocky shoreline first appeared on the peninsula and was developed into Queens Wharf. The first road of the European settlement must have been a track that followed the rise of the land from the landing place to the north, up the gentlest incline of the steep bank, to the low linear ridge where the camp was formed. This route, dictated by the original topography, evolved into Queens Wharf Road, which is the most significant surviving remnant of the original street pattern of the Moreton Bay Penal Settlement.

The topographic contours are at 5 m intervals and are derived from the 3D terrain reconstruction model and based on contemporary maps and charts.

Figure 4.1. Reconstruction of the landscape of the central Perth region at the time of British settlement, circa 1827–1831. The map view is centered on Frasers Point and the Perth Ridge, where the terrain was dominated by a linked system of groundwater-fed lakes, swamps, and marshes formed behind sand ridges (particularly the Perth Ridge). The lakes were the surface expression of the groundwater system and the water level varied seasonally. During dry periods, many of the lakes may have been dry, or reduced to marshes or swamps. All had rush margins.

The Frasers Point wetlands were the south-eastern outlet of a more extensive groundwater drainage system that extended to the northwest, to at least as far as Herdsman Lake. It probably formed where groundwater flow was deflected around the outcropping mass of Tamala Limestone forming Mount Eliza. During the wettest times of the year, east-flowing channels probably linked the lakes to form a nearly continuous surface drainage system. The outlets of the wetlands were surface creeks, such as Clause (Claise) Brook and Walters Brook, which drained directly into the Swan River, and through intermediate saltmarshes; or via springs and spring-fed runnels emanating from the steep riverbanks in areas of higher ground (such as the Perth Ridge and Mounts Bay foreshores).

The reconstruction of the main vegetation types of central Perth, c. 1827–1831 is based on the contemporary written descriptions of Charles Fraser (1827), and observations by surveyors on land survey charts in the early 1830s. The contemporary information is augmented by modern vegetation reconstructions.

Inset. The geology of central Perth (simplified after Gozzard 1986).

Figure 4.2. Reconstruction of the landscape of the Perth Settlement c. 1830, based on historic maps and 3D topographic reconstructions. This plan shows the location of the main buildings of the camp as recorded by Arrowsmith (1839). The first settlement was a temporary camp on the crest of the steep riverbank, where a gully provided easy access to the high ground of the Perth Ridge. The camp was established in the approximate position of the current government domain, in the grounds of Government House and near where the WA Supreme Court stands.

The landing place at the base of the gully became the focus of river traffic to the camp and a jetty was developed, and eventually, a large brick-built Commissariat Store (now the site of the WA Supreme Court). This became a major transport hub, and the location remains the main jetty in the city centre (moved progressively south as land reclamation of the foreshore progressed).

The street grid of the current CBD was surveyed by John Roe in 1830 and extended northwards in 1832. The lake system that dominated the landscape north of the Perth Ridge was the major influence on the location of land available for subdivision into town blocks and strongly influenced their geometry when the streetscape was extended to this area.

Key sources for this map include row Aerosmith and written descriptions by Charles Fraser, James Stirling, and others. The topographic contours are at 1 m intervals and are derived from the 3D terrain reconstruction model, and based on contemporary maps and charts.

Figure 5.1. The regional landscape context of the 1835 settlement at Melbourne, based on a compilation of historic map information (mainly Hoddle 1835 and 1837; Selwyn 1860, and Cox et al. 1866) and 3D topographic reconstructions. The settlement site was bordered to the south and west by an extensive fresh and saltwater wetland system that once dominated the Melbourne landscape. Most drainage lines passed from freshwater into fresh and then salt marsh systems before entering the lower reaches of the Yarra River (the estuary). The Elizabeth Street Creek entered the Yarra River immediately above the falls and had to be diverted by drains to the west to remove town runoff from the freshwater supply. The Moonee Moonee Chain of Ponds dissipated in a series of increasingly saline ponds and wetlands as it entered the influence of the Batman Swamp, as did much of the western drainage from the high ground on which the settlement was established.

The distribution of native vegetation types shown is a non-specific reconstruction of the density of woodland, mostly interpreted from the survey maps of Robert Hoddle, and various other early cadastral land survey maps. Where no vegetation distribution information was recorded by contemporary observers, some conjectural infill interpretations are made (dotted lines). The distribution of native vegetation changed markedly with the underlying geology, particularly between the places underlain by the young sediments of the delta and those part of the landscape formed on Silurian basement and the Older and Newer Volcanics. The geology was the dominant influence on vegetation type and density.

It is difficult to recognize the site of the modern Melbourne CBD in this reconstruction (mid upper center) because the location was not a major topographic feature in the original landscape. Only with British settlement did it become a site of significance because it was located near the beginning of freshwater in the Yarra River. The location of the original regular street grid by surveyor Robert Hoddle (1836) is shown in pale grey.

Inset. Simplified geology of the Melbourne settlement site, based on the map compilation of Neilson (1989), with additions from Vandenberg (1975), Selwyn (1860), and Neilson (1989).

Figure 5.2. Topographic reconstruction of the landscape of the Melbourne settlement site at the time of Van Diemonian (Tasmanian) settlement, circa 1836, based mainly on the historic mapping of Hoddle, (1835; 1837), and 3D topographic reconstructions.

The first unnamed settlement on the Yarra River was founded by settlers from Van Diemen's Land (Tasmania) without government authorization. The buildings of this first ramshackle village were concentrated near the landing place, on high ground above the north bank of the river. A natural rock bar prevented navigation of vessels further upriver and marked the change from brackish estuarine to fresh water. There was low-lying marshland and swamps to the south and west and the high ground here provided the first site on the Freshwater River that could be reached by small ships.

The cluster of early buildings spread along the crest of the high north bank, and along the lower slopes of the Western Hill. John Batman built his house on a southern spur of this hill and John Fawkner built his residence on a high point overlooking the landing place. Several occupants established gardens that were situated between the crest of the bank and the river. When government authority was finally established, the appointed Superintendent (later Lieutenant Governor) built his residence and offices at the western end of the settlement, facing the river, and on John Batman's death, eventually took over his house as offices.

The settlement high ground was formed on Silurian basement and Cenozoic volcanics and sediments. The western margin of this high ground was steeply bluffed, dropping sharply away to the low-lying lands of the Yarra Delta, occupied by saltwater lagoons and marshes. The sharp change in topography coincided with the geological boundary of the volcanics.

A burial ground and flagstaff were established on a high conical eminence to the north of the settlement. This site had a clear view to Hobsons Bay and became an important early site of communication between the main Hobsons Bay anchorage and the settlement.

The first street grid surveyed by Robert Hoddle in 1835 is shown. This regularized grid was imposed on the original settlement and the footprint of the first settlement was rapidly obliterated by more formalized land allocation and development.

Figure 6.1. The landscape of the Adelaide Plains circa 1836, based on the original survey plans of William Light (published by Arrowsmith 1839), with additions for Port Adelaide by Light and others. The distribution of denser stands of woodland and forest shown is based on the information recorded by William Light in various survey plans of the main settlement and the rural block surveys of the surrounding plains. Additional details are derived from Kraehenbuehl (1996).

This regional view illustrates the strong association between the distribution of dense stands of woodland and forest and the geological units underlying the landscape (see Inset). The location of denser woodland is particularly tied to the distribution of calcrete and calcretized Hallett Cove Sandstone (east of the Para Fault Scarp) and the alluvial sediments deposited as outwash and along river courses and smaller tributaries of the Torrens flowing northwestwards from the Mount Loft Ranges.

The distribution of drainage and wetland systems is shown. The Torrens River did not directly enter the sea but entered an extensive hind-dune wetland system of reed beds—the Cowandilla Wetland system—lying parallel to the coastline. The wetlands then discharged to the sea through the Port River and the Patawalonga River, which flowed to the north and south.

Inset. Simplified surface geology of the Adelaide Plain. Based on Forbes (1980), with additions from Selby (1981).

Figure 6.2. Reconstruction of the topography of the Adelaide settlement area circa 1836. The reconstruction is based on historic maps and modern topographic data, with reference to written descriptions of native vegetation and contemporary art depicting the area (chiefly from William Light).

The scarp east of the Para Fault system is the major geological control on the landforms of the Adelaide settlement. Minor variations in the undulating topography of the high ground east of the scarp, which is underlain by calcrete, controlled the distribution of denser stands of woodland. The margins of the calcrete also provide an environment for denser woodland. The course of the Torrens River, which cuts through the scarp, was marked by alluvial sediments that hosted an open river red gum forest. The boundaries of North and South Adelaide and the parklands are also shown.

William Light fitted the required number of town blocks into this topography very well and used the high ground to full effect. The earliest years of the settlement were focused on the south bank of the river, at the northwest corner of South Adelaide. The government domain was also established centrally on the south bank of the river. Fresh water came from the billabongs at the western end of the river, where the channel became more complex as it crossed the Para Fault scarp.

Sources for this map include Light's map (Arrowsmith 1839, and others). Topographic information is based on 1:50,000 scale topographic mapping, and more detailed topography from Selby and Lindsay 1982).

Figure 7.1. The topography of the Darwin region c 1869–1870. The landscape was a deeply incised plateau formed by flat lying Cretaceous sediments that unconformably overlay a deformed and metamorphosed Proterozoic basement. The steep drainage lines of the plateau were in-filled by Quaternary and recent sediments and were occupied by wetlands and mangroves along their lower courses.

The distribution of the major vegetation types is constrained by the topography and the recent geological deposits. Particularly important were the drainage lines of the margins of the plateau which were filled with Quaternary sediments and occupied by water courses that continued to be fed by freshwater springs during the dry season. These low lying and watered places supported remnant stands of monsoon rain forest. Once cleared, they provided important patches of fertile, well-watered soil that supported market gardens and recreational areas.

Inset 1. Geology. The geology is based on Pietsch (1983).

Figure 7.2. The topography of Darwin at the time of the South Australian settlement in 1869. The landscape chosen for the first permanent settlement was a drowned mesa comprised of mostly horizontally bedded Cretaceous sediment overlying a Proterozoic basement.

The most influential parts of the topography on the first European settlement were the outcrop geometry of the Cretaceous cover sequence, which controlled the mesa geometry, and the unconformity at the basement contact, which controlled the shoreline, and the sources of freshwater (springs).

This reconstruction is based on the survey plans of George Goyder (1869) with additional details from various subsequent surveys, and 3D topographic reconstructions. The contours are at 2 m intervals.

Figure 8.1. Examples of the uses of characteristic local stone in the built environment of Sydney (Panel 1), which used Hawkesbury Sandstone (and brick); Hobart (Panel 2), where sandstone, some dolerite and brick predominated; and Brisbane (Panel 3), where Brisbane Tuff (ignimbrite) and sandstone were mainly used.

Panel 1 Sydney

1A. Partly modified outcrop of Hawkesbury Sandstone on the eastern shore of Farm Cove, near the tip of Mrs Macquarie's Point. Note the sandstone walling and steps fitted into the shape of the outcrop. Many original outcrops survive in this area.

1B. Fig tree on quarried sandstone face, with ornate carved stone fence pillar. Native figs are a common feature of the exposed sandstone faces in central Sydney.

1C. View of Admiralty House, Kirribilli Point, north shore, Port Jackson. An example of a grand harbourside home built from local sandstone within the context of natural and modified sandstone outcrops and landforms. This grand house is the Sydney residence of the Governor General.

1D. Stone steps built into the quarried face of Bennelong Point, opposite the Sydney Opera House. The vertical face is known as the Tarpeian Wall. Vertical faces like this are dominant features of the shores of Bennelong Point and Millers Point. They contrast sharply with the shoreline of Mrs Macquarie's Point where many natural outcrops survive relatively intact.

1E. Quarried face in sandstone with a sandstone block retaining wall fitted neatly onto the upper surface. The sandstone has weathered to reveal details of the bedding and structure of the natural stone. Hungry Mile – Hickson Road.

1F. Former warehouses built from sandstone, with upper storeys added in brick. The Rocks, western shore of Sydney Cove. This type of sandstone building once lined the shores of Circular Quay.

1G. Sandstone Guardhouse of Hyde Park Barracks (1811–1819). The lower courses of sandstone blocks have been affected by groundwater and are more weathered. The stone used does not seem to have been of particularly high quality.

1H. Finely carved sandstone royal crest on the façade of the Land Titles Office (1912–13), Albert Road, central Sydney. Ornate stonework was a feature of many of the great public buildings of Sydney and many survive as important architectural reference points in the inner city.

1I. Sandstone portico, St James Church (1824), Sydney. The portico shows the fine classical detailing that a skilled stone carver could produce in Hawkesbury Sandstone.

Panel 2 Hobart

2A. View of Davey Street looking eastwards from Franklin Square, central Hobart. The difference in elevation between Davey Street and the high ground of the park reflects the height difference between the original sea-level shoreline and the site of the camp (though the original shoreline bluffs have been modified by quarrying). Franklin Square occupies ground that was once at the top of the cliffs. The site of the first Government House was approximately where the road intersection occurs in the middle-distance. The road level at this point has been greatly lowered to provide an easier gradient to the waterfront. The buildings in the background occupy reclaimed land.

2B. View of the city end of the Brooker Highway, on the eastern side of the Hobart CBD, looking south from the Old University Building. The valley occupied by the highway in the middle distance was formerly the course of the Park Street (Domain) Rivulet. Like the rivulet, which it follows, the Brooker Highway marks the eastern boundary of the CBD.

2C. The channelised Hobart Rivulet where it is crossed by Barrack Street. East of this point, the rivulet becomes obscured by city developments and becomes a sub-surface drain, until reappearing at the Collins Street Canal. However, the geometry of central city cadastral boundaries, buildings and roads built over this course are all strongly influenced by the course of the stream, and it is still deeply imprinted on the layout of the modern city.

2D. The steep incline in Harrington Street looking north from the intersection of Macquarie Street. The steep topography reflects the north slope of the Central Ridge, as it drops towards the bed of the rivulet (now a sub-surface drain just south of Liverpool Street).

2E. The former Governor's Private Secretary's Cottage seen from Davey Street, and now within the modern Tasmanian Museum and Art Gallery complex. The cottage once stood high above the north shore of Sullivan's Cove, within the grounds of Old Government House. Horizontally bedded Triassic sandstone and siltstone were exposed in a cutting below the cottage but were obscured in the early 2000s (Hope, 2012, 150).

2F. The 1810 Commissariat Store building, which originally stood at the shoreline of Sullivans Cove and at the head of the sand spit that joined Hunter Island to the mainland. Subsequent land reclamation has pushed the shoreline over 80 m to the southeast. This brick-built building, created from local clay sources, remains a pivotal reference point in the built environment of central Hobart.

2G. Sequence of buildings forming part of the State Treasury Department facing east across modern Franklin Square (far right = former 1858–60 Supreme Court building; center = former 1884–88 Executive Council building, left = 1894–95 Deeds Building). This complex of buildings occupies the western end of the former Georges Square, which in 1811 was the most important public square in Hobart. This complex of government buildings developed throughout the colonial era. Like many other government sites in central Hobart, it occupies land at the crest of the bluff, and overlooking the cove, which was chosen in 1804 for the temporary residences of the first government officials.

2H. Sandstone mile marker on the Harrington Street bridge (1844) across the Hobart Rivulet. Commuters daily cross this rebuilt bridge across the rivulet, linking the north and south sides of the formerly rugged banks.

2I. A rare building constructed mostly from dolerite blocks in Salamanca Square. This stone for this building was presumably sourced from the adjacent Salamanca quarry. The use of shaped blocks of dolerite is relatively rare in Hobart buildings. This stone was more commonly used for the foundations and lower courses of buildings that were completed in sandstone and brick.

2J. An example of a modern red gravel driveway to a garage in suburban Hobart. Red gravel is a product of weathered dolerite with good road making properties and there are many dirt roads and driveways constructed from this resilient material. Red gravel is one of the most distinctive and characteristic elements of urban and rural southern Tasmania.

Panel 3 Brisbane

3A. A remnant outcrop of the Neranleigh-Fernvale Beds on Queens Wharf Road, central Brisbane. The foliated metamorphic rock outcrops have been incorporated into a wall constructed from Brisbane Tuff.

3B. Detail view of the water-level outcrops of Neranleigh-Fernvale Beds at the base of the Riverside Expressway, near the site of the former Queens Wharf. The surviving exposure may be a remnant of the outcrops that spurred Henry Miller to choose this site as the main landing place for the Moreton Bay Penal Settlement. There are few other recognizable features of the original Queens Wharf.

3C. View to the east, from the Spine of Brisbane (now the site of the modern Queensland University of Technology—QUT) and across the site of the former convict farms and gardens of the penal settlement. The moderate to steep slopes of this landscape were once occupied by wetlands and rain forest and the low-lying marshy, but fertile land was a focus of early agriculture. The space is now occupied by the City Botanic Gardens.

3D. Detail view of a corner of the riverside boundary wall of the Commissariat Store (1829) of the Moreton Bay Penal Settlement. The pillar is made from local sandstone, but the store building and the bulk of the bounding wall are made from locally quarried Tuff. The tuff is sourced from the Kangaroo Point quarry and the sandstone from quarries developed in Triassic sediments of the Ipswich Basin, downriver to the east.

3E. Panoramic view of part of the Kangaroo Point Cliffs, a former stone quarry near the southern base of Kangaroo Point. The remaining vertical quarry faces are set far back from the original cliff faces that once bordered the river here. This quarry was a source for much of the tuff that was used for buildings (such as the Commissariat Store), walling and kerbing in central Brisbane.

3F. The soaring structure of St John's Anglican Cathedral, Brisbane; with external walls built largely from Brisbane Tuff, and internal walls of Helidon sandstone. It was finally completed in 2009. The scale of this enormous building shows the potential of the Brisbane Tuff as a monumental and distinctive local building stone.

3G. Retaining walls and steps built from Brisbane Tuff, leading eastwards from QUT to the low-lying ground of the City Botanic Gardens. Feature walls and steps built from this stone are common in older parts of Brisbane.

3H. The Brisbane General Post Office building (1872), built from local Triassic sandstone (quarried near Breakfast Creek) and stone from Murphy's Creek (near Toowoomba) and tuff. This most beautiful and elegant of the surviving colonial buildings of central Brisbane is constructed on a site that was once an isolated knoll to the east of the main settlement, and previously occupied by the convict Female Factory.

3I. The ANZAC Square Shrine of Remembrance (1930). It is built mainly from Helidon Sandstone and shows the remarkable structural qualities of the stone. The central location of the memorial, in front of the main railway station, reflects the original importance of the railways in Brisbane.

Figure 8.2. Examples of the uses of characteristic local stone in the built environment of Perth (limestone), Melbourne (bluestone—basalt), and Adelaide (bluestone—metamorphic rock, and calcrete).

Panel 1 Perth

1A. Face of former limestone quarry on Mounts Bay Road, on the eastern slope of Mt Eliza (Kings Park). The building complex in the foreground is a former pump house for a bore that exploited fresh groundwater, which was probably an original natural spring.

1B. The limestone rubble wall on the southern boundary of the Government House grounds, near the Supreme Court buildings in central Perth. This wall was once on the original shoreline but now lies 180 m from the river due to land reclamations. The lake in the Government House gardens to the immediate east preserves a section of the original river shoreline.

1C. Steps in the grounds of Government House, which traverse the steeply rising ground of the Swan River shoreline, to the level of St. Georges Terrace.

1D. Limestone block wall of an Edwardian era house in West Perth. Such limestone walls were (and are) a common feature of the suburban environment. Note the weathered and etched block faces that emphasize the eolian bedding of the limestone.

1E. Limestone block wall with variably weathered and etched surfaces that reveal the eolian cross bedding in some examples.

1F. Detail of Tamala Limestone base courses of a legal chambers in central Perth. Limestone was often integrated with brick as an ornamental but practical feature component in the late Victorian and Edwardian buildings of Perth.

1G. The Old Perth No 1 Fire Station (1901). Like many of the most important government buildings of the post gold rush, late nineteenth and early twentieth centuries, this intricately detailed building was mostly constructed from Tamala Limestone.

1H. Ornate limestone-built house with brick detailing, Mount Street, West Perth. An intricately detailed limestone building in which courses of red brick have been integrated as a feature. Mount Street was once the most elite residential area of Perth.

1I. Intricate and detailed limestone masonry of the first section of St Mary's Catholic Cathedral, East Perth, and first begun in 1865. Details such as wall and window arches were created from Donnybrook Sandstone. The cathedral was finally completed with modern materials in 2009.

1J. The northern façade of the 1904 Parliament House, Perth. The intricately carved and ornate decoration of the main public face of this building was created from fine-grained Donnybrook Sandstone.

Panel 2 Melbourne

2A. Modern view of "The Pond" or "Turning Basin" of the Yarra River, looking east from near the King Street Bridge. This naturally widened area of the river was formed by the effects of tides and flow over the basalt falls that once occupied the location of the Queen Street Bridge (seen in the distance). This area became the first Port of Melbourne and was a place where the small vessels that could navigate upstream could be turned for the return journey.

2B. View southwards down Elizabeth Street, central Melbourne, looking from a position just north of LaTrobe Street. This was once the course of the main drainage line of the settlement area and was prone to flooding in the early settlement, and created a significant barrier to east-west road traffic. The alignment still reverts to a flowing creek during exceptional weather events (for example, in 2011).

2C. Bluestone (basalt) paving stones of a laneway in central Melbourne. The durable bluestone was used extensively for paving and kerbing throughout Melbourne and is still occasionally used for this purpose.

2D. Intricately ornamented bluestone (basalt) being used as a base for iron railings in the central city area. Ornamental bluestone was used extensively for this purpose throughout Melbourne.

2E. Bluestone foundation stone and base courses of the Town Hall, Swanston Street, Melbourne. Bluestone was frequently used in the foundations and base courses of Victorian and Edwardian era buildings (and bridges) because of its durability and strength.

2F. Ornamental bluestone base courses, steps and a basement public lavatory entrance at the Melbourne Town Hall (facing Collins Street).

2G. Bluestone former wool store on the corner of William and Bourke Street. Bluestone was less commonly used to construct complete buildings, despite its durability, because it was considered too gloomy, and was difficult to work. Light colored stone was preferred for the facades of important public buildings, such as the Town Hall, Parliament House and the Treasury Building, though bluestone was often used for the internal walls (e.g. Parliament House). Bluestone was often favored for churches, prison complexes and warehouses, such as this example.

2H. Detail of the blue stone construction of the former Methodist Church in Lonsdale Street. Note the often conchoidal surface texture of the stone created by the fine-grained to glassy character of the basalt during the shaping of the blocks.

2I. The Wesley Uniting Church in Lonsdale Street, central Melbourne. The use of bluestone in this building is typical of many of Melbourne churches built in the nineteenth and early twentieth centuries and they survive as defining element of the built environment. The most significant is St. Mary's Catholic Cathedral.

2J. Bluestone forming the southern bridge pylon and the river embankment at the south end of Queens Bridge. This is the approximate location where walkers from the Beach would have crossed the river at the falls in the 1830s. Note the use of large bluestone pavers for the modern South Bank Yarra Promenade.

2K. Weathered and worn simple kerbing and gutters formed from both roughly shaped and cut bluestone blocks, in a side street near the Queen Victoria Markets.

Panel 3 Adelaide

3A. View of the former Hallett Cove Sandstone (known as limestone) quarry on the south bank of the Torrens River, at the boundary of the Government House grounds. This view is looking south and is taken from the Torrens Parade Ground Headquarters building. The quarry exploited calcareous sandstone and probably also calcrete and calcretized sandstone and once extended westward along the south bank of the river to the railway station.

3B. View eastward along the Torrens River from the 1879 Albert Bridge (Frome Road), which is east of the modern CBD. This section of the river shows the often-steep banks of the river in a section that has seen less modification than in the main settlement area.

3C. View of the Torrens River to the west of the main settlement area, now occupied by the weir walls that form the modern ornamental lakes. In this area, the riverbed became more complex and multi-channeled, with billabongs on side channels near the old Adelaide gaol. This is the place where William Light established his survey camp, and it was an important freshwater source for early Adelaide.

3D. Detail of the stone wall bordering the grounds of Government House on North Terrace, Adelaide. The upper courses are constructed from irregular blocks of pale-yellow calcrete ("limestone"), and some slabs of more tabular and regularly shaped bluestone (metamorphic rock). The base courses and foundations of this section of the wall are formed from bluestone.

3E. Calcrete rubble and brick wall bordering St Laurence's Catholic Church, corner Buxton and Hill Streets, North Adelaide. Bluestone (?) and brick form the base courses.

3F. Old Parliament House on North Terrace Adelaide (1843, 1857). This original legislative building of South Australia was constructed from Hallett Cove Sandstone and calcrete quarried from along the banks of the Torrens River. This building shows that significant structures could be successfully constructed by a skilled builder from what, at first appearance, seems like a poor building material. Other examples of major buildings built from this material are Holy Trinity Church, and the oldest parts of Government House.

3G. Typical double-storied and verandaed stone building constructed from shaped blocks of Adelaide bluestone, a metamorphic schistose metasedimentary rock, often referred to as slate. Front facades of such buildings were typically made from more carefully shaped stonework, while side walls were built of less regular stonework. Adelaide abounds in stone buildings, like this North Adelaide example, because good quality timber was in short supply, wooden buildings were prone to termite attack, and quality stone was plentiful.

3H. Detail view of the Allan Campbell Buildings of the Adelaide Children's Hospital, King William Road. Bluestone ("slate") forms the base courses and foundations of this building, while the upper part of the structure is built from faced sandstone (freestone) from Stirling Point.

3I. Detail of stone building façade of a corner building on Pirie Street, Adelaide. This example shows well-shaped rectangular Adelaide bluestone blocks and fine mortar pointing integrated with brick detailing of window and door arches to create a decorative effect. This building probably once had a two-story veranda.

3J. Adelaide bluestone used as the foundations, base courses, and arched entrance steps of this 1920s building on Kintore Avenue, which was formerly a Teacher Training College. Even well-into the 20th Century, Adelaide bluestone remained a key architectural character element of elegant Adelaide buildings.

Bibliography

Ahmad, M. and J. Hollis. 2013. "Pine Creek Orogen." In *Geology and Mineral Resources of the Northern Territory*. Geological Survey, Special Publication 5, edited by M. Ahmad and T. Munson. Darwin: Northern Territory Geological Survey.

Aitken, Richard, David Jones and Colleen Morris. 2006. *Adelaide Botanic Garden Conservation Study*. Prepared for the Board of the Botanic Gardens of Adelaide. Adelaide Research & Innovation Ltd: University of Adelaide.

Anderson, R. 1992. "Werribee Formation – Geology." In *Engineering Geology of Melbourne: Proceedings of the Seminar on Engineering Geology of Melbourne*, edited by W. Peck, J. Neilson, R. Olds, K. Seddon, 137–143. Rotterdam: A.A. Balkema.

Angus, M. 1984. *Simpkinson De Wesselow: Landscape Painter in Van Diemen's Land and the Port Phillip District, 1844–1848*. Hobart: Blubber Head Press.

Arrowsmith, J. 1833. "Inset Map of Perth." In *Discoveries in Western Australia: From Documents Furnished to the Colonial Office by J. S Roe*. TROVE, National Library of Australia. Accessed August 19, 2020. https://nla.gov.au/nla.obj-231891898/view.

Arrowsmith, J. 1839. *The District of Adelaide, South Australia, As Divided into Country Sections. From the Trigonometrical Surveys of Colonel Light, Late Survr Genl.* Published February 18, 1840. London: John Arrowsmith. TROVE, National Library of Australia. Accessed August 20, 2020. https://nla.gov.au/nla.obj-232669583/view.

ATSE. 2000. *Technology in Australia 1788–1988*. Online edition. Australian Academy of Technology and Engineering. http://www.austehc.unimelb.edu.au/tia/332.html.

Attenbrow, Val. 2002. "Pre-colonial Aboriginal Land and Resource Use in Centennial, Moore and Queens Parks – Assessment of Historical and Archaeological Evidence for Centennial Park and Moore Park Trust Conservation Management Plan." In *Centennial Parklands Conservation Management Plan: Volume 3*, Appendix S,

Sydney: Australian Museum. Accessed July 12, 2019. https://www.centennialparklands.com.au/getmedia/e32ae90a-e730-4c28-82c4-4b17e9e3c5e1/Appendix_S_-_Pre-colonial_Archaeology_report_Val_Attenbrow.pdf.aspx.

Attenbrow, Val. 2009. "Aboriginal Place Names around Port Jackson and Botany Bay, New South Wales, Australia: Sources and Uncertainties." In *Aboriginal Placenames, Naming and re-naming the Australian landscape*, edited by H. Koch and L. Hercus, 9–54. Aboriginal History Monograph 19. Canberra: ANU Press. Accessed July 12, 2019. http://press-files.anu.edu.au/downloads/press/p17331/pdf/ch0115.pdf.

Attenbrow, Val. 2010. *Sydney's Aboriginal Past: Investigating the Archaeological and Historical Records*. Sydney, NSW: University of New South Wales Press. Accessed August 17, 2020. ProQuest Ebook Central.

Attenbrow, Val and N. Kononenko. 2017. "Creating Sydney-Hawkesbury Rock Engravings: A Mangrove Mountain Engraving Tool." *Archaeology in Oceania* 52: 161–170.

Australian Bureau of Statistics. 2019. "Interesting Facts about Australia's 25,000,000 Population." First published August 7, 2018. Accessed November 26, 2019. https://www.abs.gov.au/websitedbs/D3310114.nsf/home/Interesting+Facts+about+Australia%E2%80%99s+population.

Australian Bureau of Meteorology. 2020. "Climate Classification Maps." Accessed August 20, 2020. http://www.bom.gov.au/jsp/ncc/climate_averages/climate-classifications/index.jsp?maptype=kpngrp#maps.

Australian Government. 2019. "State and Territory Government." Accessed August 20, 2020. https://www.australia.gov.au/about-government/how-government-works/state-and-territory-government.

Australian Stratigraphic Units Database. 2020. *Stratigraphic Unit Details: Werribee Formation*. https://asud.ga.gov.au/search-stratigraphic-units/results/29131.

Babaahmadi, A. and G. Rosenbaum. 2014. "Late Mesozoic and Cenozoic Wrench Tectonics in Eastern Australia: Insights from the North Pine Fault System (southeast Queensland)." *Journal of Geodynamics* 73: 83–99.

Bach, J. 2019. "Barker, Collet (1784–1831)." Australian Dictionary of Biography. Accessed August 22, 2019. http://adb.anu.edu.au/biography/barker-collet-1740.

Bacon, C. and M. Banks. 1989. "A History of Discovery, Study and Exploitation of Coal in Tasmania." *Papers and Proceedings of the Royal Society of Tasmania* 123: 137–189.

Bacon, C. 1986. *The New Town Coalfield*. Tasmanian Department of Mine Report 1986/07. Hobart.

Banks, M. and D. Leaman. 1999. "Charles Darwin's Field Notes on the Geology of Hobart—A Modern Appraisal." *Papers and Proceedings of the Royal Society of Tasmania* 133(1).

Barney, George. 1839. "Plan of Brisbane Town." State Library of New South Wales. Cadastral map. Accessed August 19, 2020. https://www.sl.nsw.gov.au/collection-items/map-brisbane-town-moreton-bay-g-w-barney-1839.

Bates, Daisy. 1992. *Aboriginal Perth and Bibbulmun Biographies and Legends*. Edited by P. Bridge, with an introduction by P. Bindon. Carlisle, Western Australia: Hesperian Press.

Bates, Daisy M. 2018. *Oldest Perth: The Days before the White Men Won*. Carlisle, Western Australia: Hesperian Press.

Bauman, Toni. 2006. *Aboriginal Darwin: A Guide to Exploring Important Places*. Canberra: Aboriginal Studies Press.

Beard, J. 1999. "Evolution of the River Systems of the South-West Drainage Division, Western Australia." *Journal of the Royal Society of Western Australia* 82: 147–164.

Bembrick, C. S., C. Herbert, E. Scheibner and J. Stuntz. 1973. "Structural Subdivision of the New South Wales Portion of the Sydney-Bowen Basin." *Quarterly Notes Geol. Survey of N.S.W* 11: 1–13.

Bennett, C. and M. Norbert. 2014. "Tunnelling within the Bunya Phyllite of Legacy Way, Brisbane." *Queensland 15th Australasian Tunnelling Conference 2014. Sydney, NSW, 17–19 September 2014*, 131–140.

Bennett, John. 1869. *Vocabulary of the Woolner District Dialect, Adelaide River, Northern Territory*. Annotated by Paul Foelsche, 1869. State Library of New South Wales. Accessed July 20, 2020. http://archival.sl.nsw.gov.au/Details/archive/110361550.

Benson, D. and J. Howell. 1990. *Taken for Granted: The Bushland of Sydney and Its Suburbs*. Kenthurst, N.S.W. Sydney: Kangaroo Press in association with the Royal Botanic Gardens.

Blainey, Geoffrey. 1989. *The Rush That Never Ended: A History of Australian Mining*. 3rd ed. Parkville, VIC: Melbourne University Press.

Blewett, R. ed. 2012. *Shaping a Nation: A Geology of Australia*. Canberra: Geoscience Australia and ANU E Press.

Bolt, F. 1981. "James Meehan's Survey of Hobart Town in 1811." *Papers and Proceedings of the Royal Society of Tasmania* 115: 5–18.

Bolt, F. 2004. *The Founding of Hobart: A Diary Recounting the Events on the Derwent 1803–1804*. 1st ed. Kettering, Tasmania: Peregrine Press.

Bolton, G. 1989. "Perth: A Foundling City." In *The Origins of Australia's Capital Cities*, edited by Pamela Statham, 141–155. Cambridge, UK: Cambridge University Press.

Bonar, Lynda. 2011. *Hobart Rivulet Catchment Management Plan 2011–2016*. Prepared for Hobart City Council by Lynda Bonar. Accessed July 21, 2018 https://www.hobartcity.com.au/files/assets/public/roads-and-transport/stormwater/hobart_cmp_-_final.pdf.

Bonwick, J. 1868. *John Batman, the Founder of Victoria*. 2nd ed. Melbourne: Fergusson and Moore. Internet Archive Ebook. https://archive.org/details/johnbatmanfound01bonwgoog

Bowden, B. 2016. *Direct Hit: The Bombing of Darwin Post Office*. Australia: Rosenberg Publishing Pty Ltd.

Bowes Smyth, Arthur. *A Journal of a Voyage from Portsmouth to New South Wales and China in the Lady Penrhyn, Merchantman William Cropton Sever, Commander by Arthur Bowes-Smyth, Surgeon—1787–1788–1789*. Transcript. Sydney: State Library of NSW. http://archival-classic.sl.nsw.gov.au/_transcript/2017/D36405/a1085.html.

Botanic Gardens and Parks Authority. 2019. Kings Park and Botanic Garden: Geology. Accessed January 25, 2018. https://www.bgpa.wa.gov.au/kings-park/area/bushland/geology.

Bourke, Patricia. 2005. "Identifying Aboriginal 'Contact Period' Sites Around Darwin: Long Past Due for Native Title?" *Australian Aboriginal Studies* 1: 54–65. Canberra: Aboriginal Studies Press.

Bourke, Patricia. 2014. "Goyder and the Ancient Aboriginal Shell Mounds of Virginia." *Northern Territory Historical Studies* 25: 45–56. Darwin: Historical Society of the Northern Territory.

Bowern, C. 2012. The Riddle of Tasmanian Languages. *Proceedings of the Royal Society B* 279 (1747): 4590–4595.

Boyce, James. 2011. *1835: The Founding of Melbourne and the Conquest of Australia.* Collingwood, Victoria: Black Inc.

Boyce, James. 2019. *Van Diemen's Land.* Carlton, Victoria: Black Inc.

Bradley, W. 1802a. *Sydney Cove, Port Jackson, 1788.* Drawings from his journal 'A Voyage to New South Wales,' 1802. State Library of New South Wales. Accessed January 21, 2019. https://www.sl.nsw.gov.au/collection-items/sydney-cove-port-jackson-1788.

Bradley, W. 1802b. *William Bradley Journal: A Voyage to New South Wales, December 1786—May 1792. A Voyage to New South Wales, December 1786—May 1792.* Transcript of 1802 Journal Compilation. Sydney: University of Sydney Library, Australian Digital Collections. http://adc.library.usyd.edu.au/view?docId=ozlit/xml-main-texts/brajour.xml.

Branagan, D. 2000a. "The Hawkesbury Sandstone: It's Origins and Later Life." In *Sandstone City: Sydney's Dimension Stone and Other Sandstone Geomaterials*, edited by G. McNally and B. Franklin. Sydney: Environmental Engineering and Hydrogeology Specialists Group, Geological Society of Australia Monograph No. 5, 23–38.

Branagan, D. 2000b. "Structural Geology of the Hawkesbury Sandstone, with Particular Reference to the City of Sydney and Nearby Areas." In *Sandstone City: Sydney's Dimension Stone and other Sandstone Geomaterials*, edited by G. McNally and B. Franklin. Sydney: Environmental Engineering and Hydrogeology Specialists Group, Geological Society of Australia Monograph No. 5, 39–54.

Brearley, Anne. 2005. *Ernest Hodgkin's Swanland: Estuaries and Coastal Lagoons of South-Western Australia.* Perth, Western Australia: UWA Publishing.

Briggs, Cecily. 1929. "The Brisbane Tuff." *Proceedings of the Royal Society of Queensland* 40 (12): 147–164.

Broadbent, John. 2010. *Transformations: Ecology of Pyrmont Peninsula 1788–2008.* Report prepared for the City of Sydney: Sydney. https://www.cityofsydney.nsw.gov.au/__data/assets/pdf_file/0018/228033/Transformations-Ecology-of-Pyrmont-Peninsula-17882008.pdf.

Brown, Andrew. 2006. *Geology and the Gettysburg Campaign.* 11th ed. Pennsylvania Department of Conservation and Natural Resources, Educational Series 5. Accessed July 20, 2019. http://elibrary.dcnr.pa.gov/GetDocument?docId=1751059&DocName=ES5_Geo-GettysburgCampaign.

Brumley, J. 1972. *Melbourne Underground Rail Loop. Wellington Parade and Richmond Parade South Areas.* Second Engineering Report for Railway Construction Board, March 1972. Melbourne, Victoria: Geological Survey of Victoria.

Brumley, J. 1974a. *Explanatory Notes on the 1:5,000 Stratigraphic Map of Melbourne's Central Business District.* Geological Survey of Victoria Report 23, 1974/7. Melbourne, Victoria: Geological Survey of Victoria.

Brumley, J. 1974b. "Stratigraphic Map of the Central Business District of Melbourne." Prepared by the Survey Division, Melbourne and Metropolitan Board of Works. State Library of Victoria. http://search.slv.vic.gov.au/permalink/f/1o9hq1f/SLV_VOYAGER678683.

Brunker, R. L. and G. Rose. 1967 (compilers). *New South Wales 1:500,000 Geological Sheet (Special), Sydney Basin.* Sydney: New South Wales Department of Mines, Division of Regional Geology.

Butler, Edward. 2005. "The Tail of Two Rivers in Tasmania: The Derwent and Huon Estuaries." In: *Estuaries (The Handbook of Environmental Chemistry)*, edited by Peter Wangersky. 5H: 1–49. Published online: 8 November 2005. Berlin: Springer.

Camp, C. and Maxwell Banks. 1978. "A Proterosuchian Reptile from the Early Triassic of Tasmania." *Alcheringa* 2 (2): 143–158.

Carter, Max. 1998. *No Convicts There: Thomas Harding's Colonial South Australia.* Port Melbourne, Victoria: Thames & Hudson.

Casey and Lowe Pty Ltd. 2011. *Archaeological Investigation, 710–722 George Street, Haymarket*, Volume1, Section3. Report to Inmark, June 2011. Accessed May 12, 2017. http://www.caseyandlowe.com.au/pdf/710george/volume1_section3.pdf.

Chesnut, W. 1976. *Notes on the Engineering Geology of the Sydney Area.* Sydney: NSW Department of Mines, Geological Survey Report GS1976/005.

Chesnut, W. 1979. *Sydney 1:100,000 Sheet: Engineering Geological Considerations and Environmental Geological Hazards*, NSW Department of Mineral Resources, Geological Survey Report GS1978/444.

Chesnut, W. 1983. "Engineering Geology." In *Geology of the Sydney 1:100,000 Sheet 9130*, edited by C. Herbert, 182–203. Geological Survey of New South Wales.

City of Melbourne. 2019. "Parks and Gardens History." Accessed September 23, 2019. https://www.melbourne.vic.gov.au/about-melbourne/melbourne-heritage/history/Pages/parks-gardens-history.aspx.

Clarke, P. 1991. "Adelaide as an Aboriginal Landscape." *Aboriginal History* 15: 54–72. Canberra: ANU Press. Accessed March 16, 2019. https://www.jstor.org/stable/24046402.

Clarke, P. 1996. "Early European Interaction with Aboriginal Hunters and Gatherers on Kangaroo Island, South Australia." *Aboriginal History* 20 (1): 51–81.

Clarke, Graham, Greg McNally and Graeme Nyland. 1999. "Geotechnical Characteristics of the Darwin Porcellanite." In *Proceedings 8th Australia New Zealand Conference on Geomechanics: Consolidating Knowledge*, edited by Nihal Vitharana and Randal Colman, 911–914. Australian Geomechanics Society.

Colhoun, Eric and Adrian Piper. 1982. "Stone Fish-Traps at Cooks Corner, Freycinet Peninsula, Eastern Tasmania." *Australian Archaeology* 14 (June 1982): 115–118. Taylor and Francis Ltd.

Collins, D. 1798. *An Account of the English Colony in New South Wales, Vol. 1, With Remarks on the Dispositions, Customs, Manners, Etc. Of the Native Inhabitants of That Country. To Which Are Added, Some Particulars of New Zealand.* The Project Gutenberg EBook. http://www.gutenberg.org/files/12565/12565-h/12565-h.htm

Collins, D. 1802. *An Account of the English Colony in NSW: From its First Settlement in 1788, to August 1801.* Vol. 2. A. Strahan, for T. Cadell Jun. and W. Davies. The Strand, London.

Commander, Philip. 2003. "Outline of the Geology of the Perth Region." *Australian Geomechanics Journal* 38 no 3, September 2003. The Engineering Geology of Perth, Part 1: 7–16.

Cook, James and New South Wales Government Printing Office. 1893. *A sketch of Botany Bay in New South Wales* [Cartographic Material]. TROVE, National Library of Australia. Accessed May 21, 2015. http://nla.gov.au/nla.obj-231834491.

Cook, J. 2003. *The Journals.* Edited by J. Beaglehole, P. Edwards, and the Hakluyt Society. Penguin Books.

Cook, J. 2005. *Captain Cook's Journal During the First Voyage Round the World Made in H.M. Bark "Endeavour" 1768–1771.* Accessed November 5, 2013. Project Gutenberg EBook #8106. Last Updated August 24, 2012. Accessed August 20, 2020. http://www.gutenberg.org/files/8106/8106-h/8106-h.htm.

Cosgrove Richard. 1999. "Forty-Two Degrees South: The Archaeology of Late Pleistocene Tasmania." *Journal of World Prehistory* 13 (4): 357.

Cosgrove, Richard. 2015. "Raw Material Movement and Past Tasmanian Aboriginal Interaction: Implications for Understanding Isolated Human Population Dynamics." *Archaeology in Oceania* 50 (Suppl): 69–81.

Cox, H. L., T. Bourchier, P. H. McHugh and Great Britain Hydrographic Department. 1866. Victoria-Australia, Port Phillip. Hobson Bay and River Yarra leading to Melbourne [cartographic material] / surveyed by H.L. Cox; assisted by Thos. Bourchier & P.H. McHugh, 1864; engraved by J. & C. Walker. ([1866 ed.]) [Map]. Published by the Admiralty. Accessed July 21, 2018. http://search.slv.vic.gov.au/permalink/f/1o9hq1f/SLV_VOYAGER1364707.

Craig, C. 1945. "Early Town Planning in Hobart." *Papers and Proceedings of the Royal Society of Tasmania* 1942–1944: 99–108.

Cranfield, L., R. Day and H. Schwarzbock. 1976. *Geology of the Ipswich and Brisbane 1:250,000 Sheet Areas.* Brisbane: Geological Survey of Queensland.

Cromer, B. 2015. "Reinterpreting the Coastal Geology South of Charles Darwin Cliff at Sandy Bay, Southern Tasmania March 16, 2015." Accessed July 26, 2019. http://www.williamccromer.com/reinterpreting-the-coastal-geology-south-of-charles-darwin-cliff/.

Cross, Jack. 2011. *Great Central State: The Foundation of the Northern Territory.* Kent Town, South Australia: Wakefield Press.

Crowley, F. 1962. *The History of South Perth.* Rigby Limited.

Cuffley, Peter. *Australian Houses of the '20s & '30s.* Fitzroy: Five Mile Press, 1989. 141.

Cunningham, C. 1996. *Blue Mountains Rediscovered: Beyond the Myths of Early Australian Exploration.* Kenthurst, NSW: Kangaroo Press.

Darwin Waterfront. 2018. "Things To Do–Experience: Stokes Hill Wharf." Accessed June 21, 2018. https://www.waterfront.nt.gov.au/wp-content/uploads/2018/06/History-6-Stokes-Hill-Wharf.pdf.

DASH Architects. 2018. *Adelaide Park Lands and City Layout: Issues and Opportunity Analysis for the National Heritage Listing.* Report prepared for Heritage South Australia. Adelaide: Department of Environment and Water.

DASH Architects and Peter Bell. 2018. *Heritage Assessment Report. Adelaide Park Lands and City Squares.* Issue Date: 17-May-2018.

Davidson, W. 1995. *Hydrogeology and Groundwater Resources of the Perth Region, Western Australia.* Western Australia Geological Survey, Bulletin 142.

Davidson, W. and Yu, X. 2008. *Perth Regional Aquifer Modelling System (PRAMS) Model Development: Hydrogeology and Groundwater Modelling.* Western Australia Department of Water, Hydrogeological Record Series HG 20.

Davies, M. 1989. "Establishing South Australia." In *The Origins of Australia's Capital Cities*, edited by Pamela Statham, 159–178. Cambridge UK: Cambridge University Press.

Dawes, William, John Hunter, Arthur Phillip and John Stockdale (Bookseller). 1789. *Sketch of Sydney Cove, Port Jackson in the County of Cumberland, New South Wales, July 1788.* Coastline by W. Dawes, soundings by Capt. Hunter. Thomas Medland (lithographer). Originally published in Phillip, A. 1789. *The Voyage of Governor Phillip to Botany Bay.* London: John Stockdale. TROVE, National Library of Australia. Accessed July 21, 2013. http://nla.gov.au/nla.obj-230635094.

De La Rue, K. 2014. *Ruled by Remote Control: The Commonwealth's Role in the History of Darwin, 1911–1978.* PhD thesis, Charles Darwin University.

Denholm, T. 1989. "Adelaide: A Victorian Bastide?" In *The Origins of Australia's Capital Cities*, edited by Pamela Statham, 178–198. Cambridge UK: Cambridge University Press.

DENR and David Young. 2008. *Heritage Conservation Practice Notes: Technical Note 3.6. Stone Masonry in South Australia.* Adelaide: Department of Environment and Natural Resources.

Department of Environment and Climate Change. 2008. *Best Practice Guidelines Sydney Turpentine–Ironbark Forest.* Sydney: Department of Environment and Climate Change, NSW. Accessed October 12, 2015. http://www.environment.nsw.gov.au/resources/threatenedspecies/08528tsdssydturpironforestbpg.pdf

Dix, S. 2016. "Short Report. New Aboriginal Hand Stencil Discoveries at the Freycinet National Park, Tasmania." *Australian Archaeology* 82 (1): 76–79.

Dixon, R. Circa. 1827. *Part of a Map of Port Hacking*, In Irish, Paul and Michael Ingrey. 2013. "Marking Their Footsteps: Aboriginal People and Places in Nineteenth-Century Sydney." In *Exploring Urban Identities and Histories*, edited by Christine Hansen and Kathleen Butler, 75–90. Sydney: AIATSIS Research Publications School of History.

Dixon, R. 1840. *Map of Brisbane Town by Robert Dixon 24 March 1840.* Photograph taken by Henry Mobsby, Mobsby Collection, Fryer Library, University of Queensland, Brisbane, Australia, Accessed August 19, 2020. Flickr https://www.flickr.com/photos/hwmobs/7173003438.

Dortch, C. 1975. "Recent Research in South Western Prehistory." *Australian Archaeology* 3: 13–18. Accessed Jul 4, 2018. https://dspace.flinders.edu.au/xmlui/handle/2328/279?show=full.

Dowson, John. 2011. *Fremantle Port: A Pictorial History*. Fremantle, WA: The Chart & Map Shop.

Doyle, H. 2011. *Thematic History—A History of the City of Melbourne's Urban Environment. Planning for Future Growth 2012.* Report prepared by Context Pty Ltd for the City of Melbourne.

https://www.melbourne.vic.gov.au/SiteCollectionDocuments/history-city-of-melbourne-urban-environment.pdf.

Doyle, N. 2001. *Extractive Minerals Within the Outer Darwin Area.* Northern Territory Geological Survey, Report 14. Darwin: Northern Territory Geological Survey. https://geoscience.nt.gov.au/gemis/ntgsjspui/handle/1/81550

Edwards, Robert. 1964. "A Former Aboriginal Camp-site on the Sturt River at Marion, South Australia." *Mankind* 6 (4.1): 184–188.

eMelbourne. 2019. "eMelbourne: The City Past and Present: Marvellous Melbourne." Published July 2008, University of Melbourne. Accessed July 23, 2019. http://www.emelbourne.net.au/biogs/EM00906b.htm.

Engineers Australia. 2012. *Recognition of Perth's first public water supply scheme for an Engineering Heritage Marker.* Perth, W.A. Engineers Australia Western Australian Division. Accessed April 21, 2017. https://www.engineersaustralia.org.au/sites/default/files/content-files/2016-12/perths_first_water_supply.pdf.

Eriksen, R., L. Koehnken, A. Brooks and D. Ray. 2011. *Derwent Catchment Review. Part 1 Introduction and Background.* Prepared for Derwent Catchment Review Steering Committee, June, 2011. August 21, 2019. Accessed https://www.derwentestuary.org.au/assets/Derwent_Catchment_Review_2011_Part1.pdf.

Etheridge, R., T. Edgeworth David and J. Grimshaw. 1896. "On the Occurrence of a Submerged Forest with Remains of the Dugong, at Shea's Creek, near Sydney." *Journal and Proceedings of the Royal Society of New South Wales*, XXX: 158–185. https://archive.org/details/mobot31753002624135.

Facer, R., and P. Carr. 1979. "K-Ar dating of Permian and Tertiary Igneous Activity in the Southeastern Sydney Basin, New South Wales." *Journal of the Geological Society of Australia* 26 (1–2): 73–79.

Fannie Bay History and Heritage Society. 2016. *Know Where You Stand: Fannie Bay and Surrounds Darwin's Industrial Heartland, 1870–1950.* Darwin: Fannie Bay History and Heritage Society. Accessed May 02, 2016. http://fanniebayhistory.net.au/wp-content/uploads/2016/06/KWYS-sample.pdf.

Fenner, C. "Adelaide, South Australia: A Study in Human Geography." *Transactions and Proceedings of the Royal Society of South Australia* vLI: 193–256. Adelaide: Royal Society of South Australia.

Fetherston, J. 2006. "Tamala Limestone—a Significant Resource for the Australian Building Block Industry." In *Geological Survey of Western Australia Annual Review 2005–2006*, 26–35. Perth, Western Australia: Geological Survey of Western.

Fetherston, J. 2007. *Dimension Stone in Western Australia: Volume 1—Industry Review and Dimension Stones of the Southwest Region.* Mineral Resources Bulletin 23. Perth, Western Australia: Geological Survey of Western.

Flannery, T. 1995. *The Future Eaters: An Ecological History of the Australasian Lands and People*. Port Melbourne: Reed Books.

Flannery, T., ed. 1999. *The Birth of Sydney.* Melbourne: The Text Publishing Company.

Flannery, T., ed. 2006. *Terra Australis: Matthew Flinders' Great Adventures in the Circumnavigation of Australia.* Melbourne: The Text Publishing Company.

Flannery, T., ed. 2007. *The Birth of Melbourne*. Melbourne: The Text Publishing Company.

Flemming, James. 1802. "The Voyage of His Majesty's Colonial Schooner "Cumberland" from Sydney to King Island and Port Phillip in 1802–3. A Journal of the Explorations of Charles Grimes, Acting Surveyor-General of New South Wales. Kept by James Flemming." In *Historical Records of Port Phillip: The First Annals of the Colony of Victoria,* edited by John Shillinglaw. Project Gutenberg of Australia, Accessed September 22, 2019. eBook no.1304081h.html, first posted July 2013. http://gutenberg.net.au /ebooks13/1304081h.html #ch03.

Flinders, Matthew. 1802. "Matthew Flinders Journal - Port Phillip Bay - 26th April to 3rd May 1802." POI Australia. Accessed August 10, 2019. https://poi-australia .com.au/matthew-flinders-journal-port-phillip-bay-26th-april-to-3rd-may-1802/.

Flinders, Matthew. 1814. *A Voyage to Terra Australis Vol 1.* Transcript from facsimile edition, 1966. Project Gutenberg Australia Ebook. Accessed January 22, 2019. http://gutenberg.net.au/ebooks/e00049.html.

Flood, Josephine. 1990. *The Riches of Ancient Australia*. Queensland: Queensland University Press.

Flood, Josephine. 1997. *Rock Art of the Dreamtime: Images of Ancient Australia.* Sydney: Angus & Robertson.

Fowkes, F., R. Cribb and S. Neele. 1789. *Sketch & description of the settlement at Sydney Cove Port Jackson in the County of Cumberland*, [cartographic material]. London. TROVE, National Library of Australia. Accessed January 02, 2015. http:// nla.gov.au/nla.obj-230578175.

Forsyth, S. M. and M. J. Clarke (compilers) 1999. *Digital Geological Atlas 1:25 000 Scale Series. Sheet 5225 Hobart*. Hobart: Mineral Resources Tasmania.

Friend, Mary Ann. 1829. *Mary Ann Friend, Typescript of Journal, 1829–1831*. State Library of Western Australia. Accessed January 22, 2019. https://purl.slwa.wa.gov .au/download/slwa_b1761196_1.pdf.

Frost, Alan. 2001. *Green Gold: A History of Donnybrook, 1842–1974*. Facsimile third edition. Donnybrook, Western Australia: A.C. Frost and Donnybrook-Balingup Shire Council.

Gaffney, L. and J. Stockton. 1980. "Results of the Jordan River Midden Excavation." *Australian Archaeology* 10: 68–78.

Gapps, Stephan. 2018. "Contested Waterways—Aboriginal Resistance in Early Colonial Sydney." Posted on June 19, 2018. Australian National Maritime Museum https://www.sea.museum/2018/06/19/contested-waterways.

Geological Survey of Victoria. 1967. *Geology of the Melbourne District, Victoria*. Geological Survey of Victoria, Bulletin 59. Victoria: Mines Department, http://earthresources.efirst.com.au/product.asp?pID=721&cID=56.

Geoscience Australia. 2019. "Area of Australia - States and Territories." Accessed August 18, 2020. http://www.ga.gov.au/scientific-topics/national-location-information/dimensions/area-of-australia-states-and-territories.

Geoscience Australia. 2019. "Perth Basin." Accessed October 17, 2019. http://www.ga.gov.au/scientific-topics/energy/province-sedimentary-basin-geology/petroleum/offshore-southwest-australia/perth-basin.

Gibbs, Martin. 2011. "An Aboriginal Fish Trap on the Swan Coastal Plain: The Barragup Mungah." In *Fire and Hearth Forty Years on: Essays in Honour of Sylvia J. Hallam*, edited by Caroline Bird and R. Webb. Records of the Western Australian Museum 79: 4–15.

Gilligan, Ian. 2007. "Resisting the Cold in Ice Age Tasmania: Thermal Environment and Settlement Strategies." *Antiquity* 81 (313). Cambridge University Press.

Glover, J. and A. Cockbain. 1971. "Transported Aboriginal Artifact Material, Perth Basin, Western Australia." *Nature* 234: 545–546.

Golder Associates. 2016. *Melbourne Metro Rail Project. Regional Groundwater Numerical Modelling EES Summary Report*. Submitted to AJM Joint Venture. Accessed April 2019. https://metrotunnel.vic.gov.au/__data/assets/pdf_file/0009/51111/MMRP-Technical-Appendix-O-Groundwater-Part-6.pdf.

Goodman, J. 2005. *The Rattlesnake: A Voyage of Discovery to the Coral Sea*. London: Faber and Faber Limited.

Gostin, Victor. 2014. "Peeling Back the Layers under Adelaide." *Bibliophile* 13 (4), December 2014: 3–8. https://digital.library.adelaide.edu.au/dspace/bitstream/2440/99014/3/hdl_99014.pdf.

Goyder, G. 1869. "Survey of Northern Territory: Ordered by the House of Assembly to be printed, 23rd November 1869. Copy of Surveyor-General's Report on Survey of Northern Territory."

Goyder, G. W. 1869b. "Northern Territory Survey Progress Reports." Ordered by the Legislative Council and House of Assembly to be printed, July 30, 1869.

Goyder, G. W. 1869c. "Northern Territory Survey Expedition, 1868–1870 Diary kept by the Surveyor General (G.W. Goyder) Jan 1 to Sep 18 1869." Presented by Estate of Mrs Andrew Smith (Per Sir Keith Smith) Archives, Public Library of South Australia

Gozzard, J. 1983. *Fremantle Part Sheet 2033 I and 2033 IV*. Perth Metropolitan Region Environmental Geology Series. Perth: Geological Survey of Western Australia.

Gozzard, J. 1986. *Perth Sheet 2034 II and part 2034 III and 2134 III*. Perth Metropolitan Region Environmental Geology Series. Perth: Geological Survey of Western Australia.

Gozzard, J. 2007. "The Guildford Formation Re-evaluated." *Australian Geomechanics Journal* 42, no 3: 59–80.

Gray, N., J. Sofoulis, L. de la Hunty and G. Low. 1951. *Clay Loam and Shale Resources, Perth and Environs Sheet 2*, In Perth and Environs Geological Maps, GeoDocs. Perth: Geological Survey of Western Australia. Accessed June 23, 2019. https://geodocs.dmirs.wa.gov.au/Web/documentlist/3/Combined/PO20.

Griffin, R. 1963. *The Botany Basin.* Geological Survey NSW, Bulletin 18. Sydney: Department of Mines.

Grimes, K., L. Cranfield, P. Donchak, L. Hutton, M. Jones. 1986. *Brisbane, Queensland, 1:100 000 Geological Series Map. Sheet 9543.* 1st ed. Brisbane: Geological Survey of Queensland.

Grubb, Kenneth. 1989. "Engineering Geology of the Central Business District of Brisbane." Master of Applied Science thesis, Queensland University of Technology.

Hall, James. 2018. *The Thermochronological Evolution of the Northern Gawler Craton and Northern Adelaide Rift Complex.* PhD thesis, University of Adelaide.

Hallam, Sylvia. 1991. "Aboriginal Women as Providers: The 1830s on the Swan." *Aboriginal History* 15: 38–53. https://search.informit.com.au/fullText;dn=037744120792144;res=IELAPA.

Harris, George. 1804. *Map of part of the Freshwater River at Hobart Town, River Derwent, Island of Van Diemen, 1804.* Pen, Ink and Watercolour on Paper, Department of Primary Industries, Water and Environment Hobart 10. Accessed July 2013. https://stors.tas.gov.au/AI/AF394-1-9.

Harrison, T. 1865. "On the Geology of Hobart Town." *Geological Magazine* 2(17): 491–498.

Harvey, Nick and Brian Caton. 2010. *Coastal Management in Australia.* [Electronic Resource]. Adelaide: University of Adelaide Press. Accessed August 17, 2020. https://digital.library.adelaide.edu.au/dspace/handle/2440/87913.

Hateley, Ronald. 2010. *The Victorian Bush: Its 'Original and Natural' Condition.* South Melbourne, Victoria: Polybractea Press.

Haworth, R. 2003. "The Shaping of Sydney by Its Urban Geology." *Quaternary International* 103: 41–55.

Haworth et al. 2004. "A 6000 Year Old Fossil Dugong from Botany Bay: Inferences about Changes in Sydney's Climate, Sea Levels and Waterways." *Australian Geographical Studies* 42 (1) 46–59.

Haworth, R., Baker, R. and Flood, P. 2004. "A 6000 Year-old Fossil Dugong from Botany Bay: Inferences about Changes in Sydney's Climate, Sea Levels and Waterways." *Australian Geographical Studies* 42 (1): 46–59.

Haworth, R. 2010. "Changes in Mangrove/Salt Marsh Distribution in the Georges River Estuary, Southern Sydney, 1930–1970." *Wetlands* 20 (2): 80–103.

Hay, J., ed. 1906. *The Visit by Charles Fraser (Colonial Botanist for New South Wales) To the Swan River in 1827, with his Opinion on the Suitableness of the District for a Settlement.* Paper read before the West Australian Historical Society. Perth: J. G. Hay. http://handle.slv.vic.gov.au/10381/90951.

Heeres, J. and C. Coote eds. 1898. *Abel Janszoon Tasman's Journal.* Project Gutenberg eBook Accessed August 20, 2019. http://gutenberg.net.au/ebooks06/0600571h.html.

Heirisson, F. 1801. "Swan River Chart, from Direct Observation after His Journey by Longboat along the Swan River from 17–22 June 1801." Western Australia: Battye Library https://slwa.wa.gov.au/treasures/freycinet/swanriver.htm.

Heiss, Anita and Melodie-Jane Gibson. 2019. "Barani: Sydney's Aboriginal History. Aboriginal People and Place." Accessed August 21, 2019. https://www.sydneybarani.com.au/sites/aboriginal-people-and-place/.

Herbert, C. 1983. *Sydney 1:100,000 Geological Sheet 9130.* 1st ed. Sydney: NSW Department of Mineral Resources. Geological Survey of New South Wales.

Herbert, C. and R. Helby. 1980. *A Guide to the Sydney Basin.* Bulletin No 26 Geological Survey of New South Wales. Sydney: Department of Mineral Resources.

Higginbotham, E. 1986. *Report on Archaeological Watching Brief: Gateway Plaza, Circular Quay.* Sydney: Gateway Plaza Pty Ltd., N.S.W.

Higinbotham and Robinson. 1888. *North Botany. Parish of Botany.* Sydney: Lithographed & published by Higinbotham & Robinson. TROVE, National Library of Australia. Accessed August 22, 2020. http://nla.gov.au/nla.obj-229916137.

Hill, E. 1941. "The Silurian Rocks of the Studley Park District." *Proceedings of the Royal Society of Victoria* 53(NS) Part 1: 167–191.

Hillman, M., G. Cocks and J. Ameratunga. 2003. "Guildford Formation." *The Engineering Geology of Perth - Part 2, Australian Geomechanics Journal* 38 (4): 31–39.

Hofto, P., D. Sloane and B. Weldon. 1991. *Engineering Geology of the Greater Hobart Area.* Urban Geological mapping project Report 1. Hobart: Tasmania Department of Resources and Energy.

Hoddle, R. 1837a. "Town of Melbourne 1837." Historic Plan Collection, Public Records Office of Victoria, VPRS 8168/P2, unit 6166. https://ictedservices.typepad.com/.a/6a00e5537796fd883401b8d2c11ff2970c-pi.

Hoddle, R. 1837b. "Plan of the Settlement: Port Phillip." East Melbourne Historical Society. Date digitized, November 18, 2009. Accessed May 22, 2019. https://emhs.org.au/catalogue/eml0003.

Holcombe, R. 1977. "Structure and Tectonic History of the Brisbane Metamorphics in the Brisbane Area." *Journal of the Geological Society of Australia* 24 (7–8): 475–489.

Holdgate, G., B. Geurin, M. Wallace and S. Gallagher. 2001. "Marine Geology of Port Phillip, Victoria." *Australian Journal of Earth Sciences* 48 (3): 439–455. Sydney: Geological Society of Australia.

Holdgate, G. and M. Norvick. 2017. "Geological Evolution of the Holocene Yarra Delta and Its Relationship with Port Phillip Bay." *Australian Journal of Earth Sciences* 64 (3): 301–318. Sydney: Geological Society of Australia.

Holdgate, G., B. Wagstaff and S. Gallagher. 2011. "Did Port Phillip Bay Nearly Dry Up Between ~2800 and 1000 cal. yr BP? Bay Floor Channelling Evidence, Seismic and Core Dating." *Australian Journal of Earth Sciences* 58 (2): 157–175. Sydney: Geological Society of Australia.

Hordern, Marsden. 1989. *Mariners Be Warned! John Lort Stokes and H.M.S. Beagle in Australia 1837–1843.* Ebook, Victoria, Carleton: Melbourne University Press.

Hoskins, W. G. 1985. *The Making of the English Landscape*. Harmondsworth: Penguin.

Houston, Beverley, R. Tucker and Queensland Mines Department. 1965. *Explanatory Notes to Accompany a Geological Map of the City of Brisbane*. Brisbane: Geological Survey of Queensland.

Howchin, Walter. 1918. *The Geology of South Australia (In Two Divisions)*. Adelaide: Education Department.

Historical Records of Port Phillip: The First Annals of the Colony of Victoria, edited by John Shillinglaw. Project Gutenberg of Australia, Accessed September 22, 2019 eBook no. 1304081h.html, first posted July 2013. http://gutenberg.net.au/ebooks13/1304081h.html#ch03.

Hughes, Robert. 1988. *The Fatal Shore: A History of the Transportation of Convicts to Australia, 1787–1868*. London: Collins Harvill.

Hughes-Hallett, Debra. 2010. *Indigenous History of the Swan and Canning Rivers*. Student work placement: Curtin University for the swan River Trust. https://parks.dpaw.wa.gov.au/sites/default/files/downloads/parks/Indigenous%20history%20of%20the%20Swan%20and%20Canning%20rivers.pdf.

Hunter, J. 2003/1793. *An Historical Journal of the Transactions at Port Jackson and Norfolk Island*. Sydney: University of Sydney Library, Australian Digital Collections. Accessed May 01, 2014. http://adc.library.usyd.edu.au/view?docId=ozlit/xml-main-texts/hunhist.xml.

Irish, Paul and Tamika Goward. 2019. "Barani: Sydney's Aboriginal History. Dawes Point/Tar-Ra Barani." Accessed September 01, 2019. https://www.sydneybarani.com.au/sites/dawes-point-tar-ra/.

Jack, R. 1923. *The Building Stones of South Australia*. Adelaide: Geological Survey of South Australia.

Jenkin, Graham. 1995. *Conquest of the Ngarrindjeri: A Story of the Lower Murray Lakes Tribes*. Point McLeay, South Australia: Raukkan Publishers.

Johnson, B., A. Albani, P. Rickwood and J. Tayton. 1977. "The Bedrock Topography of the Botany Basin, New South Wales." *Journal of the Geological Society of Australia* 24 (7–8): 403–408.

Johnson, Murray and Ian McFarlane. 2015. *Van Diemen's Land: An Aboriginal History*. Sydney: New South Publishing.

Johnston, W. and Helen Gregory. 1989. "Brisbane: Making It Work." In *The Origins of Australia's Capital Cities*, edited by Pamela Statham, 252–268. Cambridge UK: Cambridge University Press.

Justice, R. 2000. "Putting the Geo into Geotechnical – the Role of the Engineering Geologist During Construction of the Eastern Distributor Tunnel." *Australian Geomechanics Journal*. Sept 2000: 63–75. Accessed August 17, 2020. http://australiangeomechanics.org/wp- content/uploads/2015/03/35_3_7.pdf.

Karskens, G. 1999. *Inside the Rocks: The Archaeology of a Neighbourhood*. Sydney: Hale and Iremonger.

Karskens, G. 2010. *The Colony: A History of Early Sydney*. Sydney: Allen & Unwin.

Kass, T., C. Liston, J. McClymont. 1996. *Parramatta: A Past Revealed*. Parramatta: Parramatta City Council.

Keneally, T. 2006. *The Commonwealth of Thieves: The Sydney Experiment.* Milsons Point: Random House Australia Pty Ltd.

Kerrison, A. R. and M. A. Binns. 1984. "A Midden Excavation - Royal Tasmanian Botanical Gardens." *Papers and Proceedings of the Royal Society of Tasmania* 118: 53–64.

Kerkhove, Ray. 2015. *Aboriginal Camp Sites of Greater Brisbane: An Historical Guide.* Salisbury, Queensland: Boolarong Press.

King, Philip Gidley. 1786–1792. "Collection 06 Vol. 1: Philip Gidley King Private Journal, in Two Volumes, 1786–1792. Vol. 1 'Remarks & Journal Kept on the Expedition to Form a Colony in His Majestys Territory of New South Wales . . . is Majesty's Ship Sirius . . . .'" 24 October 1786–12 January 1789. https://primo-slnsw.hosted.exlibrisgroup.com/permalink/f/1ocrdrt/ADLIB110316474.

Knopwood, Robert. 1977. *The Diary of the Rev. Robert Knopwood 1805–1838.* Edited by Mary Nicholls, Hobart, Tasmanian Historical Research Association.

Kraehenbuehl, D. 1996. *Pre-European Vegetation of Adelaide: A Survey from the Gawler River to Hallett Cove.* Adelaide: Nature Conservation Society of South Australia.

Krause, F. 1883. *Report on the New Town Carbonaceous Deposits.* Parliamentary Papers Tasmania 1884(59). Mineral Resources Tasmania, OS/051. http://www.mrt.tas.gov.au/mrtdoc/dominfo/download/OS_051/OS_051.pdf.

Laing, A. 1975. *History of Melbourne's Water Supply System.* Geological Survey of Victoria Unpublished Report 1975/77. Victoria: Mines Department. http://gsv.vic.gov.au/searchAssistant/document.php?q=parent_id:5221.

Lampert, R. J. 1980. "Aboriginal Site." In *The Archaeological Investigation of the Bond Store, Moore's Wharf,* edited by R. Lampert and M Truscott. Appendix 1. Accessed August 18, 2020. http://dx.doi.org/10.4227/11/50481E92CFC2E.

Land and Property Management Authority. 2011a. Botany Bay, 1:25000 GeoPDF Topographic Map 9130–3S.

Land and Property Management Authority. 2011b. Parramatta River, 1:25000 GeoPDF Topographic Map 9130–3N.

Lands Department. 1838. *Perth 18a. Plan of Perth Township showing Lots, Streets & Lakes.* Unsigned, probably Hillman, 1838. https://archive.sro.wa.gov.au/uploads/r/srowa/f/9/f90a411f11420be1cc067cf3037127a36a903090f661c1e2e36eadfa02710fe5/cons_3868_item_289.jpg.

Land Registry Services. Undated (probably 1830s). *Parish of Sutherland.* Historical Lands Records Viewer. Historical Parish Maps. Sheet 6. Filename 14062001.jp2. last modified August 12, 2011. Sydney: NSW Land Registry Services. https://hlrv.nswlrs.com.au/.

Land Registry Services. 1868–74. *Parish of Sutherland.* Historical Lands Records Viewer. Historical Parish Maps. Sheet 7. Filename 14066401.jp2. last modified 2011-08-12. Sydney: NSW Land Registry Services. https://hlrv.nswlrs.com.au/.

Larrakia Development Corporation. 2019. "Larrakia." Accessed August 20, 2020. https://www.larrakia.com.au/.

Latinovic, M., A. Waite, C. Calver and S. Forsyth. 2001. *An Investigation of Land Stability in the Taroona Area.* Tasmanian Geological Survey Record 2001/01.

Leaman, D. 1975. "Form, Mechanism, and Control of Dolerite Intrusion near Hobart, Tasmania." *Journal of the Geological Society of Australia* 22:2: 175–186.

Leaman, D. 1976. *Geological Atlas 1:50,000 Series. Sheet 82 (83125). Hobart.* Explanatory Report, Tasmania: Department of Mines.

Leaman, D. 1977. *The River Derwent: Elwick Bay to Macquarie Point. Geology: facts, deductions and problems.* Tasmanian Department of Mines Technical Report TR20 85–101.

Leaman, D. 1978. *Bowen's Landing Historic Site, Risdon. Geology, Building Materials, Survey Factors and Recommendations for Geophysical Work.* Tasmanian Mines Department Report UR 1978/27.

Leaman, D. 1999. *Walk into History in Southern Tasmania.* Hobart, Tasmania: Leaman Geophysics.

Leaman, D. 2002. *The Rock Which Makes Tasmania.* Hobart, Tasmania: Leaman Geophysics.

Leaman, D. E. 2009a. "Darwin's Sullivans Cove." In *Charles Darwin in Hobart Town*, edited by M. Davies. Tasmania: Royal Society of Tasmania.

Leaman, D. E. 2009b. "Darwin's Mount Wellington." In *Charles Darwin in Hobart Town*, edited by M. Davies. Tasmania: Royal Society of Tasmania.

Leaman, D. and R. Richardson. 1981. *Revised Interpretation: Gravity Survey of the Henry Jones Building, Old Wharf, Hobart.* Tasmanian Mines Department unpublished report UR1981/16.

Lee, Ida. 1906. *The Coming of the British to Australia, 1788 to 1829.* London: Longman Green and Co.

Lewis, Miles. 2020. "6.04 Brick Veneer." In Miles Davis, *Australian Building: A Cultural Investigation.* [dynamic web publication], Accessed August 20, 2020. https://www.mileslewis.net/australian-building/pdf/06-bricks-tiles/6.04-brick-veneer.pdf.

Leyland, L. 2012. *Reinterpretation of the Hydrogeology of the Leederville Aquifer Gnangara Groundwater System: Looking after All Our Water Needs.* Hydrogeological Record Series Report no 59. Western Australia: Department of Water. https://www.water.wa.gov.au/_data/assets/pdf_file/0015/4713/103063.pdf.

Light, William. 1984. *William Light's Brief Journal and Australian Diaries: With an Introduction and Notes by David Elder.* Adelaide: Wakefield Press.

Ling, T. 2011. *Commonwealth Government Records about the Northern Territory.* Canberra: National Archives of Australia. Accessed June 21, 2019. https://www.naa.gov.au/help-your-research/research-guides/commonwealth-government-records-about-northern-territory.

Lister, A. 1979. *Archaeology of Risdon Cove Historic Site 1978/79 Part III.* Unpublished Report. Tasmania: National Parks and Wildlife Service.

Kingston, G. and E. Stephens. 1842. *Kingston Map.* London: Edward J. Wheeler & Co. State Library of South Australia, SLSA: B141879. http://adelaidia.sa.gov.au/sites/default/files/images/maps/b1414879_kingston_map.png.

McArthur, W. and E. Bettenay. 1960. *Development and Distribution of Soils on the Swan Coastal Plain.* C.S.I.R.O. Soil Publication 16. Western Australia.

McDonald, Jo. 2004. *Archaeological Survey for an Aboriginal Heritage Assessment, University of Sydney, NSW. June 2004.* Report prepared for Capital Insight Pty Ltd by Jo McDonald Cultural Heritage Management Pty Ltd.

McIntosh, P., K. Kiernan and D. Price. 2004. "An Aeolian Sediment Pulse at c. 28 kyr BP in Southern Tasmania." *Journal of the Royal Society of New Zealand* 34 (4): 369–379.

McNally, G. 1995. "Engineering Characteristics and Uses of Duricrusts in Australia." *Australian Journal of Earth Sciences* 42: 535–547.

McNally, G., G. Clarke and B. Weber. 2000. "Porcellanite and the Urban Geology of Darwin, Northern Territory." *Australian Journal of Earth Sciences* 47 (1): 35–44.

McNally, G. and B. Franklin, eds. 2000. *Sandstone City: Sydney's Dimension Stone and other Sandstone Geomaterials.* Sydney: Environmental Engineering and Hydrogeology Specialists Group, Geological Society of Australia Monograph No. 5.

McPherson, K. 2004. *Selected Industries and their Impact on the Aboriginal Landscape of Van Diemen's Land from Invasion to 1830.* MSc thesis, University of Tasmania.

Macphail, M.K. 1999. "A Hidden Cultural Landscape: Colonial Sydney's Plant Microfossil Record." *Australasian Historical Archaeology* 17: 79–105.

Marsden, Cheramie. 1995. *Sustainable Hillside Planning for Glenorchy.* Master of Town Planning thesis, University of Tasmania.

Martens, J. 2011. *Government House and Western Australian Society 1829–2010.* Crawley, Western Australia: UWA Publishing.

Martin, Ged. 1989. "The Founding of New South Wales." In *The Origins of Australia's Capital Cities*, edited by Pamela Statham, 37–51. Cambridge UK: Cambridge University Press.

Matthews, W. and R. Donaldson. 1994. *A Review of Sand Resources in the Hobart Area.* Mineral Resources Tasmania UR1994/10.

Mayer, Wolf. 2008. "Early Geological Investigations of the Pleistocene Tamala Limestone, Western Australia." In *History of Geomorphology and Quaternary Geology*, edited by R. Grapes, D. Oldroyd and A. Grigelis. Geological Society, London: Special Publications, 301, 279–293.

Mayne, S., E. Nicholas, A. Bigg-Wither, J. Rasidi, M. Raine. 1974. *The Geology of the Sydney Basin—A Review.* Canberra: Department of Minerals and Energy, Bureau of Mineral Resources Geology and Geophysics, Australian Government Publishing Service.

Mazengarb, C., G. J. Dickens and C. R. Calver. 2007. *A Compilation of Historical Accounts of the 1872 Glenorchy Landslide.* Tasmanian Geological Survey Record 2007/01.

Meehan, James. 1804. *Plan of the Settlement at the River Derwent as examined by James Meehan, Assistant to the Surveyor General between the 16th of October 1803 and March 1804.* Accessed January 2019. https://stors.tas.gov.au/AI/AF396-1-206.

Millar, J. 1860. "Design for Ship Canal of "Tidal Harbor" and Docks for the Port of Melbourne." [cartographic material]. Melbourne: Lithographed at the Office of Lands and Survey, December 21, 1860.

Mitchell, T., J. W. Lowry, H. Nicholl, and W. Boone (Firm). 1853. *Trigonometrical survey of Port Jackson* [cartographic material]. Engraved by Lowry, J. W. London. T. & W. Boone, New Bond Street. TROVE, National Library of Australia. Accessed August 22, 2018. http://nla.gov.au/nla.obj-231443454.

Mulder, J. A., R. F. Berry, J. A. Halpin, S. Meffre and J. L. Everard. 2018. "Depositional Age and Correlation of the Oonah Formation: Refining the Timing of Neoproterozoic Basin Formation in Tasmania." *Australian Journal of Earth Sciences* 65 (3): 391–407.

Murray, John. 2013. *H.M. Survey Vessel Lady Nelson and the Discovery of Port Phillip.* Edited and introduced by John Currey. Malvern, Victoria: Banks Society Publications.

Murray, P. 1985. "Ichthyosaurs from Cretaceous Mullaman Beds near Darwin, Northern Territory." *Beagle* (1): 39–55.

Nayton, G., ed. 2011. *The Archaeology of Market Capitalism: A Western Australian Perspective.* 11 Contributions to Global Historical Archaeology, Springer.

Neilson, J. 1967. "The Physiography of the Melbourne Area." In *Geology of the Melbourne District, Victoria*, 12–18. Geological Survey of Victoria Bulletin 59. Melbourne, Victoria. http://earthresources.efirst.com.au/product.asp?pID=721&cID=56.

Neilson, J. and J. Jenkin. 1967. "Quaternary." In *Geology of the Melbourne District, Victoria*, Bulletin 59. Geological Survey of Victoria. Melbourne: Mines Department.

Neilson, J. 1989. "Engineering Geology Map, Central Melbourne Area." 1:5 000. Melbourne: Geological Survey of Victoria. http://gsv.vic.gov.au/searchAssistant/document.php?q=parent_id:36624.

Neilson, J. 1992. "Geology of the Yarra Delta." In *Engineering Geology of Melbourne: proceedings of the Seminar on Engineering Geology of Melbourne*, edited by W. Peck, J. Neilson, R. Olds and K. Seddon, 223–243. Proceedings of the seminar on the engineering geology of Melbourne, 16 September 1992. Rotterdam: A.A. Balkema.

Nicholls, A. 1930. "The Structural Features of the Silurian Rocks in the Melbourne District." *Proceedings of the Royal Society of Victoria* 42 (NS) Part 2: 129–134.

Noetling, F. 1910. "Red Ochre and Its Use by the Aborigines of Tasmania." *Papers and Proceedings of the Royal Society of Tasmania for 1909*: 30–38.

Nott, J. 1994. "Long-term Landscape Evolution in the Darwin Region and Its Implications for the Origin of Landsurfaces in the North of the Northern Territory." *Australian Journal of Earth Sciences* 41 (5): 407–15.

Nott, J. 2003. "The Urban Geology of Darwin Australia." *Quaternary International* 103: 83–90.

NSWLAPMA. 2011a. 1:25000 Topographic and Orthophotomap, 9130-3S BOTANY BAY, NSW Land and Property Management Authority, 2011, downloaded April 18, 2016.

NSWLAPMA. 2011b. 1:25000 Topographic and Orthophotomap, 9130-3N PARRAMATTA RIVER NSW Land and Property Management Authority, 2011, downloaded April 18, 2016.

NSWLAPMA. 2011c. 1:25000 Topographic and Orthophotomap, 9129-4N PORT HACKING NSW Land and Property Management Authority, 2011, downloaded April 18, 2016.

NT Chinese Museum. 2020. "Short History." https://www.chungwahnt.asn.au/wp-content/uploads/2018/08/themuseumproject.pdf.

NT Place Names Register. 2019a. "Place Names Register Extract, Garramilla Boulevard." https://www.ntlis.nt.gov.au/placenames/view.jsp?id=24955.

NT Place Names Register. 2019b. "Place Names Register Extract, Freds Pass." https://www.ntlis.nt.gov.au/placenames/view.jsp?id=12980.

Och, D., R. Offler, H. Zwingmann, J. Braybrooke and I. Graham. 2009. "Timing of Brittle Faulting and Thermal Events, Sydney Region: Association with the Early Stages of Extension of East Gondwana." *Australian Journal of Earth Sciences* 56, no. 7: 873–887.

Och, D., R. Offler and J. Braybrooke. 2007. "Implications of K-Ar Dating of Fault Gouges in NNE Trending Faults, Sydney Region." *ASEG Extended Abstracts* 2006.

Och, D., P. Pells and J. Braybrooke. 2004. "Geological Faults and Dykes in Sydney CBD." In *The Engineering Geology of the Sydney Region*. Sydney Chapter Mini-Symposium, October 13, 2004. Sydney: Australian Geomechanics Society.

O'Connor, R., C. Bodney and L. Little. 1985. *Preliminary Report the Survey of Aboriginal Areas of significance in the Perth Metropolitan and Murray River Regions*. Unpublished report, Department of Aboriginal Sites. Perth: Western Australia.

Officer, I. 1980. *Survey of Derwent River Aboriginal midden and quarry sites*. B.Ed. Dissertation Tasmanian College of Advanced Education.

Owen, John. 2010. *Risdon Cove, 3 May 1804*. Sydney: Macleay Press.

Owen, Timothy and Michael Macphail. 2018. "What was Growing Along the Tank Stream Valley, Sydney Cove, in 1788?" *Australasian Historical Archaeology* 36: 16–28.

Owen, Timothy and F. Pate. 2014. "A Kaurna Burial, Salisbury, South Australia: Further Evidence for Complex Late Holocene Aboriginal Social Systems in the Adelaide Region." *Australian Archaeology* 79: 45–53.

Parkhouse, T. 1895. "Native Tribes of Port Darwin and Its Neighbourhood." *Australasian Association for the Advancement of Science* 6: 638–647.

Parks Victoria. 2002. *1803 Collins Settlement Site: Future Directions Management Plan: Final Report.* Report prepared for Parks Victoria and Mornington Peninsula Shire. Victoria: Parks Victoria. http://search.slv.vic.gov.au/permalink/f/1cl35st/SLV_VOYAGER3713309.

Pastmasters. 2019. "A Notional Port Darwin Heritage Cruise: Stokes Hill Wharf." https://www.pastmasters.net/darwin-harbour.html.

Patison, N. L., R. F. Berry, G. J. Davidson, B. P. Taylor, R. S. Bottrill, B. Manzi, J. Ryba and R. E. Shepherd. 2001. "Regional Metamorphism of the Mathinna Group, Northeast Tasmania." *Australian Journal of Earth Sciences* 48: 281–292.

Pells, P., ed. 1985. *Engineering Geology of the Sydney Region*. Published on behalf of the Australian Geomechanics Society. Rotterdam: A.A. Balkema.

Pells, P. 1993. "A Thumbnail Engineering Geology of the Triassic Rocks of the Sydney Area." Unpublished Part 3 of the 1993 EH Davis Memorial Lecture. Part 1 published in the Australian Geomechanics Journal. This part was handed out at the time of the lecture but not formally published. Accessed January 23, 2013. http://www.pellsconsulting.com.au/downloads/engineeringGeologyOfTheTriassicRocksOfTheSydneyArea.pdf.

Pells, P., J. Braybrooke and D. Och. 2004. *Map and Selected Details of Near Vertical Structural Features in the Sydney CBD*. Sydney: Australian Geomechanics Society.

Peron, Francois. 2012. *A Voyage of Discovery to the Southern Hemisphere*. A Project Gutenberg Australia eBook. Date first posted: September 2012. http://gutenberg.net.au/ebooks12/1203691h.html.

Phillip, A. 1789. *The Voyage of Governor Phillip to Botany Bay*. London: Printed for John Stockdale, Piccadilly. Project Gutenberg Australia. Accessed February 22, 2013. http://gutenberg.net.au/ebooks/e00101.html.

Pietsch, B. 1983. *Darwin 1:100,000 Geology Map Series Explanatory Notes*. Darwin: Northern Territory Geological Survey.

Pike, Glenville. 1959. "Early Attempts at Settlement of the Northern Territory (1824–1870)." *Journal of the Royal Historical Society of Queensland* 6, issue 2: 324–341. Brisbane: Royal Historical Society of Queensland. UQ Espace. Accessed April 21, 2019. https://espace.library.uq.edu.au//view/UQ:212586.

Pillans, B. and P. Gibbard. 2012. "The Quaternary Period." In *The Geologic Time Scale*, edited by Felix Gradstein, James Ogg, Mark Schmitz and Gabi Ogg, 979–1010. Elsevier.

Pittman, E. F. 1903. *Geological Sketch Map of the Country in the Vicinity of Sydney*. Sydney. Geological Survey of New South Wales. Accessed July 12, 2019. https://search.geoscience.nsw.gov.au/product/488.

Playford, P., A. Cockbain and G. Low. 1976. *Geology of the Perth Basin, Western Australia*. Western Australia Geological Survey, Bulletin 124. Perth: Western Australia.

Plomley, N. J. B. and Caroline Goodall. 1990. *Tasmanian Aboriginal Place Names*. Occasional Paper No. 3., Launceston: Queen Victoria Museum and Art Gallery.

Powell, Alan. 1986. "John Stokes and the Men of the Beagle: Discoverers of Port Darwin." *Occasional Papers (Northern Territory Library Service)* no. 1. Darwin: Library Services of the Northern Territory.

Powell, C., P. Baillie and A. VandenBerg. 2003. "Silurian to Mid-Devonian Basin Development of the Melbourne Zone, Lachlan Fold Belt, Southeastern Australia." *Tectonophysics* 375: 9–36.

Prangnell Jonathan and Hannah Craig-Ward. 2017. "Domestic Archaeology of 1 William Street, Brisbane City." *Australasian Historical Archaeology* 35: 71–78.

Preiss, W. 2000. "The Adelaide Geosyncline of South Australia and Its Significance in Neoproterozoic continental reconstruction." *Precambrian Research* 100 (2000): 21–63

Presland, Gary. 1985. *The Land of the Kulin: Discovering the Lost Landscape and the First People of Port Phillip*. Melbourne: McPhee Gribble.

Presland, Gary. 1993. *An Archaeological Survey of the Melbourne Metropolitan Area*. Reprinted 1993. Victoria Archaeological Survey, Ministry for Planning and Environment. Albert Park: Victoria.

Presland, Gary. 1994. *Aboriginal Melbourne: The Lost Land of the Kulin People*. Melbourne: McPhee Gribble.

Presland, Gary. 2009. *The Place for a Village: How Nature Has Shaped the City of Melbourne*. Melbourne: Museum Victoria Publishing.

Presland, Gary. 2010. *First People: The Eastern Kulin of Melbourne, Port Phillip & Central Victoria*. Melbourne: Museum Victoria Publishing.

Presser, J., M. Stoneking and A. Redd. 2002. "Tasmanian Aborigines and DNA." *Papers and Proceedings of the Royal Society of Tasmania* 136: 35–38.

Price, R., I. Nicholls and C. Grey. 2003. "Cainozoic Igneous Activity." In *Geology of Victoria*, edited by William Birch, 361–375. Geological Society of Australia Special Publication 23. Melbourne: Geological Society of Australia (Victoria Division).

Purcell, D. and G. Trand. 1978. *Melbourne Cable Tunnels Geological Investigations 1977–78*. Bureau of Mineral Resources Australia Record 1978/16 (unpubl). Canberra: Bureau of Mineral Resources. https://d28rz98at9flks.cloudfront.net/13612/Rec1978_016.pdf.

Pugh, Derek. 2017. *The British in North Australia 1824–29: Fort Dundas*. Northern Territory: Rapid Creek.

Pugh, Derek. 2018a. *Escape Cliffs: The First Northern Territory Expedition, 1864–66*. Northern Territory: Rapid Creek.

Pugh, Derek. 2018b. *Darwin 1869: The Second Northern Territory Expedition*. Northern Territory: Rapid Creek.

Pugh, Derek. 2019. *Darwin: Origin of a City*. Northern Territory: Rapid Creek.

Q-Album. 2020. *Charlotte Street, flood of 1864*.Accessed September 2019. https://qalbum.archives.qld.gov.au/qsa/charlotte-street-flood-1864.

Qualischefski, David. 2001. "Excavations in the Brisbane CBD." *Australian Geomechanics Journal*. September 2001. Accessed August 19, 2020. https://australiangeomechanics.org/wp-content/uploads/2015/03/36_3_5.pdf.

Rands, W. 1887. *Report Accompanying the Geological Map of Brisbane*. Geological Survey of Queensland Publication 34.

Raper, George and John Hunter. 1791. Plan of Botany Bay, New South Wales [cartographic material]. TROVE, National Library of Australia. Accessed August 17, 2020. http://nla.gov.au/nla.obj-232280436

Rawlinson, N., H. Tkalčić and A. Reading. 2010. "Structure of the Tasmanian Lithosphere from 3-D Seismic Tomography." *Australian Journal of Earth Sciences* 57: 381–394.

Reece, Robert. 1989. "Palmerston (Darwin): Four Expeditions in Search of a Capital." In *The Origins of Australia's Capital Cities*, edited by Pamela Statham, 292–313. Cambridge UK: Cambridge University Press.

Richards, H. and W. Bryan. 1933. "The Problem of Brisbane Tuff." *Proceedings of the Royal Society of Queensland* 45 (11): 50–65. Brisbane.

Robson, L. 1989. "Settling Van Diemens Land." In *The Origins of Australia's Capital Cities*, edited by Pamela Statham, 80–96. Cambridge, UK: Cambridge University Press.

Rocky Bay Heritage Trail. 1988. *Pamphlet Created for the Rocky Bay Heritage Trail for the Bicentennial Project*. Perth: Western Australia. http://whadjukwalkingtrails.org.au/brochures-and-maps/heritage-trails/Rocky-Bay-Hertiage-Trail.pdf.

Roe, J. 1829. *Item 134—J.S. Roe—Eye Sketch of the Swan River, October-November 1829*. State records office of WA. Rhttps://archive.sro.wa.gov.au/index.php/j-s-roe-eye-sketch-of-the-swan-river-october-november-1829–134.

Russell, Robert. 1837. *Map Shewing the Site of Melbourne and the Position of the Huts & Buildings Previous to the Foundation of the Township by Sir Richard Bourke in 1837*. State Library of Victoria. http://handle.slv.vic.gov.au/10381/74744

Ryan, L. 1996. *The Aboriginal Tasmanians*. 2nd ed. St Leonards, Australia: Allen & Unwin Pty Ltd.

Sagona, A. 1994. *Bruising the Red Earth: Ochre Mining and Ritual in Aboriginal Tasmania*. Melbourne University Press.

Saintilan, Neil and Robert Williams. 1999. "Mangrove Transgression into Saltmarsh Environments in South-East Australia." *Global Ecology and Biogeography* 8, no. 2: 117–24. Accessed July 30, 2020. www.jstor.org/stable/2997853.

Sala, G. 1885. "The Land of the Golden Fleece." *Argus,* Saturday 8 August 1885, page 5. Melbourne, Victoria.

Sanderson, E. and M. Boyer. 2009. *Mannahatta: A Natural History of New York City*. New York: Abrams.

Savours, Ann and A. McConnell. "The History of the Rossbank Observatory, Tasmania." *Annals of Science* 82, 39 (6): 527–564.

Selby, J. and J. Lindsay. 1982. *Engineering Geology of the Adelaide City Area*. South Australia: Geological Survey of South Australia.

Select Committee. 1885. *Western Australia Parliament Legislative Council Select Committee Appointed to Consider the Report of a Commission upon the Sanitary Condition of, and as to a Water Supply for, the City of Perth and the town of Fremantle*. Perth, WA: Government Printer.

Selwyn, A. 1860 (?). "Geological Quarter Sheet No 1." Geological Survey of Victoria. Melbourne.

Scanlon, Andrew, Graham Fish, and Murray Yaxley. 1990. *Behind the Scenery: Tasmania's Landforms and Geology*. Tasmania: Department of Education and the Arts.

Schofield, A. 2014. *The Engineering Geology of a Brecciated Sub-Unit in the Newer Volcanics of Melbourne and the Implications for Construction*. MSc thesis, University of Canterbury. http://hdl.handle.net/10092/10064.

Selwyn, A. 1855. "Report on the Geological Relations of Some of the Coal Seams of Van Diemens Land, their Probable Extent and Relative Economic Value." *Papers and Proceedings of the Royal Society of VD.L.* III (I): 116–141.

Semmens, T. 1988. "Food and Agriculture in the New Colony of Van Diemen's Land, 1803 to 1810." *Papers and Proceedings of the Royal Society of Tasmania* 122 (2): 19–29.

Shepherd, R. 2001. "Regional Metamorphism of the Mathinna Group, Northeast Tasmania." *Australian Journal of Earth Sciences* 48 (2): 281–292.

Sigleo, Wayne. 1978. *A Study of Late Quaternary Environment and Man from Four Sites in Southeastern Tasmania*. Thesis, University of Tasmania.

SIXMaps. New South Wales Spatial Services. https://maps.six.nsw.gov.au/.

Smith, M., K. Fontaine and S. Lewis. 2015. *Regional Hydrogeological Characterisation of the St Vincent Basin, South Australia: Technical Report for the National Collaboration Framework Regional Hydrogeology Project*. Record 2015/16. Canberra: Geoscience Australia. http://dx.doi.org/10.11636/Record.2015.016.

Smyth, A. 1789. *A Journal of a Voyage from Portsmouth to New South Wales and China—in the Lady Penrhyn, Merchantman—William Cropton Sever, Commander*. Transcription. Accessed Febuary 23, 2013. http://acms.sl.nsw.gov.au/album/albumView.aspx?acmsID=412747&itemID=823394.

Sobel, Dava. 1996. *Longitude: The True Story of a Lone Genius Who Solved the Greatest Scientific Problem of His Time*. London: Fourth Estate.

Solomon, R. 1967. "Sprent's Hobart, circa 1845." *Papers and Proceedings of the Royal Society of Tasmania* 101: 49–66.

Specht, R. and A. Specht. 2007. "Pre-Settlement Tree Density in the Eucalypt Open-Forest on the Brisbane Tuff." *Proceedings of the Royal Society of Queensland* 113: 9–16. Brisbane.

Spillett, Peter. 1972. *Forsaken Settlement: An Illustrated History of the Settlement of Victoria, Port Essington, North Australia, 1838–1849*. Dee Why, New South Wales: Lansdowne.

Sprent, James 1841–1846. *Sprent's Maps (AF393), the survey by Surveyor-General James Sprent of Hobart commencing on 25 February 1841*. Tasmania: Lands and Survey Department. Accessed August 19, 2020. https://stors.tas.gov.au/AI/AF393.

Stacey, A. R. and R. F. Berry. 2004. "The Structural History of Tasmania: A Review for Petroleum Explorers." *PESA Eastern Australasian Basins Symposium II*, Adelaide, September 19–22, 2004.

Stanbury, Peter, John Clegg and David Campbell. 1990. *A Field Guide to Aboriginal Rock Engravings: With Special Reference to Those Around Sydney*. South Melbourne: Sydney University Press in Association with Oxford University Press.

Statham, Pamela, ed. 1989. *The Origins of Australia's Capital Cities*. Cambridge, UK: Cambridge University Press.

Statham, Pamela. 1989. "Western Australia Becomes British." In *The Origins of Australia's Capital Cities*, edited by Pamela Statham-Drew. Cambridge, UK: Cambridge University Press.

Statham-Drew, Pamela. 2003. *James Stirling: Admiral and Founding Governor of Western Australia*. Crawley, Western Australia: University of Western Australia Press.

State Library of Western Australia. 2019. *Stereograph of Claise Brook, Perth*. slwa_b1923967_1. State Library of Western Australia. Accessed October 2019. https://catalogue.slwa.wa.gov.au/search/,?b1923967.

Steele, John. 1972. *The Explorers of the Moreton Bay District, 1770–1830*. St. Lucia: University of Queensland Press.

Steele, John. 1984. *Aboriginal Pathways in Southeast Queensland and the Richmond River*. St. Lucia: University of Queensland Press.

Steele, John. 1987. *Brisbane Town in Convict Days: 1824–1842*. Reprinted 1987. St. Lucia: University of Queensland Press.

Stevenson, P. 1969. *Site investigation at Wrest Point, Sandy Bay*. Tasmania Department of Mines Technical Report TR14: 115–117.

Stevenson, M., R. Woolley and C. Mazengarb. 2016. *Historical Assessment of the 1872 Glenorchy Debris Flow: A Basis for Modelling the Large Debris Flow Hazard from the Wellington Range, Hobart*. Tasmania: Department of State Growth, Mineral Resources Tasmania.

Stewart, Doug. 2003. "Geotechnical Borehole Database for the Perth CBD." *Australian Geomechanics Journal* 38 (4): 157–160. https://australiangeomechanics.org/papers/geotechnical-borehole-database-for-the-perth-cbd/.

Stirling, James. 1829a. "Chart of Swan & Canning Rivers on the Western Coast of Australia." Accessed July 16, 2020. https://catalogue.slwa.wa.gov.au/record=b1804831~S2

Stirling, James. 1829b. *Chart of Swan River from a Survey (1827)*. TROVE, National Library of Australia. Accessed July 16, 2020. http://nla.gov.au/nla.obj-678184652

Stokes, J. 1846. *Discoveries in Australia; with an Account of the Coasts and Rivers Explored and Surveyed During the Voyage of HMS Beagle in the Years 1837-38-39-40-41-42-43*. Vol. 2. Ebook transcript. http://gutenberg.net.au/ebooks/e00039.html.

Stockton, J. 1977. "Preliminary Notes on an Aboriginal Stone Alignment and Associated Features." *Papers and Proceedings of the Royal Society of Tasmania* 111: 181–184.

Stockton, J. 1977. "A Tasmanian Painting Site (?)." *Papers and Proceedings of the Royal Society of Tasmania* III: 185–189.

Stockton, J. and A. Wallace. 1979. "Towards a Human Prehistory in the Lower Derwent River Area, South-eastern Tasmania, Australia." *Papers and Proceedings of the Royal Society of Tasmania* 113: 81–84.

Stroud, W., H. Sherwin, H. Ray, C. Baker and K. Smart. 1985. *Port Hacking 1:100,000 Geological Sheet 9130*. 1st ed. Sydney: Geological Survey of New South Wales.

Sumerling, Patricia. 2011. *The Adelaide Park Lands: A Social History*. Kent Town, South Australia: Wakefield Press.

Surveyor General. 1911. "Township of Palmerston." Digitized map noting name change to Darwin on March 03, 1911. Northern Territory: Department of Infrastructure, Planning and Logisitics. Northern Territory. http://www.ntlis.nt.gov.au/imfPublic/sites/historic_map_index/displayLargeMap.jsp?mapId=00989.

Sutherland, F. 1972. "The Classification, Distribution, Analysis and Sources of Materials in Flaked Stone Implements of Tasmanian Aborigines." *Records of the Queen Victoria Museum* 42. Launceston, Tasmania.

Sydney City Council. 1947. "City of Sydney—Civic Survey, 1938–1950." Archives & History Resources, Historical Atlas. https://archives.cityofsydney.nsw.gov.au/pages/historicalatlas.

Sydney Mail Wednesday 17 October 1934, page 19. TROVE, National Library of Australia. https://trove.nla.gov.au/newspaper/rendition/nla.newsarticle166106230.3.pdf?followup=41306c6daaa06d6f00a297f0c6538b16.

Tardif, Phillip. 2003. *John Bowen's Hobart: The Beginning of European Settlement in Tasmania*. Hobart: Tasmanian Historical Research Association.

Tench, W. 1789. *A Narrative of the Expedition to Botany Bay*. A Project Gutenberg Australia eBook. Accessed January 24, 2013. http://gutenberg.net.au/ebooks/e00083.html.

Tench, W. 1789. *A Narrative of the Expedition to Botany Bay*. Tench, Watkin (1759–1833). Transcript 1998. Sydney: University of Sydney Library, Australian Digital Collections. http://adc.library.usyd.edu.au/data-2/p00039.pdf.

Tench, W. 1793. *A Complete Account of the Settlement at Port Jackson, Including an Accurate Description of the Situation of the Colony; of the Natives; and of Its Natural Productions*. Sydney: University of Sydney Library, Australian digital collections 1998. Accessed August 18, 2020. http://setis.library.usyd.edu.au/ozlit/pdf/p00044.pdf.

Thomas, Lan. 1991. *The Holocene Archaeology and Palaeoecology of Northeastern Tasmania, Australia*. PhD thesis, University of Tasmania.

Thureau, G. 1883. "The Carboniferous Deposits near New Town." *House of Assembly Papers Tasmania 1883(3)*.

Tindale, Norman. 1974. *Aboriginal Tribes of Australia: Their terrain, Environmental Controls, Distribution, Limits and Proper Names*. Republished in digital form (1991). Canberra: The ANU Press. Accessed May 21, 2019. https://openresearch-repository.anu.edu.au/handle/1885/114913.

Tovey, J. 1911. "Some Notes on Coode Island and its Flora." Melbourne: Field Naturalists' Club of Victoria. State Library of Victoria. http://handle.slv.vic.gov.au/10381/113195.

Trade, Business and Innovation. 2019. "Population." Accessed August 20, 2020. https://business.nt.gov.au/business-and-economic-data/business/nt-key-business-statistics/population

Trezise, D. 1981. "Minor Industrial Minerals in the Brisbane-Ipswich-Southport Mining Districts." In *Field Conference. Brisbane-Ipswich Area 6–8 June 1981*, edited by G. Hofman, 53–54.

Tuckey, James. 1805. *A Voyage to Establish a Colony at Port Philip in Bass's Strait. 1805*. University of Sydney Australian Digital Collections. Accessed August 19, 2020. http://adc.library.usyd.edu.au/view?docId=ozlit/xml-main-texts/tucacco.xml.

Turnbull, C., ed. 2006. *Following their Footsteps: Exploring Adventure Bay*. Bruny Island, Tasmania: Friends of Adventure Bay.

Twidale, C., M. Tyler and B. Webb, eds. 1976. *Natural History of the Adelaide Region*. Adelaide: Royal Society of South Australia.

Twidale, C. 1987. "Sinkholes (Dolines) in Lateritised Sediments, Western Sturt Plateau, Northern Territory, Australia." *Geomorphology* 1 (1): 33–52.

Vandenberg, A. 1974. "Melbourne, Geological Survey of Victoria, 1:63,360, Geological Map." Victoria: Department of Mines. http://earthresources.efirst.com.au/product.asp?pID=377&cID=33&c=20448.

VandenBerg, A. 1997a. "Queenscliff SJ 55-9 Edition 2, 1:250 000 Geological Map Series." Victoria: Geological Survey of Victoria.

VandenBerg, A. 1997b. "Melbourne 1:250000 Geological Map, SJ55-5, Edition 2, May 1997." Victoria: Geological Survey of Victoria.

Vanderwal, R. 1977. "The Shag Bay Rock Shelter Tasmania." *The Artefact* 2 (4): 161–70.

Van Oosterzee, Penny, Ian Morris, Diane Lucas, N. Preece. 2014. *A Natural History and Field Guide to Australia's Top End*. Marleston, South Australia: Gecko Books.

Vicary, M. and C. Mazengarb. 2019. *Field Notes for the Australian Geomechanics Society Excursion to the Midlands*. August 15, 2019. Hobart: Australian Geomechanics Society and Mineral Resources Tasmania.

Vines, G. 2016. *Fishermans Bend Heritage Study. Revised report for Fisherman's Bend Taskforce*. Melbourne: Biosis Pty Ltd. https://www.fishermansbend.vic.gov.au/__data/assets/pdf_file/0022/31675/Heritage-Study_Biosis_November-2016.pdf.

von Gnielinski, F. and J. Siemon. 2012. *Self-Guided Walking Tour Featuring Building Stones Through Brisbane CBD*. Pamphlet produced for 34th International Geological Congress, August 5–10, 2012, Brisbane, Australia.

Wade, H. 1844. *Map of the Environs of Brisbane Town, Situated in the County of Stanley*. Brisbane: Office of the Surveyor General, Queensland 1848. Accessed August 19, 2020 through flickr https://www.flickr.com/photos/queenslandstatearchives/41791195410.

Waight, Sarah. 2015. "Hobart Brick Heritage." *Proceedings of Fabric: The Threads of Conservation. Australia ICOMOS Conference, 5–8 November 2015. Adelaide*. Accessed August 19, 2020. https://www.aicomos.com/wp-content/uploads/Hobart-brick-heritage-full-paper.pdf.

Walker, James. 1889. "The Founding of Hobart by Lieutenant Governor Collins." *Papers & Proceedings of the Royal Society of Tasmania*, 223–248. Hobart: Government Printer. https://eprints.utas.edu.au/15834/1/walker-founding-hobart-1889.pdf.

Walpole, B., P. Dunn, M. Randal, S. Skwarko and J. Hays. 1968. *Geology of the Katherine-Darwin Region, Northern Territory*, Volume 1. Bureau of Mineral Resources, Geology and Geophysics, Bulletin 82. Canberra. Bureau of Mineral Resources. https://d28rz98at9flks.cloudfront.net/146/Bull_082_Vol_1.pdf.

Watson, Frederick, ed. 1923. *Official Papers Relating to the Settlement at Swan River, West Australia. December 1826—January, 1830*. Produced by Ned Overton. Project Gutenberg of Australia eBook. http://gutenberg.net.au/ebooks14/1402751h.html.

Webster, A. 2016. *Geology, Landscape and Brisbane in the 1820s: The South Brisbane Reach*. Guidebook produced for the World Science Festival walking tour. St Lucia: University of Queensland.

Wedge, J. 1835. "Map of Port Phillip from the survey of Mr. Wedge and others." State Library of Victoria. https://viewer.slv.vic.gov.au/?entity=IE7115425&mode=browse.

Wedge, J. 1836. *On the Country Around Port Philip, South Australia.* TROVE, National Library of Australia. Accessed July 20, 2020. http://nla.gov.au/nla.obj-250889930.

Wells, Samantha, 2003. *Negotiating Place in Colonial Darwin: Interactions Between Aborigines and Whites, 1869–1911.* PhD thesis, University of Technology, Sydney.

Willan, T. L. 1925. *Geological Map of the Sydney District* (DOO4951230). NSW Geological Survey Heritage Map HOO66. Sydney: NSW Department of Planning and Environment, Resources and Geosciences. Accessed August 13, 2017 https://search.geoscience.nsw.gov.au/report/R00030426?q=willan&sort=score%20desc&t=all&a=true&p=false&s=false.

Williams, B. 2005. "The Archaeological Potential of Colonial Prison Hulks: The Tasmanian Case Study." *Bulletin of the Australasian Institute for Maritime Archaeology* 29: 77–86.

Willis, J. 1859. *Plan of Portions 1 to 11 at Shea's Creek, Cook's River, Parish of Petersham, Co Cumberland.* Accessed August 22, 2020. TROVE, National Library of Australia. http://nla.gov.au/nla.obj-229983273.

Willmott, W. ed. 2012. *Rocks and Landscapes of Brisbane and Ipswich.* Brisbane: Geological Society of Australia, Queensland Division.

Wintle, S. 1864. "A Sketch of the Principal Geological Features of Hobart, Tasmania." *Quarterly Journal of the Geological Society* 20: 465–469.

Worgan, G. 1788. *Journal of a First Fleet Surgeon.* Sydney: University of Sydney Library, Australian digital collections 2003. Accessed January 12, 2013. http://adc.library.usyd.edu.au/view?docId=ozlit/xml-main-texts/worjour.xml

World Population Review. 2021. "World City Populations 2021." Accessed September 27, 2021 https://worldpopulationreview.com/world-cities.

Work Cover Authority of NSW. 2006. *Report Pursuant to Section 88 of the Occupational Health and Safety Act 2000: Lane Cove Tunnel Collapse and Subsidence 2 November 2005.* Gosford: Work Cover Authority of NSW.

Wrigley, John and Murray Fagg. 2010. *Eucalypts: A Celebration.* Sydney: Allen & Unwin.

Woodhouse, C. "Melbourne in 1838, from the Yarra Yarra." 1888. Color lithograph. (Melbourne: State Library of Victoria). http://handle.slv.vic.gov.au/10381/101254.

Zoellner, Don. 2017. *Vocational Education and Training: The Northern Territory's History of Public Philanthropy.* Acton, ACT: ANU Press. Accessed June 22, 2020. https://www.jstor.org/stable/j.ctt1pwtd3q.2.

Index

Note: Italic page numbers refer to figures.

About the Author

Anthony (Tony) Webster is a Tasmania-based consulting geologist with a deep interest in early Australian history. His particular interest is in how the geology of Australian places acted through their landscapes to influence the first years of European settlement. He is also an adjunct senior lecturer at the Centre for Ore Deposits and Earth Sciences (CODES) at the University of Tasmania and an honorary senior research fellow at the W. H. Bryan Mining and Geology Research Centre (BRC) at the University of Queensland. He is a fellow of the Geological Society (London), the Australian Institute of Geoscientists, the Society of Economic Geologists, and the Australasian Institute of Mining and Metallurgy. His previous book, *The Geology of the Broken Hill Lead-Zinc-Silver Deposit, NSW, Australia*, was published in 2006.

www.ingramcontent.com/pod-product-compliance
Ingram Content Group UK Ltd.
Pitfield, Milton Keynes, MK11 3LW, UK
UKHW020838290926
15470UKWH00004B/1429

9 781498 597951